Prentice Hall Multimedia Series in Automotive Technology

Automotive Steering, Suspension, and Alignment

Second Edition

Prentice Hall Multimedia Series in Automotive Technology

Other books by James D. Halderman and Chase D. Mitchell, Jr., in the
Prentice Hall Multimedia Series in Automotive Technology include:

Automotive Technology: Principles, Diagnosis, and Service, 0-13-359969-8.

Advanced Engine Performance Diagnosis, 0-13-576570-6.

Automotive Brake Systems, Second Edition, 0-13-080041-4.

Automotive Chassis Systems, Second Edition, 0-13-079970-X.

Prentice Hall Multimedia Series in Automotive Technology

Automotive Steering, Suspension, and Alignment

Second Edition

James D. Halderman   Sinclair Community College
Chase D. Mitchell, Jr.   Utah Valley State College

Merrill
Prentice Hall

Upper Saddle River, New Jersey   Columbus, Ohio

Library of Congress Cataloging-in-Publication Data

Halderman, James D.,
    Automotive steering, suspension, and alignment / James D.
Halderman, Chase D. Mitchell, Jr. --2nd ed.
      p.   cm. -- (Prentice Hall multimedia series in automotive
technology)
    Includes index.
    ISBN 0-13-799719-1
    1. Automobiles--Steering-gear. 2. Automobiles--Springs and
suspension. 3. Automobiles--Wheels--Alignment. I. Mitchell, Chase
D. II. Title. III. Series.
TL259.H35   2000   99-37065
629.2'47--dc21   CIP

Editor: Ed Francis
Production Editor: Stephen C. Robb
Design Coordinator: Karrie M. Converse-Jones
Cover Designer: Jason Moore
Cover Photo: James D. Halderman
Production Manager: Patricia A. Tonneman
Illustrations: Carlisle Communications, Ltd.
Production Supervision: Kelli Jauron, Carlisle Publishers Services
Marketing Manager: Chris Bracken

This book was set in Century Book and Gill Sans by Carlisle
Communications, Ltd., and was printed and bound by
Courier Kendallville. The cover was printed by Phoenix Color Corp.

Printed in the United States of America

10 9 8 7 6 5 4 3

ISBN: 0-13-799719-1

Prentice-Hall International (UK) Limited, *London*
Prentice-Hall of Australia Pty. Limited, *Sydney*
Prentice-Hall Canada, Inc., *Toronto*
Prentice-Hall Hispanoamericana, S. A., *Mexico*
Prentice-Hall of India Private Limited, *New Delhi*
Prentice-Hall of Japan, Inc., *Tokyo*
Prentice-Hall (Singapore) Pte. Ltd., *Singapore*
Editora Prentice-Hall do Brasil, Ltda., *Rio de Janeiro*

Contents

5

Drive Axle Shafts and CV Joints   96

6

Drive Axle Shaft and CV Joint Service   117

7

Steering System Components and Operation   138

8

Steering System Diagnosis and Service   166

9

Suspension System Components and Operation   212

10

Suspension System Diagnosis and Service   247

11

Wheel Alignment Principles 283

12

Alignment Diagnosis and Service 303

13

Vibration and Noise Diagnosis and Correction 343

APPENDIXES

Tech Tips, Frequently Asked Questions, Diagnostic Stories, and High-Performance Tips

Photo Sequences

Preface

Diagnostic Approach

The primary focus of this textbook is to satisfy the need for problem diagnosis. Time after time, the authors have heard that automotive technicians need more training in diagnostic procedures and skill development. To meet this need, diagnostic stories are included throughout the text to help illustrate how real problems are solved. Each new topic covers the parts involved, their purpose, function, and operation, as well as how to test and diagnose each system.

Other important features in the second edition are photo sequences, tech tips, and answers to frequently asked questions (FAQs).

ASE Content Approach

This comprehensive textbook covers the material necessary for the suspension and steering (A4) area of certification as specified by the National Institute for Automotive Service Excellence (ASE) and the National Automotive Technicians Education Foundation (NATEF). The book also includes information on drive axle shafts, CV joints, and rear axles which will help the technician study for the manual drive train and axles (A3) ASE test.

Multimedia System Approach

Twelve photo sequences are included in the text. These photo sequences are included on a videotape for instructors available upon adoption of the textbook. A multimedia CD-ROM that accompanies the textbook makes learning more fun for students. The CD includes sound, animation, color photo sequences with narration, a glossary of automotive terms, sample ASE test questions, sample worksheets, and the ASE (NATEF) task list for the Automotive Steering and Suspension ASE area.

Internet (World Wide Web) Approach

Included with the book is a coupon that entitles the owner to free access to an ASE test preparation Web site for an extended time period. Students can use the Web site to practice and take the ASE certification tests with confidence. Included at this Web site are ASE-type questions for the suspension and steering ASE automotive area. The questions are presented twenty at a time, then graded (marked). The correct answer is then given as students scroll back through the questions. This feature allows students to study at their own pace.

Worktext Approach

A worktext is also included with the book. The worksheets help the instructor and students apply the material presented to everyday-type activities and typical service and testing procedures. Also included are typical results and a listing of what could be defective if the test results are not within the acceptable range. These worksheets help students build diagnostic and testing skills.

Acknowledgments

Many people and organizations have cooperated in providing the reference material and technical information used in this text. The authors express sincere thanks to the following organizations for their special contributions:

Allied Signal Automotive Aftermarket

Arrow Automotive

ASE

Automotion, Inc.

Automotive Parts Rebuilders Association (APRA)

Bear Automotive

Bendix

British Petroleum (BP)

Cooper Automotive Company

CR Services

DaimlerChrysler

Dana Corporation

Fluke Corporation

Ford Motor Company

FMC Corporation

General Motors Corporation Service Technology Group

Hennessy Industries

Hunter Engineering Company

Lee Manufacturing Company

John Bean Company

Monroe Shock Absorbers

MOOG Automotive Inc.

Northstar Manufacturing Company, Inc.

Oldsmobile Division, GMC

Perfect Hofmann—USA

Reynolds and Reynolds Company

Shimco International, Inc.

SKF USA, Inc.

Society of Automotive Engineers (SAE)

Specialty Products Company

Tire and Rim Association, Inc.

Toyota Motor Sales, USA, Inc.

TRW Inc.

Wurth USA, Inc.

Portions of materials contained herein have been reprinted with the permission of General Motors Corporation, Service Operations.

Technical and Content Reviewers

The authors gratefully acknowledge the following people who reviewed the manuscript before production and checked it for technical accuracy and clarity of presentation. Their suggestions and recommendations were included in the final draft of the manuscript.

Victor Bridges
Umpqua Community College

Dr. Roger Donovan
Illinois Central College

A. C. Durdin
Moraine Park Technical College

Herbert Ellinger
Western Michigan University

Al Engledahl
College of Dupage

Oldrick Hajzler
Red River Community College

Betsy Hoffman
Vermont Technical College

Carlton H. Mabe, Sr.
Virginia Western Community College

Kerry Meier
San Juan College

Fritz Peacock
Indiana Vocational Technical College

Dennis Peter
NAIT (Canada)

Mitchell Walker
St. Louis Community College at Forest Park

Photo Sequences

The authors thank Rick Henry, who photographed all of the photo sequences. Many of the sequences were taken in automotive service facilities while live work was being performed. Special thanks to all who helped, including:

B P ProCare
Dayton, Ohio
    Tom Brummitt
    Jeff Stueve
    John Daily
    Bob Babal
    Brian Addock
    Jason Brown
    Don Patton
    Dan Kanapp

Rodney Cobb Chevrolet
Eaton, Ohio
    Clint Brubacker

Dare Automotive Specialists
Centerville, Ohio
    David Schneider
    Eric Archdeacon
    Jim Anderson

Foreign Car Service
Huber Heights, Ohio
    Mike McCarthy
    George Thielen
    Ellen Finke
    Greg Hawk
    Bob Massie

Genuine Auto Parts Machine Shop
Dayton, Ohio
    Freddy Cochran
    Tom Berger

Import Engine and Transmission
Dayton, Ohio
    Elias Daoud
    James Brown
    Robert Riddle
    Felipe Delemos
    Mike Pence

J and B Transmission Service
Dayton, Ohio
    Robert E. Smith
    Ray L. Smith
    Jerry Morgan
    Scott Smith
    Daryl Williams
    George Timitirou

Saturn of Orem
Orem, Utah

    We also thank the faculty and students at Sinclair Community College in Dayton, Ohio, and Utah Valley State College in Orem, Utah, for their ideas and suggestions. Most of all, we thank Michelle Halderman for her assistance in all phases of manuscript preparation.

James D. Halderman
Chase D. Mitchell, Jr.

Chassis Design, Materials, Fasteners, and Safety

Objectives: After studying Chapter 1, the reader should be able to:

1. Discuss the various methods of vehicle construction.
2. List the various types of steels and their application in the automotive chassis system.
3. Explain the safe use of tools and automotive chemicals.
4. Describe safe methods of lifting or hoisting a vehicle.
5. Discuss hazardous materials and how to handle them.

The chassis is the framework of any vehicle. The suspension, steering, and drivetrain components are mounted to the chassis. The chassis has to be a strong and rigid platform to support the suspension components. The suspension system allows the wheels and tires to follow the contour of the road. The connections between the chassis, the suspension, and the drivetrain must be made of rubber to dampen noise, vibration, and harshness (NVH). The construction of today's vehicles requires the use of many different materials.

■ CHASSIS DESIGN

A typical dictionary definition of *chassis* usually includes terms such as *frame and machinery of a motor vehicle on which the body is supported.* There are three basic designs used today: frame, unit-body, and space frame construction.

Frame Construction

Frame construction usually consists of channel-shaped steel beams welded and/or fastened together. The frame of a vehicle supports all the "running gear" of the vehicle, including the engine, transmission, rear axle assembly (if rear-wheel drive), and all suspension components.

This frame construction, referred to as *full frame*, is so complete that most vehicles can usually be driven without the body. Most trucks and larger rear-wheel-drive cars use a full frame.

There are many terms used to label or describe the frame of a vehicle, including:

Ladder Frame This is a common name for a type of perimeter frame where the transverse (lateral) connecting members are straight across, as in Figure 1–1. When viewed with the body removed, the frame resembles a ladder. Most pickup trucks are constructed with a ladder-type frame.

Perimeter Frame A perimeter frame consists of welded or riveted frame members around the entire perimeter of the body. This means that the frame members provide support underneath the sides as well as for the suspension and suspension components (see Figure 1–2).

Stub-Type Frames A stub frame is a partial frame often used on unit-body vehicles to support the power

Figure 1–1 Typical frame of a vehicle.

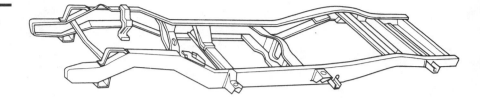

Figure 1–2 Perimeter frame.

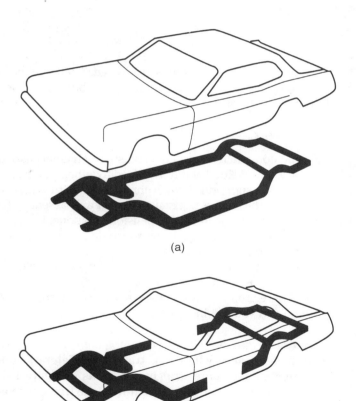

(a)

(b)

Figure 1–3 (a) Separate body and frame construction. (b) Unitized construction; the small frame members are for support of the engine and suspension components. Many vehicles attach the suspension components directly to the reinforced sections of the body and do not require the rear frame section.

Frequently Asked Question ???

What Does GVW Mean?

GVW, *gross vehicle weight,* is the weight of the vehicle, plus the weight of all passengers the vehicle is designed to carry [× 150 lb (68 kg) each], plus the maximum allowable payload or luggage load. *Curb* weight is the weight of a vehicle *wet,* meaning with a full tank of fuel and all fluids filled, but without passengers or cargo (luggage). *Model* weight is the weight of a vehicle wet and with passengers.

The GVW is found stamped on a plate fastened to the doorjamb of most vehicles. A high GVW rating does not mean that the vehicle itself weighs a lot more than other vehicles. For example, a light truck with a GVW of 6000 lb (2700 kg) will not ride like an old 6000-lb luxury car. In fact, a high GVW rating usually requires stiff springs to support the payload; these stiff springs result in a harsh ride. Often technicians are asked to correct a harsh-riding truck that has a high GVW rating. The technician can only check that everything in the suspension is satisfactory and then try to convince the owner that a harsher than normal ride is the result of a higher GVW rating.

---

train and suspension components. It is also called a *cradle* on many front-wheel-drive vehicles (see Figure 1–3).

Unit-Body Construction

Unit-body construction (sometimes called *unibody*) is a design that combines the body with the structure of the frame. The body itself supports the engine and driveline components, as well as the suspension and steering components. The body is composed of many individual stamped steel panels welded together.

The strength of this type of construction lies in the *shape* of the assembly. The typical vehicle uses 300 separate and different stamped steel panels that are spot-welded to form a vehicle's body (see Figure 1–4).

NOTE: A typical vehicle contains about 10,000 individual parts.

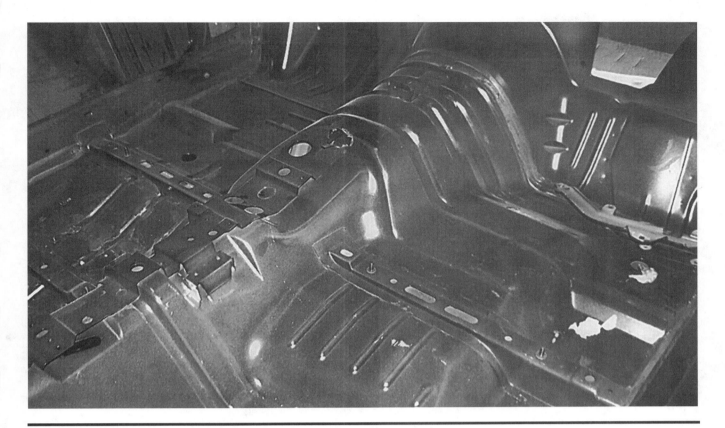

Figure 1–4 Note the ribbing and the many different pieces of sheet metal used in the construction of this body.

Space Frame Construction

Space frame construction consists of formed sheet steel used to construct a framework for the entire vehicle. The vehicle is driveable without the body, which uses plastic or steel panels to cover the steel framework (see Figure 1–5).

■ PLATFORMS

The platform of any vehicle is its basic size and shape. Various vehicles of different makes can share the same platform and, therefore, many of the same drivetrain (engine, transmission, and final drive components) and suspension and steering components.

A platform of a unit-body vehicle includes all major sheet-metal components that form the load-bearing structure of the vehicle, including the front suspension and engine supporting sections. The area separating the engine compartment from the passenger compartment is variously called **bulkhead, cowl panel, dash panel,** or **firewall.** The height and location of this bulkhead panel to a large degree determine the shape of the rest of the vehicle.

Other components of vehicle platform design that affect handling and ride are the track and wheelbase of the vehicle. *The track of a vehicle is the distance between the*

wheels, as viewed from the front or rear. A wide-track vehicle is a vehicle with a wide wheel stance; this increases the stability of the vehicle especially when cornering. *The wheelbase of a vehicle is the distance between the center of the front wheel and the center of the rear wheel, as viewed from the side.* A vehicle with a long wheelbase tends to ride smoother than a vehicle with a short one.

Examples of common platforms include:

1. Chevrolet Lumina, Buick Regal, and Pontiac Grand Prix
2. Pontiac Grand Am, Olds Alero, and Chevrolet Malibu
3. Dodge Intrepid, Chrysler Concord, and Eagle Vision
4. Olds 88, Buick LeSabre, and Pontiac Bonneville
5. Toyota Camry and Lexus ES300
6. Ford Crown Victoria and Mercury Grand Marquis

TECH TIP

Hollander Interchange Manual

Most salvage businesses that deal with wrecked vehicles use a reference book, the *Hollander Interchange Manual.* In this yearly publication, every vehicle part is given a number. If a part from another vehicle has the same Hollander number, then the parts are interchangeable (see Figure 1–6).

Figure 1–5 Space frame for a GM van, showing it without the exterior body panels. The framework surrounding the vehicle is a three-dimensional measuring system capable of accurate measurement of the vehicle.

Figure 1–6 A *Hollander Interchange Manual* is available for both domestic and imported vehicles.

■ CHASSIS MATERIALS

Many automotive component parts are made of cast iron, including brake drums and rotors, spindles, engine blocks, and many other components and fasteners. Just as each component requires different strengths and characteristics from the material, there are different types of steel. The amount of carbon in steel is very important to strength, hardness and machining characteristics.

Galvanized Steel

Galvanized steel is steel with a zinc coating that protects the steel from corrosion (rust). Another type of rust-resistant steel includes zincrometal, which is a two-coat bake-on system using chromium oxide and zinc.

> **CAUTION:** Galvanized (zinc-coated) steels should not be welded unless proper ventilation and precautions are taken. The vapors from zinc are poisonous and can cause serious injury or death.

High-Strength Steel

Since the mid-1970s, many car and light truck parts have been built with high-strength steel (HSS). HSS is commonly used in the sill area under the doors where strength, yet light weight, is needed. Bumper supports and impact beams in doors use HSS. HSS is very hard; heating causes it to lose much of its strength. High-strength steel is a low-carbon alloy steel with various amounts of silicon, phosphorus, and manganese. Body repair technicians should always follow manufacturers' recommended procedures to avoid weakening the integrity of the body.

■ THREADED FASTENERS

Most of the threaded fasteners used on engines are cap screws. They are called **cap screws** when they are threaded into a casting. Automotive service technicians usually refer to these fasteners as bolts, regardless of how they are used. In this chapter, they are called *bolts*. Sometimes, studs are used for threaded fasteners. A **stud** is a short rod with threads on both ends. Often, a stud will have coarse threads on one end and fine threads on the other end. The end of the stud with coarse threads is screwed into the casting. A nut is used on the opposite end to hold the engine parts together (see Figure 1–7).

The fastener threads must match the threads in the casting or nut. The threads may be either customary measure in fractions of an inch (called *fractional*) or metric. First, the threads must be the same size. The size is measured across the outside of the threads. This is called the **crest** of the thread.

Fractional threads are either coarse or fine. Coarse threads are called *Unified National Coarse (UNC)*, and the fine threads are called *Unified National Fine (UNF)*. Standard combinations of sizes and number of threads per inch (called **pitch**) are used. Pitch can be measured with a thread gauge (Figure 1–8).

Fractional threads are specified by giving the diameter in fractions of an inch and the number of threads per inch. Typical UNC thread sizes would be 5/16–18 and 1/2–13. Similar UNF thread sizes would be 5/16–24 and 1/2–20.

Metric threads used in automotive engines are coarse threads. The size of a metric bolt is specified by the letter *M*, followed by the diameter in millimeters across the outside (crest) of the threads. Typical metric sizes would be M8 and M12. Fine metric threads are specified by the thread diameter followed by an × and the distance between the threads measured in millimeters (M8 × 1.5).

Both fractional and metric threads have tolerance. A loose-fitting thread has a lot of tolerance. A very close-fitting thread has very little tolerance. Bolts are identified by their diameter and length as measured from below the head, as shown in Figure 1–9.

Bolts are made from many different types of steel. For this reason, some are stronger than others. Their strength is called the **grade,** or classification, of the bolt. The bolt heads are marked to indicate their grade strength. Fractional bolts have lines on the head to indicate the grade (Figure 1–10).

The actual grade of these bolts is two more than the number of lines on the bolt head. Metric bolts have a decimal number to indicate the grade. More lines or a higher grade number indicates a stronger bolt. In some cases, nuts and machine screws have similar grade markings.

When installing or replacing threaded fasteners, technicians should check that all the following are correct:

1. Fractional or metric threads
2. Diameter
3. Thread pitch
4. Bolt length
5. Length of the threads on the bolt
6. Grade of the bolt

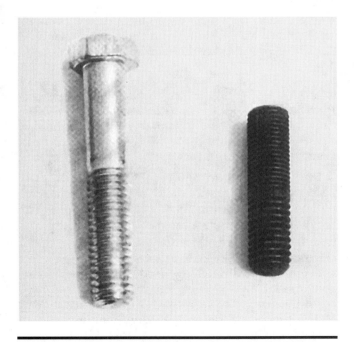

Figure 1–7 Typical bolt on the left and stud on the right. Note the different thread pitch on the top and bottom portions of the stud.

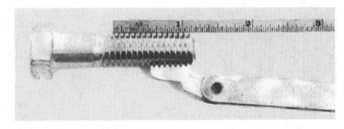

Figure 1–8 Thread pitch gauge used to measure the pitch of the thread. This is a 1/2″-diameter bolt with 13 threads to the inch (1/2 × 13).

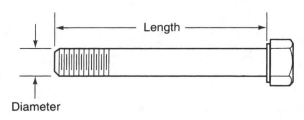

Figure 1–9 Bolt size identification.

Figure 1–10 Typical bolt (cap screw) grade markings and approximate strength.

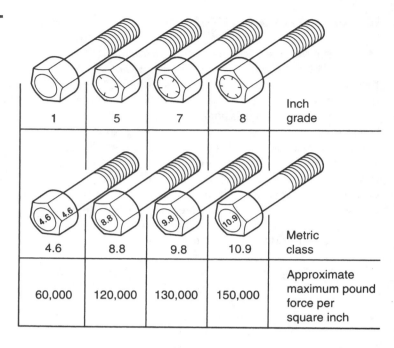

	1	5	7	8	Inch grade
	4.6	8.8	9.8	10.9	Metric class
	60,000	120,000	130,000	150,000	Approximate maximum pound force per square inch

---

> **CAUTION:** *Never* use hardware store (nongraded) bolts, studs, or nuts on any vehicle steering, suspension, or brake component. Always use the exact size and grade of hardware that is specified and used by the vehicle manufacturer.

TECH TIP

A 1/2″ Wrench Does Not Fit a 1/2″ Bolt

A common mistake made by persons new to the automotive field is that they think the size of a bolt or nut is the size of the head. The size (outside diameter of threads) is usually smaller than the size of the wrench or socket that fits the head of the bolt or nut. For example:

Wrench Size	Thread Size
7/16″	1/4″
1/2″	5/16″
9/16″	3/8″
5/8″	7/16″
3/4″	1/2″
10 mm	6 mm
12 mm	8 mm
14 mm	10 mm

Nuts

Most nuts used on cap screws have the same hex size as the cap screw head. In some specialized cases, when clearance is a problem, the next smaller hex size is used. Some inexpensive nuts use a hex size larger than the cap screw head. Metric nuts are often marked with dimples to show their strength. More dimples indicate stronger nuts. Many nuts used on cap screws in automotive work have a smooth ring on the fastener side. Machine screw nuts are the same on both sides.

TECH TIP

It Just Takes a Second

Whenever removing any automotive component, it is wise to put the bolts back into the holes a couple of threads with your hand. This ensures that the right bolt will be used in its original location when the component or part is put back on the vehicle. Often the same diameter fastener is used on a component, but the length of the bolt may vary. Spending just a couple of seconds to put the bolts and nuts back where they belong when the part is removed can save a lot of time when the part is being installed. Besides making certain that the right fastener is being installed in the right place, this method helps prevent bolts and nuts from being lost or kicked away. How much time have you wasted looking for that lost bolt or nut?

Figure 1–11 Types of lock nuts: on the left, a nylon ring; in the center, a distorted shape; and on the right, a castle for use with a cotter key.

> **NOTE:** Most of these locking nuts are grouped together and are commonly referred to as *prevailing torque* nuts. This means that the nut will hold its tightness or torque and not loosen with movement or vibration. Many prevailing torque nuts should be replaced whenever removed to be assured that the nut will not loosen during service. Always follow manufacturers' recommendations.

Anaerobic sealers, such as LOCTITE, are used on the threads where the nut or cap screw must be both locked or sealed.

Washers

Washers are often used under cap screw heads and under nuts. Plain, flat washers are used to provide an even clamping load around the fastener. Lock washers are added to prevent accidental loosening. Heavy parts use split lock washers; light parts use star lock washers. In some accessories, the washers are locked onto the nut to provide easy assembly.

In many applications, it is necessary to make sure a nut or cap screw does not accidentally loosen. In the past, this was done by drilling the head of the cap screws, then fastening two or more together with a wire going through the holes. This wire tying took a great deal of time and is still used in the aircraft industry. Chassis nuts were locked with cotter pins.

Lock washers with sharp edges used under the heads or nuts also are used to keep the nut or cap screw from accidentally loosening. Some nuts and cap screws use interference fit threads to keep them from accidentally loosening. This is done by slightly distorting the shape of the nut or by deforming part of the threads. Nuts can also be kept from loosening with a nylon washer fastened in the nut, or with a nylon patch or strip on the threads (see Figure 1–11).

■ BASIC TOOL LIST

Hand tools are used to turn fasteners (bolts, nuts, and screws). The following is a list of hand tools every automotive technician should possess. Specialty tools are not included. See Figures 1–12 through 1–27 on pages 7–11.

Figure 1–12 Combination wrench. The openings are the same size at both ends. Notice the angle of the open end to permit use in close spaces.

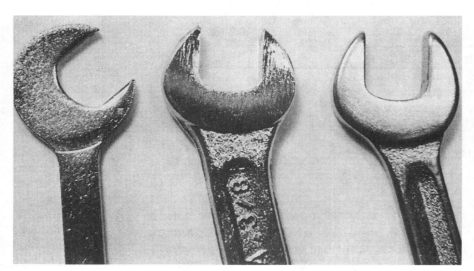

Figure 1–13 Three different qualities of open-end wrenches. The cheap wrench on the left is made from weaker steel and is thicker and less accurately machined than the standard in the center. The wrench on the right is of professional quality (and price).

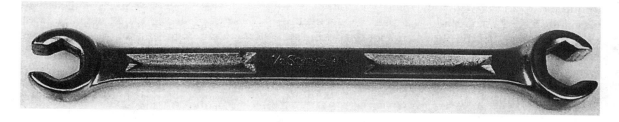

Figure I–14 Flare-nut wrench. Also known as a *line wrench, fitting wrench,* or *tube-nut wrench.* This style of wrench is designed to grasp most of the flats of a six-sided (hex) tubing fitting to provide the most grip without damage to the fitting.

Figure I–15 Box-end wrench. Recommended to loosen or tighten a bolt or nut where a socket will not fit. A box-end wrench has a different size at each end and is better to use than an open-end wrench because it touches the bolt or nut around the entire head instead of at just two places.

Figure I–16 Open-end wrench. Each end has a different-sized opening and is recommended for general usage. Do not attempt to loosen or tighten bolts or nuts from or to full torque with an open-end wrench because it could round the flats of the fastener.

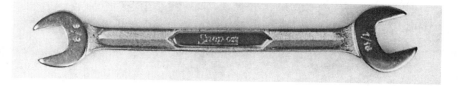

Figure I–17 A flat-blade (or straight-blade) screwdriver (*on the left*) is specified by the length of the screwdriver and width of the blade. The width of the blade should match the width of the screw slot of the fastener. A Phillips-head screwdriver (*on the right*) is specified by the length of the handle and the size of the point at the tip. A #1 is a sharp point, #2 is most common (as shown), and a #3 Phillips is blunt and is only used for larger sizes of Phillips-head fasteners.

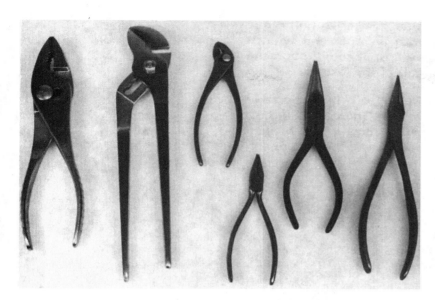

Figure 1–18 Assortment of pliers. Slip-joint pliers (*far left*) are often confused with water pump pliers (*second from left*).

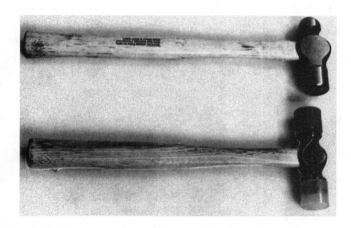

Figure 1–19 A ball-peen hammer (*top*) is purchased according to weight (usually in ounces) of the head of the hammer. At bottom is a soft-faced (plastic) hammer. Always use a hammer that is softer than the material being driven. Use a block of wood or similar material between a steel hammer and steel or iron parts to prevent damage.

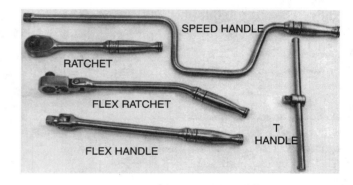

Figure 1–20 Typical drive handles for sockets.

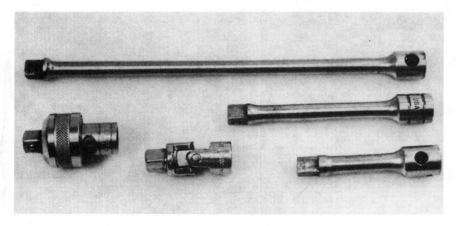

Figure 1–21 Various socket extensions. The universal joint (U-joint) in the center (*bottom*) is useful for gaining access in tight areas.

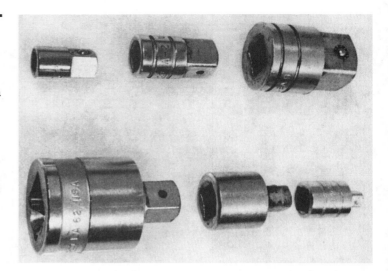

Figure 1–22 Socket drive adapters. These adapters permit the use of a 3/8″ drive ratchet with 1/2″ drive sockets, or other combinations as the various adapters permit. Adapters should *not* be used where a larger tool used with excessive force could break or damage a smaller-sized socket.

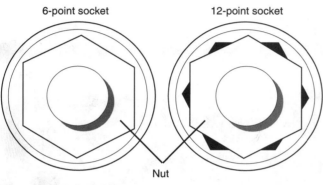

Figure 1–23 A six-point socket fits the head of the bolt or nut on all sides. A twelve-point socket can round off the head of a bolt or nut if a lot of force is applied.

Figure 1–25 An inexpensive muffin tin can be used to keep small parts separated.

Figure 1–24 Standard twelve-point short socket (*left*), universal joint socket (*center*), and deep-well socket (*right*). Both the universal and deep well are six-point sockets.

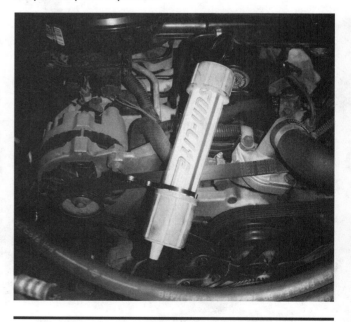

Figure 1–26 A good fluorescent trouble light is essential. A fluorescent light operates cooler than an incandescent light and does not pose a fire hazard if gasoline were accidentally dropped on an unprotected incandescent bulb used in some trouble lights.

Tool Chest

1/4″ drive socket set
1/4″ drive ratchet
1/4″ drive 2″ extension
1/4″ drive 6″ extension
1/4″ drive handle
3/8″ drive socket set
3/8″ drive Torx set
3/8″ drive 13/16″ plug socket
3/8″ drive 5/8″ plug socket
3/8″ drive ratchet
3/8″ drive 1 1/2″ extension
3/8″drive 3″ extension
3/8″ drive 6″ extension
3/8″ drive 18″ extension
3/8″ drive universal
1/2″ drive socket set
1/2″ drive ratchet
1/2″ drive breaker bar
1/2″ drive 5″ extension
1/2″ drive 10″ extension
3/8″ to 1/4″ adapter
1/2″ to 3/8″ adapter
3/8″ to 1/2″ adapter
3/8″ through 1″ combo wrench set
10 mm through 19 mm combo wrench set
1/16″ through 1/4″ hex wrench set
2 mm through 12 mm hex wrench set
3/8″ hex socket
13 mm to 14 mm flare nut wrench
15 mm to 17 mm flare nut wrench

5/16″ to 3/8″ flare nut wrench
7/16″ to 1/2″ flare nut wrench
1/2″ to 9/16″ flare nut wrench
Diagonal pliers
Needle pliers
Adjustable-jaw pliers
Locking pliers
Snap-ring pliers
Stripping or crimping pliers
Ball-peen hammer
Rubber hammer
Dead-blow hammer
Five-piece standard screwdriver set
Four-piece Phillips screwdriver set
#15 Torx screwdriver
#20 Torx screwdriver
Crowfoot set (fractional inch)
Crowfoot set (metric)
Awl
Mill file
Center punch
Pin punches (assorted sizes)
Chisel
Utility knife
Valve core tool
Coolant tester
Filter wrench (large filters)
Filter wrench (smaller filters)
Safety glasses
Circuit tester
Feeler gauge

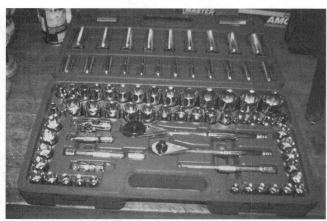

(a)

(b)

Figure 1–27 (a) A beginning technician can start with some simple basic hand tools. (b) An experienced, serious technician often spends several thousand dollars a year for tools such as found in this large (and expensive) toolbox.

TECH TIP

Japanese = Evens; Europeans = Odds

Japanese-brand vehicles usually require even-sized metric wrenches and sockets. Therefore, grab a 10-, 12-, and 14-mm wrench from your toolbox, and you will be able to disassemble and reassemble almost everything on a Japanese vehicle.

European-brand vehicles usually require odd-sized metric wrenches and sockets. Therefore, grab an 11-, 13-, and 15-mm wrench from your toolbox when working on a VW, Audi, Mercedes, BMW, etc.

American vehicles usually use metric fasteners of all sizes, so grab your entire toolbox when working on American-brand vehicles.

TECH TIP

Need to Borrow a Tool More Than Twice?—Buy It!

Most service technicians agree that it is okay for a beginning technician to borrow a tool occasionally. However, if you borrow a tool more than twice, then be sure to purchase it as soon as possible. Also, whenever borrowing, be sure that you return the tool clean and show the technician you borrowed the tool from that you are returning the tool. These actions will help in any future dealings with other technicians.

■ BRAND NAME VERSUS PROPER TERM

Technicians often use slang or brand names of tools rather than the proper term. This results in some confusion for new technicians. Some examples are given in the following table.

Brand Name	Proper Term	Slang Name
Crescent wrench	Adjustable wrench	Monkey wrench
Vise grips	Locking pliers	
Channel locks	Water pump pliers or multigroove adjustable pliers	Pump pliers
	Diagonal cutting pliers	Dikes or side cuts

■ SAFETY TIPS FOR USING HAND TOOLS

1. Always *pull* a wrench toward you for best control and safety. Never push a wrench. If a bolt or nut loosens, your entire weight is used to propel your hand(s) forward. This usually results in cuts, bruises, or other painful injury.
2. Keep wrenches and all hand tools clean to help prevent rust and for a better, firmer grip.
3. Always use a six-point socket or a box-end wrench to break loose a tight bolt or nut.
4. Use a box-end wrench for torque and an open-end wrench for speed.

5. Never use a pipe extension or other types of "cheater bars" on a wrench or ratchet handle. If more force is required, use a larger tool or use penetrating oil and/or heat on the frozen fastener. (If heat is used on a bolt or nut to remove it, always replace it with a new part.)
6. Always use the proper tool for the job. If a specialized tool is required, use the proper tool and do not try to use another tool improperly.
7. Never expose any tool to excessive heat. High temperatures can reduce the strength ("draw the temper") of metal tools.
8. Never use a hammer on any wrench or socket handle unless you are using a special "staking-face" wrench designed to be used with a hammer.
9. Replace any tools that are damaged or worn.

■ SAFETY TIPS FOR TECHNICIANS

Safety is not just a buzz word on a poster in the work area. Safe work habits can reduce accidents and injuries, ease the workload, and keep employees pain-free. Suggested safety tips include:

1. *Safety glasses should be worn at all times while servicing any vehicle.*
2. Watch your toes. Always keep your toes protected with steel-toed safety shoes. If safety shoes are not available, then leather-topped shoes offer more protection than canvas or cloth.
3. Wear gloves to protect your hands from rough or sharp surfaces. Thin rubber gloves are recommended to be worn when working around automotive liquids such as engine oil, antifreeze, transmission fluid, or any other liquids that may be hazardous.

TECH TIP ✔

Pound with Something Softer

If you must pound on something, be sure to use a tool that is softer than what you are about to pound on to avoid damage. Examples are given in the following table.

The Material Being Pounded	What to Pound with
Steel or cast iron	Brass or aluminum hammer or punch
Aluminum	Plastic or rawhide mallet or plastic-covered dead-blow hammer
Plastic	Rawhide mallet or plastic dead-blow hammer

4. Service technicians working under a vehicle should wear a **bump cap** to protect their head against under-vehicle objects and the pads of the lift.
5. Remove jewelry that may get caught on something or act as a conductor to an exposed electrical circuit.
6. Avoid loose or dangling clothing.
7. When lifting any object, get a secure grip with solid footing. Keep the load close to your body to minimize the strain. Lift with your legs and arms, not your back.
8. Do not twist your body when carrying a load. Instead, pivot your feet to help prevent strain to the spine.
9. Ask for help when moving or lifting heavy objects.
10. Push a heavy object rather than pull it. (This is the opposite of how you should work with tools—never push a wrench!)
11. Work with objects, parts, and tools that are between chest high and waist high while standing. If seated, work at tasks that are at elbow height.
12. Always connect an exhaust hose to the tailpipe of any running vehicle to help prevent the buildup of carbon monoxide inside a closed garage space (see Figure 1–28).
13. Store all flammable liquids in an approved fire safety cabinet as shown in Figure 1–29.

■ SAFETY IN LIFTING (HOISTING) A VEHICLE

Many chassis and under-body service procedures require that the vehicle be hoisted or lifted off the ground.

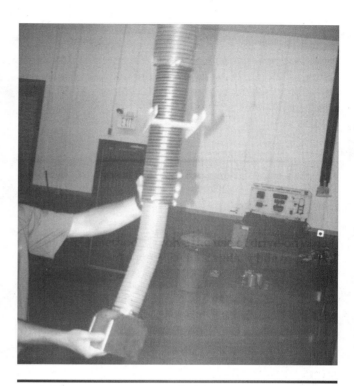

Figure 1–28 Always connect an exhaust hose to the tailpipe of the engine of a vehicle to be run inside a building.

Figure 1–29 Typical fireproof flammable storage cabinet.

The simplest methods involve the use of drive-on ramps or a floor jack and safety (jack) stands, while in-ground or surface-mounted lifts provide greater access.

Setting the pads is a critical part of the hoist lifting procedure. All automobile and light truck service manuals include recommended locations to be used when hoisting (lifting) a vehicle. Newer vehicles are marked with a triangle marking with a decal on the driver's door, indicating the recommended lift points. The recommended standard for the lift points is in SAE Standard JRP-2184. These recommendations typically include the following points:

1. The vehicle should be centered on the lift or hoist so as not to overload one side or put too much force either too far forward or too far rearward (see Figure 1–30).
2. The pads of the lift should be spread as far apart as possible to provide a stable platform.
3. The pad should be placed under a portion of the vehicle that is strong and capable of supporting the weight of the vehicle.
   a. Pinch welds at the bottom edge of the body are generally considered to be strong.

> **CAUTION:** Even though pinch weld seams are the recommended location for hoisting many vehicles with unitized bodies, care should be taken not to place the pad(s) too far forward or rearward. Incorrect placement of the vehicle on the lift could cause the imbalance of the vehicle on the lift, and the vehicle could fall. This is exactly what happened to the vehicle in Figure 1–31.

   b. Boxed areas of the body are the best places to place the pads on a vehicle without a frame. Be careful to note whether the arms of the lift may come in contact with other parts of the vehicle before the pad touches the intended location. Commonly damaged areas include:
      • Rocker panel moldings (see Figures 1–32 and 1–33)
      • Exhaust system (including the catalytic converter)
4. Before using a hoist for the first time, always check with an instructor or other responsible person for directions and/or suggestions.
5. The vehicle should be raised about 1 ft (30 cm) off the floor, then stopped. The vehicle should be shaken to check for stability. If the vehicle seems to be stable

(a)

(b)

Figure 1–30 (a) Tall safety stands can be used to provide additional support for a vehicle while on a hoist. (b) A block of wood should be used to avoid the possibility of doing damage to components supported by the stand.

Figure 1–31 This vehicle fell from the hoist because the pads were not set correctly. No one was hurt, but the vehicle was a total loss.

(a)

(b)

Figure 1–32 (a) An assortment of hoist pad adapters that are often necessary to use to safely hoist many pickup trucks, vans, and sport utility vehicles. (b) A view from underneath a Chevrolet pickup truck showing how the pad extensions are used to attach the hoist lifting pad to contact the frame.

(a)

(b)

Figure 1–33 (a) In this photo the pad arm is just contacting the rocker panel of the vehicle. (b) This photo shows what can occur if the technician places the pad too far inward underneath the vehicle. The arm of the hoist has dented in the rocker panel.

when checked a short distance from the floor, then continue to raise the vehicle and view it until it has reached the desired height.

CAUTION: Do not look away from the vehicle while it is being raised (or lowered) on a hoist. One side or one end of the hoist can stop or fail, resulting in the vehicle being slanted enough to slip or fall. This creates physical damage not only to the vehicle and/or hoist but also to the technician or others who may be near.

HINT: Most hoists can be safely placed at any desired height. For ease while working, the area that you are working on should be at chest level. When working on brakes or suspension components, raise the hoist so that the components are at chest level.

6. Before lowering the hoist, the safety latch(es) must be released, and the direction of the controls reversed, to lower the hoist. The speed downward

is often adjusted to as slow as possible for additional safety.

■ HAZARDOUS MATERIALS

Hazardous materials include used oils and transmission fluids, as well as other chemicals or products that can harm people or the environment. The Environmental Protection Agency (EPA) regulates and controls the handling of hazardous materials in the United States. A material is considered hazardous if it meets one or more of the following conditions:

1. It contains over 1000 parts per million (PPM) of halogenated compounds. (Halogenated compounds are chemicals containing chlorine, fluorine, bromine, or iodine.) Common items that contain these solvents include carburetor cleaner, silicone spray, aerosols, adhesives, Stoddard solvent, trichloromethane, gear oils, brake cleaner, A/C compressor oils, floor cleaners, and anything else that contains "chlor" or "fluor" in its ingredient name.
2. It has a flash point below 140°F (60°C).
3. It is corrosive (a pH of 2 or less, or 12.5 or higher).
4. It contains toxic metals or organic compounds. Volatile organic compounds (VOCs) must also be limited and controlled. This classification greatly affects the painting and finishing aspects of the automobile industry.

Always follow recommended procedures for handling of any chemicals and dispose of all used engine oil and other possible waste products according to local, regional, state, or federal laws.

To help safeguard workers and the environment, the following are recommended.

1. A technician's hands should always be washed thoroughly after touching used engine oils, transmission fluids, and greases. These used products contain combustion by-products and other elements that may cause personal harm. *Rubber gloves should be worn by any technician changing engine oil.*
2. Dispose of all waste oil according to established standards and laws in your area.

> **NOTE:** The EPA current standard permits fewer than 1000 PPM of total halogens (chlorinated solvents) that used engine oil can contain and still be recycled. Oil containing greater amounts of halogens must be considered hazardous waste (see Figure 1–34).

Do not pour used engine oil down sewers or onto the ground to kill weeds or control dust; the toxic chemicals can seep into drinking water miles

Figure 1–34 All solvents and other hazardous waste should be disposed of properly.

away. Oil floats on water and will continue to spread out over the surface until the oil is only one molecule thick!

3. Asbestos and products that contain asbestos are known cancer-causing agents. Even though brake linings and clutch facing materials no longer contain asbestos from the factory, millions of vehicles are being serviced every day that *may* contain asbestos. The general procedure for handling asbestos is to put the used parts into a sealed plastic bag and return the parts as cores for rebuilding, or to be disposed of according to current laws and regulations.

4. Eye wash stations should be readily accessible near the work area or near where solvents or other contaminants could get into the eyes (see Figure 1–35).

■ MATERIAL SAFETY DATA SHEETS

Businesses and schools in the United States are required to make available to all employees a list of the materials that a person may be exposed to in the area and provide a detailed data sheet on each chemical or material. These sheets of information on each of the materials that *may* be harmful are called *Material Safety Data Sheets*, or *MSDS*.

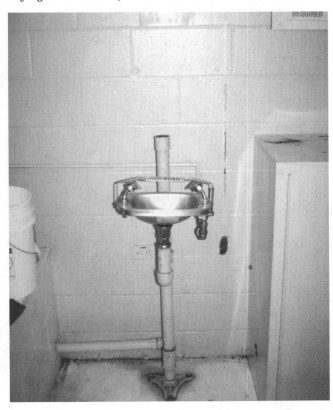

Figure 1–35 An eye wash station should be centrally located and near where solvent may be splashed.

TECH TIP

The Binder Clip Trick

It is important to use fender covers whenever working on an engine. The problem is few covers remain in place and they often become more of a hindrance than a help. A binder clip, available at most office supply stores, can be easily used to hold fender covers to the lip of the fender of most vehicles (see Figure 1–36). When clipped over the lip, the cover is securely attached and cannot even be pulled loose. This method works with both cloth and vinyl covers.

■ TECHNICIAN CERTIFICATION

Even though individual franchises and companies often certify their own technicians, there is a nationally recognized certificate organization, the **National Institute for Automotive Service Excellence**, better known by its abbreviation, **ASE**. See Figure 1–37.

The eight automotive certifications include:

1. Engine repair
2. Automatic transmission/transaxles
3. Manual drivetrain and axles
4. Suspension and steering
5. Brakes
6. Electrical systems
7. Heating and air-conditioning
8. Engine performance

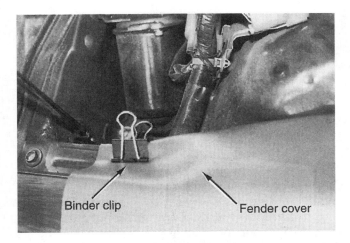

Binder clip   Fender cover

Figure 1–36 It is very important to use fender covers to protect the paint of the vehicle from being splashed with brake fluid. Use a binder clip available at local office supply stores to clip the fender cover to the lip of the fender, preventing the fender cover from slipping.

**WE SUPPORT
VOLUNTARY TECHNICIAN
CERTIFICATION THROUGH**

National Institute for
**AUTOMOTIVE
SERVICE
EXCELLENCE**

Figure 1–37 The ASE logo. ASE is an abbreviation for the National Institute for Automotive Service Excellence.

To become certified by ASE, the service technician must have two years of experience and pass a test in each area. If a technician passes all eight automotive certification tests, then the technician is considered a master certified automobile service technician. Tests are administered twice a year, in May and again in November. Registration and payment are required to be sent in early April for the May test and in early October for the November test.

■ CERTIFICATION IN CANADA

In Canada, in most provinces and territories, an Inter-Provincial (IP) Certificate is required (excluding Quebec and British Columbia). An apprenticeship program is in place that takes a minimum of four years, combining ten months in a shop and about two months in school training in each of the four years. Most apprentices must undergo 7200 hours of training before they can complete the IP examination. ASE certifications are currently used on a voluntary basis since 1993; however, an IP Certificate is still required. Other licensing of automotive technicians may be required in some cases, such as environmental substances, liquefied petroleum gas, or steam operators.

NOTE: A valid driver's license and a good driving record are important for any automotive service technician.

TECH TIP

Work Habit Hints

The following statements reflect the expectations of service managers for their technicians:

1. Report to work every day on time. Being several minutes early every day is an easy way to show your service manager and fellow technicians that you are serious about your job and career.
2. If you *must* be late or absent, call your service manager as soon as possible.
3. Keep busy. If not assigned to a specific job, ask what activities the service manager or supervisor wants you to do.
4. Report any mistakes or accidents *immediately* to your supervisor or team leader. *Never* allow a customer to be the first to discover a mistake.
5. Never lie to your employer or to a customer.
6. Always return any borrowed tools as soon as you are done with them (in *clean* condition). *Show* the person whom you borrowed the tools from that you are returning them to the toolbox or workbench.
7. Keep your work area neat and orderly.
8. Always use fender covers when working under the hood.
9. Double-check your work to be sure that everything is correct.
   a. Remember: "If you are forcing something, you are probably doing something wrong."
   b. Ask for help if unclear as to what to do or how to do it.
10. Do not smoke in a customer's vehicle.
11. Avoid profanity.
12. DO NOT TOUCH THE RADIO! If the radio is turned on and prevents you from hearing noises, turn the volume down. Try to return the vehicle to the owner with the radio at the same volume as originally set.

NOTE: Some shops have a policy that requires employees to turn the radio off.

13. Keep yourself neatly groomed including:
    a. Shirt tail tucked into pants
    b. Daily bathing and use of deodorant
    c. Clean hair, regular haircuts, and hair tied back if long
    d. Men: daily shave or keep beard and/or mustache neatly trimmed
    e. Women: makeup and jewelry kept to a minimum

PHOTO SEQUENCE   How to Set Pads and Safely Hoist a Vehicle

PS1–1 The first step in hoisting a vehicle is to properly align the vehicle in the center of the stall.

PS1–2 Most vehicles will be correctly positioned when the left front tire is centered on the tire pad.

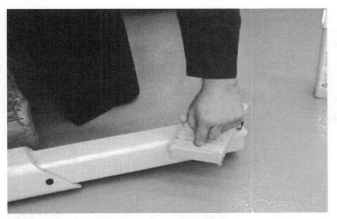

PS1–3 Most pads at the end of the hoist arms can be rotated to allow for many different types of vehicle construction.

PS1–4 The arms of the lifts can be retracted or extended to accommodate vehicles of many different lengths.

PS1–5 Most lifts are equipped with short pad extensions that are often necessary to use to allow the pad to contact the frame of a vehicle without causing the arm of the lift to hit and damage parts of the body.

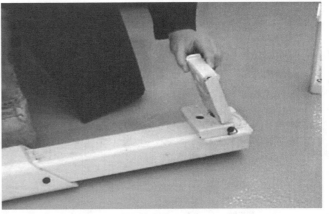

PS1–6 Tall pad extensions can also be used to gain access to the frame of a vehicle. This position is needed to safely hoist many pickup trucks, vans, and sport utility vehicles (SUVs).

How to Set Pads and Safely Hoist a Vehicle—continued

PS1–7 An additional extension may be necessary to hoist a truck or van equipped with running boards to give the necessary clearance.

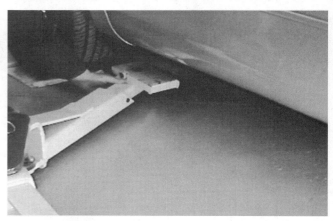

PS1–8 Position the front hoist pads under the recommended locations as specified in the owner's manual and/or service information for the vehicle being serviced.

PS1–9 Position the rear pads under the vehicle under the recommended locations.

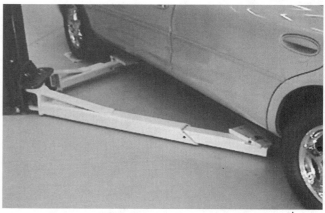

PS1–10 This photo shows an asymmetrical lift where the front arms are shorter than the rear arms. This design is best used for passenger cars and allows the driver to exit the vehicle easier because the door can be opened wide without it hitting the vertical support column.

PS1–11 After being sure all pads are correctly positioned, use the electromechanical controls to raise the vehicle.

PS1–12 Raise the vehicle about 1 ft (30 cm) and stop to double-check that all pads contact the body or frame in the correct positions.

PS1–13 With the vehicle raised about 1 ft off the ground, push down on the vehicle to check to see if it is stable on the pads. If the vehicle rocks, lower the vehicle and reset the pads. If the vehicle is stable, the vehicle can be raised to any desired working level. Be sure the safety is engaged before working on or under the vehicle.

PS1–15 Where additional clearance is necessary for the arms to clear the rest of the body, the pads can be raised and placed under the pinch weld area as shown.

PS1–17 After lowering the vehicle, be sure all arms of the lift are moved out of the way before driving the vehicle out of the work stall.

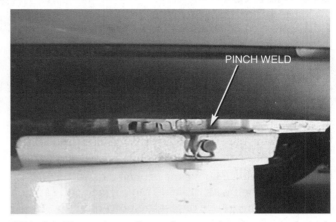

PS1–14 The pads set flat and contacting the vertical pinch welds of the body. This method spreads the load over the entire length of the pad and is less likely to dent or damage the pinch weld area.

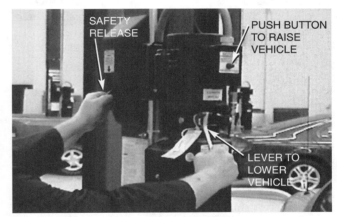

PS1–16 When the service work is completed, the hoist should be raised slightly and the safety released before using the hydraulic lever to lower the vehicle.

PS1–18 Carefully back the vehicle out of the stall. Notice that all of the lift arms have been neatly moved out of the way to provide clearance so that the tires will not contact the arms when the vehicle is driven out of the stall.

■ SUMMARY

1. Vehicle chassis designs include frame, unit-body, and space frame construction. A full-frame vehicle is often stronger and quieter, and permits the towing of heavier loads. Unit-body and space frame construction is often lighter and more fuel efficient.

2. Bolts, studs, and nuts are commonly used as fasteners in the chassis. Sizes for fractional and metric threads are different and are not interchangeable. Grade is the rating of the strength of a fastener.

3. Lifting a vehicle above the ground can be accomplished with drive-on ramps, jack and safety stands, or hydraulic or electric-powered lifts.

4. Whenever a vehicle is raised above the ground, it must be supported at a substantial section of the body or frame.

5. Hazardous materials include common automotive chemicals, liquids, and lubricants, especially those whose ingredients contain "chlor" or "fluor" in their names. Asbestos fibers should be avoided and removed according to current laws and regulations.

6. Certification and a valid driver's license are important for any service technician.

■ REVIEW QUESTIONS

1. List two advantages and two disadvantages of frame and unit-body chassis construction.

2. List three necessary precautions whenever hoisting (lifting) a vehicle.

3. List five common automotive chemicals or products that may be considered hazardous materials.

4. List five precautions every technician should take when working with automotive products and chemicals.

■ ASE CERTIFICATION-TYPE QUESTIONS

1. Two technicians are discussing the hoisting of a vehicle. Technician A says to put the pads of a lift under a notch at the pinch weld of a unit-body vehicle. Technician B says to place the pads on four corners of the frame of a full-frame vehicle. Which technician is correct?
   a. Technician A only
   b. Technician B only
   c. Both Technician A and B
   d. Neither Technician A nor B

2. The correct location for hoisting or jacking the vehicle can often be found in the
   a. Service manual
   b. Shop manual
   c. Owner's manual
   d. All of the above

3. What does the "24" mean in a bolt labeled size 5/16–24?
   a. Diameter in millimeters
   b. Length in millimeters
   c. Number of threads per inch
   d. Distance between the threads crests in millimeters

4. To add strength, yet save weight, many chassis and body parts are made from
   a. Zinc
   b. HSS
   c. Galvanized steel
   d. Zincrometal

5. For best heavy-duty towing, the tow vehicle should have what type construction?
   a. Space frame
   b. Unit-body
   c. Frame
   d. Body

6. For the best working position, the work should be at
   a. Neck or head level
   b. Knee or ankle level
   c. Overhead about 1 ft
   d. Chest or elbow level

7. When working with hand tools, always
   a. Push the wrench—don't pull toward you
   b. Pull a wrench—don't push a wrench

8. A high-strength bolt is identified by
   a. A UNC symbol
   b. Lines on the head
   c. Strength letter codes
   d. The coarse threads

9. The size of a bolt is determined by its length and the distance across
   a. The head of the bolt
   b. The threaded portion of the bolt

10. To determine if a shop material is hazardous, the service technician should check the
    a. Vehicle service manual
    b. Vehicle owner's manual
    c. MSDS
    d. Label

Tires and Wheels

Objectives: After studying Chapter 2, the reader should be able to:

1. List the various parts that make up a tire.
2. Explain how a tire is manufactured.
3. Discuss tire sizes and ratings.
4. Describe tire purchasing considerations and maintenance.
5. Explain the construction and sizing of steel and alloy wheels.
6. Define sprung and unsprung weight.
7. Demonstrate the correct lug nut tightening procedure and torque.

The *friction (traction) between the tire and the road determines the handling characteristics of any vehicle.* Think about this statement for a second. The compounding, construction, and condition of tires is one of the most important aspects of the steering, suspension, alignment, and braking systems of any vehicle! A vehicle that handles poorly or that pulls, darts, jumps, or steers "funny" may be suffering from defective or worn tires. Understanding the construction of a tire is important for the technician to be able to identify tire failure or vehicle handling problems.

Tires are mounted on wheels that are bolted to the vehicle to provide:

1. Shock absorber action when driving over rough surfaces
2. Friction (traction) between the wheels and the road

All tires are assembled by hand from many different component parts consisting of various rubber compounds, steel, and various types of fabric material. Tires are also available in many different designs and sizes.

■ PARTS OF A TIRE

Tread

Tread refers to the part of the tire that contacts the ground. *Tread rubber* is chemically different from other rubber parts of a tire and is compounded for a combination of traction and tire wear. *Tread depth* is usually 11/32″ deep on new tires (this could vary, depending on manufacturer, from 9/32″ to 15/32″). Figure 2–1 shows a tread depth gauge.

NOTE: A tread depth is always expressed in 1/32s of an inch, even if the fraction can be reduced to 1/16s or 1/8s, etc.

Wear indicators are also called *wear bars*. When tread depth is down to the legal limit of 2/32″, bald strips appear across the tread (see Figure 2–2).

Tie bars are molded into the tread of most all-season-rated tires. These rubber reinforcement bars are placed between tread blocks on the outer tread rows to prevent unusual wear and to reduce tread noise. As the tire wears normally, the tie bars will gradually appear. This should not be mistaken for an indication of excess outer edge wear. A tire tread with what appears to be a solid band across the entire width of the tread is what the service technician should consider the wear bar indicator.

Grooves are large, deep recesses molded in the tread and separating the tread blocks. These grooves are called *circumferential grooves* or *kerfs*. Grooves running sideways across the tread of a tire are called *lateral grooves* (see Figure 2–3).

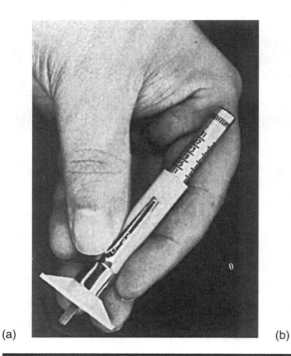
(a)

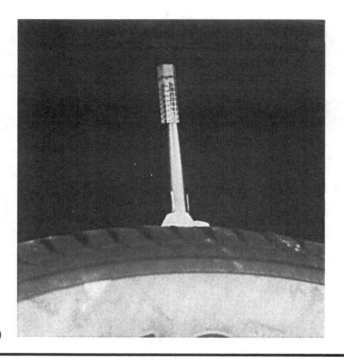
(b)

Figure 2–1 Tread depth gauge. The center of the gauge is pushed down into the groove of the tire and the depth is read at the top edge of the sleeve. Tread depth is usually expressed in 1/32s of an inch.

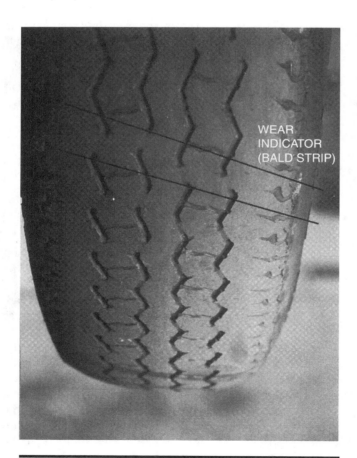

WEAR INDICATOR (BALD STRIP)

Figure 2–2 Wear indicators (wear bars) are strips of bald tread that show when the tread depth is down to 2/32″, the legal limit in many states.

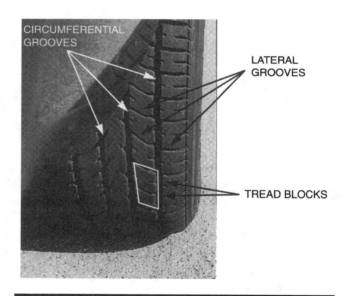

CIRCUMFERENTIAL GROOVES

LATERAL GROOVES

TREAD BLOCKS

Figure 2–3 Circumferential grooves and lateral grooves are the main water channels that help evacuate water from under the tire preventing hydroplaning.

Grooves in both directions are necessary for wet traction. The trapped water can actually cause the tires to ride up on a layer of water and lose contact with the ground, as shown in Figure 2–4. This is called **hydroplaning.** With worn tires, hydroplaning can occur at speeds as low as 30 mph on wet roads. Stopping and cornering is impossible when hydroplaning occurs. **Sipes** are small slits in the tread area to increase wet and dry traction (see Figure 2–5).

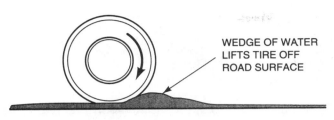

WEDGE OF WATER
LIFTS TIRE OFF
ROAD SURFACE

ROAD SURFACE

Figure 2–4 Hydroplaning can occur at speeds as low as 30 mph (48 km/h). If the water is deep enough and the tire tread cannot evacuate water through its grooves fast enough, the tire can be lifted off the road surface by a layer of water. Hydroplaning occurs at lower speeds as the tire becomes worn.

Sidewall

The sidewall is that part of the tire between the tread and the wheel. The sidewall contains all the size and construction details of the tire.

Some tires turn brown on the sidewalls after a short time. This is due to ozone (atmosphere) damage which actually causes the rubber to oxidize. Premium quality tires contain an antioxidizing chemical additive blended with the sidewall rubber to prevent this discoloration.

White Sidewall When pneumatic tires were first constructed in the early 1900s, only natural rubber was used. The entire tire was white. When it was discovered that **carbon black** greatly increased the toughness of a tire, it was used. The public did not like the change from white tires to black, so tire manufacturers put the carbon black (lamp black) only in the rubber that was to be used for the tread portion of the tire. This tire lasted a lot longer because the black rubber tread that touched the ground was stronger and tougher; it sold well because the sidewalls were white. White sidewall tires are still being manufactured and sold as a styling theme.

White sidewall tires actually contain a strip of white rubber under the black sidewall. This is ground off at the factory to reveal the white rubber. Only tires that are to be whitewalls contain this expensive white rubber. Various widths of whitewalls are made possible simply by changing the width of the grinding wheel.

Bead

The bead is the foundation of the tire and is located where the tire grips the inside of the wheel rim.

1. The bead is constructed of many turns of copper- or bronze-coated steel wire.
2. The main body plies (layers of material) are wrapped around the bead.

> **CAUTION:** If the bead of a tire is cut or damaged, the tire must be replaced!

3. Most radial ply and all truck tires wrap the bead with additional material to add strength.

Body Ply

A tire gets its strength from the layers of material wrapped around both beads under the tread and sidewall rubber. This creates the main framework, or "carcass," of the tire; these body plies are often called **carcass plies.** A four-ply tire has four separate layers of material. If the body plies overlap at an angle (bias), the tire is called a *bias ply* tire. If only one or two body plies are used and they do not cross at an angle, but lie directly from bead to bead, then the tire is called **radial ply** (see Figure 2–6).

1. **Rayon** is a body ply material used in many tires because it provides a very smooth ride. A major disadvantage of rayon is that it rots if exposed to moisture.
2. **Nylon** is a strong body ply material. Still used in some tires, it tends to "flat-spot" after sitting overnight.

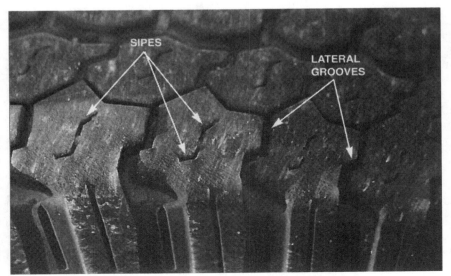

SIPES

LATERAL
GROOVES

Figure 2–5 Sipes are small slits molded into the tread of the tire to aid traction.

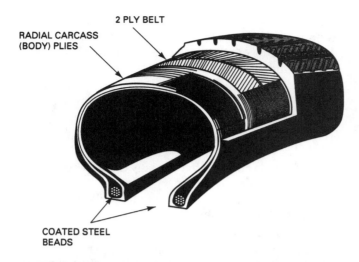

RADIAL CARCASS (BODY) PLIES

2 PLY BELT

COATED STEEL BEADS

Figure 2–6 Typical construction of a radial tire. Some tires have only one body ply, and some tires use more than two belt plies.

3. **Aramid** is the generic name for aromatic polyamide fibers developed in 1972. It is several times stronger than steel (pound for pound) and is used in high-performance tire construction. **Kevlar** is the DuPont brand name for aramid and a registered trademark of E. I. DuPont de Nemours and Co.
4. **Polyester** is the most commonly used tire material because it provides the smooth ride characteristics of rayon with the rot resistance and strength of nylon.

Belt

As shown in Figure 2–6, a tire belt is two or more layers of material applied over the body plies and under the tread area only, to stabilize the tread and increase tread life and handling.

1. Belt material can be:
   a. Steel mesh
   b. Nylon
   c. Rayon

Figure 2–7 The major splice of a tire can often be seen and felt on the inside of the tire. The person who assembles (builds) the tire usually places a sticker near the major splice as a means of identification for quality control.

   d. Fiberglass
   e. Aramid
2. *All radial tires are belted.*
3. A bias-ply tire that has a belt under the tread is called a **bias-belted tire.**

> **NOTE:** Most tires rated for high speed use a nylon "overlay" or "cap belt" between the two-ply belt and the tread of the tire. This overlay helps stabilize the belt package and helps hold the tire together at high speeds, when centrifugal force acts to tear a tire apart.

Inner Liner

The inner liner is the soft rubber lining (usually a butyl rubber compound) on the inside of the tire which protects the body plies and helps provide for self-sealing of small punctures.

Major Splice

When the tire is assembled by a craftsman on a tire-building machine, the body plies, belts, and tread rubber are spliced together. The fabric is overlapped approximately five threads. The point where the majority of these overlaps occur is called the *major splice*, which represents the stiffest part of the tire. This major splice is visible on most tires on the inside, as shown in Figure 2–7.

> **NOTE:** On most new vehicles and/or new tires, the tire manufacturer paints a dot on the sidewall near the bead, indicating the largest diameter of the tire. The largest diameter of the tire usually is near the major splice. The wheel manufacturer either marks the wheel or drills the valve core hole at the smallest diameter of the wheel. Therefore, the dot should be aligned with the valve core or marked for best balance and minimum radial runout.

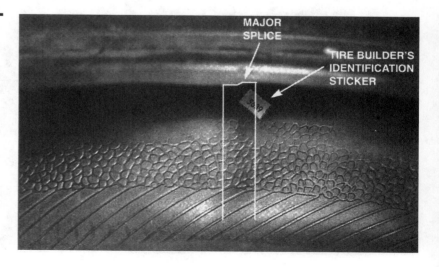

MAJOR SPLICE

TIRE BUILDER'S IDENTIFICATION STICKER

Frequently Asked Question ???

Why Do I Get Shocked by Static Electricity When I Drive a Certain Vehicle?

Static electricity builds up in insulators due to friction of the tires with the road. Newer tires use silica and contain less carbon black in the rubber, which makes the tires electrically conductive. Because the tires cannot conduct the static electricity to the ground, static electricity builds up inside the vehicle and is discharged through the body of the driver and/or passenger whenever the metal door handle is touched.

NOTE: Tollbooth operators report being shocked by many drivers as money is being passed between the driver and the tollbooth operator.

Newer tire sidewall designs that use silica usually incorporate carbon sections that are used to discharge the static electricity to ground. To help reduce the static charge buildup, spray the upholstery with an antistatic spray available at discount and grocery stores.

■ TIRE MOLDING

After the tire has been assembled by the tire builder, it is called a **green tire** (see Figure 2–8). At this stage in construction, the rubber can be returned and reused because it has not been changed chemically. The completed green tire is then placed in a mold where its shape, tread design, and all sidewall markings are formed (see Figure 2–9).

While in the mold, a steam bladder fills the inside of the tire and forces the tire against the outside of the mold. After approximately 30 minutes at 300°F (150°C), the heat changes the chemistry of the rubber. The tire is no longer called a green tire but a *cured tire;* and after inspection and cleaning, it is ready for shipment.

■ TIRE VALVES

All tires use a tire valve, called a **Schrader valve,** to hold air in the tire. The Schrader valve was invented in New York in 1844 by August Schrader for the Goodyear Brothers: Charles, Henry, and Nelson. Today, Schrader valves are used not only as valves in tires but also on fuel injection systems, air-conditioning systems, and air shock (ride control) systems. Most tire experts agree that the valve stem (which includes the Schrader valve) should be replaced whenever tires are replaced: tires can last four or more years, and in that time the valve stem can become brittle and crack. A defective or leaking valve

Figure 2–8 Tire construction is performed by assembling the many parts of a tire together on a tire-building machine.

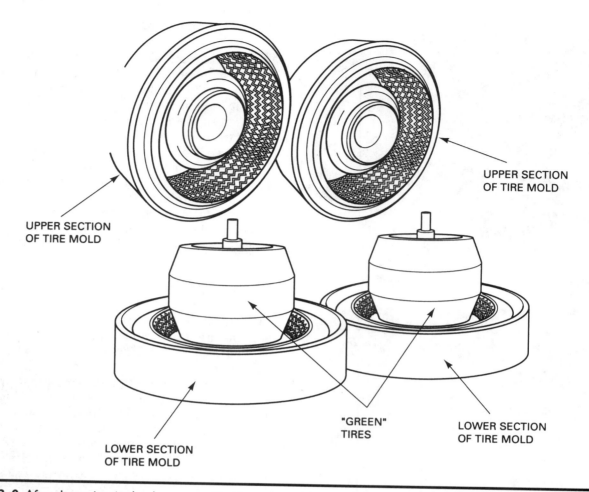

UPPER SECTION
OF TIRE MOLD

UPPER SECTION
OF TIRE MOLD

LOWER SECTION
OF TIRE MOLD

"GREEN"
TIRES

LOWER SECTION
OF TIRE MOLD

Figure 2–9 After the entire tire has been assembled into a completed "green" tire, it is placed into a tire molding machine where the tire is molded into shape and the rubber is changed chemically by the heat. This nonreversible chemical reaction is called vulcanization.

stem is a major cause of air loss. Low tire pressure can cause the tire to become overheated. Replacement valve stems are, therefore, a wise investment whenever purchasing new tires. Aluminum (alloy) wheels often require special metal valve stems that use a rubber washer and are actually bolted to the wheel (see Figure 2–10).

■ OLDER TIRE SIZE DESIGNATION

Older sizes use a numbering method which is still being used on some truck tires, such as the 7.00 × 14 designation:

7.00—the tire is 7″ wide (at its widest cross section)
 14—the tire fits a 14″-diameter wheel

■ METRIC DESIGNATION

European and Japanese tires use metric designations. For example, 185SR × 14:

185—the tire is 185 millimeters (mm) wide (cross-sectional width)
 S—the speed rating (see page 33 or 34 for speed ratings)
 R—radial design
 14—fits a 14″-diameter wheel

The European size indicates the exact physical size (width) of the tire and the speed ratings. Because of the lack of speed limits in many countries, this information is important. Because of tire design changes needed for an H- and V-rated tire, their cost is usually much higher. European sizes also include the tire's aspect ratio, for example, 185/70SR × 14. If the aspect ratio of a European-sized tire is not indicated, it is generally 83 percent for most radials.

American Metric Tire Size Designations

After 1980, American tires were also designated using the metric system. For example, P205/75R × 14:

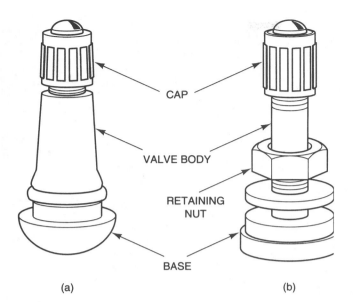

CAP

VALVE BODY

RETAINING
NUT

BASE

(a)   (b)

Figure 2–10 (a) A rubber snap-in style tire valve assembly. (b) A metal clamp-type tire valve assembly used on most high-pressure (over 60 psi) tire applications such as is found on many trucks, RVs, and trailers. The internal Schrader valve threads into the valve itself and can be replaced individually, but most experts recommend replacing the entire valve assembly to help prevent air loss every time the tires are replaced.

HIGH-PERFORMANCE TIP

How Much Bigger Can I Go?

Many owners think they can improve their vehicle by upgrading the tire size over the size that comes from the factory to make their vehicle look sportier and ride and handle better. When changing tire size, there are many factors to consider:

1. The tire should be the same outside diameter as the original to maintain the proper suspension, steering, and ride height specifications.
2. Tire size affects vehicle speed sensor values, ABS brake wheel sensor values that can change automatic transmission operation, and ABS operation.
3. The tire should not be so wide as to contact the inner wheel well or suspension components.
4. Generally, a tire that is 10 mm wider is acceptable. For example, an original equipment tire size 205/75 × 15 (outside diameter = 27.1") can be changed to 215/75 × 15 (outside diameter = 27.6"). This much change is less than 1/2" in width and increases the outside diameter by 1/2".

> **NOTE:** Outside diameter is calculated by adding the wheel diameter to the cross-sectional height of the tire, multiplied by 2 (see Figure 2–11).

5. Whenever changing tires, make sure that the load capacity is the same or greater than that of the original tires.
6. If wider tires are desired, a lower aspect ratio is required to maintain the same, or close to the same, overall outside diameter of the tire.

Old	New
P205/75 × 15	P215/70 × 15
205 × 0.75 = 154 mm	215 × 0.70 = 151 mm

Notice that the overall sidewall height is generally maintained.

If even larger tires are needed, then 225/60 × 15s may be okay—let's check the math:

$$225 \times 0.60 = 135 \text{ mm}$$

Notice that this is much too short a sidewall height when compared with the original tire (see 6).

7. Use the "plus 1" or "plus 2" concept. When specifying wider tires, the sidewall height must be reduced to maintain the same or close to the same as original equipment specifications. The "plus 1" concept involves replacing the wheels with wheels 1 inch larger in diameter to compensate for the lower aspect of wider tires.

Original	Plus 1
205/75 × 15	225/60 × 16

The overall difference in outside diameter is only 0.5≤, even though the tire width has increased from 205 mm to 225 mm and the wheel diameter has increased by 1 inch. If money is no object and all-out performance is the goal, a "plus 2" concept can also be used (use a P245/50 3 17 tire and change to 17≤-diameter wheels). (See Figure 2–11.)

Here the overall diameter is within 1/2≤ of the original tire/wheel combination, yet the tire width is 1.6" (40 mm) wider than the original tire. Refer to the section entitled "Wheels" later in this chapter for proper wheel back spacing and offset when purchasing replacement wheels.

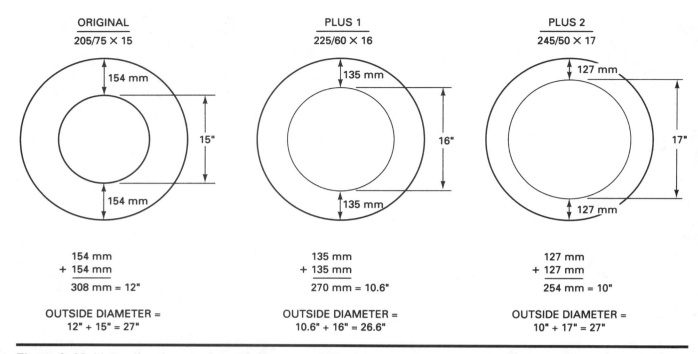

Figure 2–11 Notice that the overall outside diameter of the tire remains almost the same and at the same time the aspect ratio is decreased and the rim diameter is increased.

Frequently Asked Question   **???**

How Much Does Tire Pressure Change with a Change in Temperature?

As the temperature of a tire increases, the pressure inside the tire also increases. The general amount of pressure gain (when temperatures increase) or loss (when temperatures decrease) is as follows:

10°F increase causes 1 psi increase

10°F decrease causes 1 psi decrease

For example, if a tire is correctly inflated to 35 psi when cold and then driven on a highway, the tire pressure may increase 5 psi or more.

CAUTION: DO NOT LET AIR OUT OF A HOT TIRE! If air is released from a hot tire to bring the pressure down to specifications, the tire will be underinflated when the tire has cooled. The tire pressure specification is for a cold tire.

Always check the tire pressures on a vehicle that has been driven fewer than 2 miles (3.2 km).

Air pressure in the tires also affects fuel economy. If all four tires are underinflated (low on air pressure), fuel economy is reduced about **0.1 mile per gallon (mpg) for each 1 psi low.** For example, if all four tires were inflated to 25 psi instead of 35 psi, not only is tire life affected but fuel economy is reduced by about 1 mile per gallon (10 × 0.1 = 1 mpg).

P—passenger vehicle

205—205-mm cross-sectional width

75—75 percent aspect ratio. The height of the tire (from the wheel to the tread) is 75 percent as great as its cross-sectional width (the width measured across its widest part). *This percentage ratio of height to width is called the* **aspect ratio.** (A 60 series tire is 60 percent as high as it is wide.)

R—radial

14—14″-diameter wheel

If a tire is constructed as a bias ply tire only, then its size designation uses the letter *D* to indicate *diagonal:* P205/75D × 14.

NOTE: Many "temporary use only" spare tires are constructed with diagonal (bias) plies; the size designation is T for temporary.

If a tire is constructed as a bias ply with a belt of additional material under the tread area, its size designation uses the letter *B* to indicate *belted:* P205/75B × 14. Some tires use letters at the end of the tire size (suffixes) to indicate special applications including:

LT—light truck

ML—mining and logging

MH—mobile home

ST—special trailer

TR—truck

■ SERVICE DESCRIPTION

Tires built after 1990 use a "service description" method of sidewall information in accordance with ISO 4000 (International Standards Organization) that includes size, load, and speed rating together in one easy-to-read format (see Figure 2–12).

P-Metric Designation	Service Description
P205/75HR × 15	**205/75R × 15 92H**
P passenger vehicle	205 cross-sectional width in mm
205 cross-sectional	75 aspect ratio width in mm
75 aspect ratio	R radial construction
H speed rating (130 mph/210 km/h)	15 rim diameter in inches
R radial construction	92 load index
15 rim diameter in inches	H speed rating (130 mph/210 km/h)

■ HIGH-FLOTATION TIRE SIZES

High-flotation light truck tires are designed to give improved off-road performance on sand, mud, and soft soil and still provide acceptable hard-road surface performance. These tires are usually larger than conventional tires and usually require a wider than normal wheel width. High-flotation tires have a size designation such as 33 × 12.50R × 15LT:

33—approximate overall tire diameter in inches
12.50—approximate cross-sectional width in inches
R—radial-type construction
15—rim diameter in inches
LT—light truck designation

Frequently Asked Question　　　???

If I Have an Older Vehicle, What Size Tires Should I Use?

Newer radial tires can be used on older model vehicles if the size of the tires is selected that best matches the original tires. See the following cross-reference chart. (This chart does not imply complete interchangeability.)

NOTE: Vehicles designed for older bias ply tires may drive differently when equipped with radial tires.

Pre-1964	'65 to '72	80 Series Metric	Alphanumeric 78 Series	P-Metric 75 Series Radial	P-Metric 70 Series Radial
590-13	600-13	165-13	A78-13	P165/75R13	P175/70R13
640-13	650-13	175-13	B78-13	P175/75R13	P185/70R13
725-13	700-13	185-13	D78-13	P185/75R13	P205/70R13
590-14	645-14	155-14	B78-14	P175/75R14	P185/70R14
650-14	695-14	175-14	C78-14	P185/75R14	P195/70R14
700-14	735-14	185-14	E78-14	P195/75R14	P205/70R14
750-14	775-14	195-14	F78-14	P205/75R14	P215/70R14
800-14	825-14	205-14	G78-14	P215/75R14	P225/70R14
850-14	855-14	215-14	H78-14	P225/75R14	P235/70R14
590-15	600-15	165-15	A78-15	P165/75R15	P175/70R15
650-15	685-15	175-15	C78-15	P175/75R15	P185/70R15
640-15	735-15	185-15	E78-15	P195/75R15	P205/70R15
670-15	775-15	195-15	F78-15	P205/75R15	P215/70R15
710-15	815-15	205-15	G78-15	P215/75R15	P225/70R15
760-15	855-15	215-15	H78-15	P225/75R15	P235/70R15
800-15	885-15	230-15	J78-15	P225/75R15	P235/70R15
820-15	900-15	235-15	L78-15	P235/75R15	P255/70R15

Figure 2–12 Cross-sectional view of a typical tire showing the terminology.

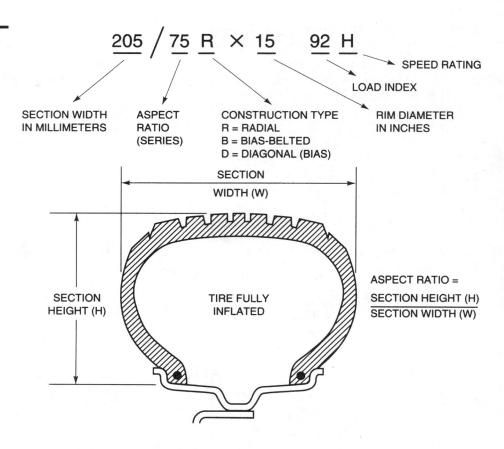

LOAD INDEX AND EQUIVALENT LOADS

The load index, as shown in Figure 2–13, is an abbreviated method to indicate the load-carrying capabilities of

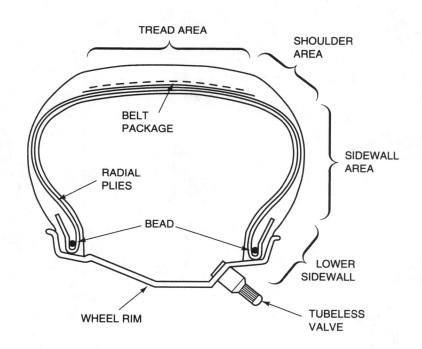

a tire. The weights listed in the chart represent the weight that *each tire* can safely support. Multiply this amount by four to get the maximum that the vehicle should weigh fully loaded with cargo and passengers.

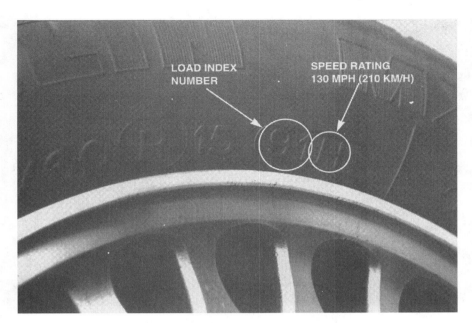

Figure 2–13 Typical sidewall markings for load index and speed rating following the tire size.

Load Index	Load (kg)	Load (lb)
75	387	853
76	400	882
77	412	908
78	425	937
79	437	963
80	450	992
81	462	1019
82	475	1047
83	487	1074
84	500	1102
85	515	1135
86	530	1168
87	545	1201
88	560	1235
89	580	1279
90	600	1323
91	615	1356
92	630	1389
93	650	1433
94	670	1477
95	690	1521
96	710	1565
97	730	1609
98	750	1653
99	775	1709
100	800	1764
101	825	1819
102	850	1874
103	875	1929
104	900	1934
105	925	2039
106	950	2094
107	975	2149
108	1000	2205

Load Index	Load (kg)	Load (lb)
109	1030	2271
110	1060	2337
111	1090	2403
112	1120	2469
113	1150	2535
114	1180	2601
115	1215	2679

■ SPEED RATINGS

Tires are rated according to the maximum *sustained* speed. A vehicle should never be driven faster than the speed rating of the tires.

> **CAUTION:** A high-speed rating does not guarantee that the tires will not fail, even at speeds much lower than the rating. Tire condition, inflation, and vehicle loading also affect tire performance.

As the speed rating of a tire increases, fewer compromises exist for driver comfort and low noise level. The higher speed rating does not mean a better tire. To survive, a high-speed tire must be built with stiff tread compounds, reinforced body (carcass) construction, and fabric angles that favor high speed and high performance over other considerations. For example, a V-rated tire often has less tread depth than a similar tire with an H speed rating and, therefore, will often not give as long a service life. Since the speed ratings were first developed in Europe, the letters correspond to metric speed in kilometers per hour, with a conversion to miles per hour.

Frequently Asked Question ???

What Effect Does Tire Size Have on Overall Gear Ratio?

Customers often ask what effect changing tire size has on fuel economy and speedometer readings. If larger (or smaller) tires are installed on a vehicle, many other factors change also. These include:

1. *Speedometer reading.* **If larger diameter tires are used, the speedometer will read slower** than you are actually traveling. This can result in speeding tickets!
2. *Odometer reading.* Even though larger tires are said to give better fuel economy, just the opposite can be calculated! Since a larger diameter tire travels farther than a smaller diameter tire, the larger tire will cause the odometer to read a shorter distance than the vehicle actually travels. For example, if the odometer reads 100 miles traveled

on tires that are 10 percent oversized in circumference, then the actual distance traveled is 110 miles.

3. *Fuel economy.* If fuel economy is calculated on miles traveled, the result will be *lower* fuel economy than for the same vehicle with the original tires.

$$\text{Calculation: mph} = \frac{\text{rpm} \times \text{diameter} \times 3.14}{\text{gear ratio}}$$

$$\text{rpm} = \frac{\text{mph} \times \text{gear ratio}}{\text{diameter} \times 3.14}$$

$$\text{gear ratio} = \frac{\text{rpm} \times \text{diameter} \times 3.14}{\text{mph}}$$

Letter	Maximum Rated Speed
L	120 km/h (75 mph)
M	130 km/h (81 mph)
N	140 km/h (87 mph)
P	150 km/h (93 mph)
Q	160 km/h (99 mph)
R	170 km/h (106 mph)
S	180 km/h (112 mph)
T	190 km/h (118 mph)
U	200 km/h (124 mph)
H	210 km/h (130 mph)
V	240 km/h (149 mph)
W	270 km/h (168 mph)
Y	300 km/h (186 mph)
Z	open-ended*

*The exact speed rating for a particular Z-rated tire is determined by the tire manufacturer and may vary according to size. For example, not all Brand X Z-rated tires are rated at 170 mph, even though one size may be capable of these speeds.

■ TIRE PRESSURE AND TRACTION

All tires should be inflated to the specifications given by the vehicle manufacturer. Most vehicles have recommended tire inflation figures written in the owner's manual or on a placard or sticker on the door post or glove compartment (see Figure 2–14).

The pressure number molded into the sidewall of a tire should be considered the maximum pressure when the tire is cold. (Pressures higher than that stated on the sidewall may be measured on a hot tire.)

■ TIRE CONICITY AND PLY STEER

Tire conicity can occur during the construction of any radial or belted tire when the parts of the tire are badly positioned, causing the tire to be smaller in diameter on one side. When this tire is installed on a vehicle, it can

Figure 2–14 The proper air pressure is often printed on a decal or placard on the edge of the driver's door or post. The owner's manual also includes the recommended tire pressure.

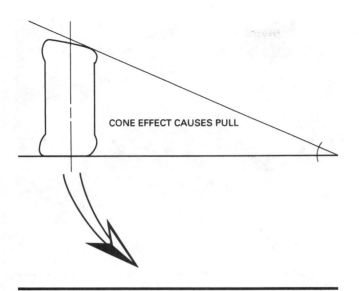

CONE EFFECT CAUSES PULL

Figure 2–15 Conicity is a fault in the tire that can cause the vehicle to pull to one side due to the cone effect (shape) of the tire.

cause the vehicle to pull to one side of the road due to the cone shape of the tire (as shown in Figure 2–15).

Since the cause of conicity is due to the construction of the tire itself, there is nothing the service technician can do to correct the condition. The exact cause of the conicity is generally due to the slight movement of the belt and tread in the mold during inflation. If a vehicle pulls to one side of the road, the service technician should switch tires left to right (left-side tires to the right side and the right-side tires to the left side of the vehicle). If this swap of tires corrects the pulling condition, then tire conicity was the possible cause.

> **NOTE:** *Radial pull* or *radial tire pull* are other terms often used to describe tire conicity.

Ply steer is another term that describes a slight pulling force on a vehicle due to tire construction. Ply steer is due to the angle of the cords in the belt layers, as seen in Figure 2–16, and not in the ply layer of the carcass (body), as the name implies. Ply steer will cause a slight drift regardless of its direction of rotation. Switching tires left to right will *not* correct a ply steer condition.

> **NOTE:** Whenever a wheel and tire assembly are switched from one side of a vehicle to the other, the tire revolves in the opposite direction.

Ply steer is built into the tire during construction. There is nothing a service technician can do to correct ply steer, except to compensate for it with alignment angles. (See Chapter 12 for details on alignment diagnosis and correction.)

■ VEHICLE HANDLING AND TIRE SLIP ANGLE

The tire surface contact area or tire patch size is about one-half the area of one page of this book. All accelerating, braking, and cornering forces of a vehicle are transferred to the pavement at just four spots. The combined area of these four spots is about equal to the size of this opened book.

Think about this whenever you are braking or cornering. As a vehicle is turned, the wheels are moved while the tires remain in contact with the road. These actions "twist" the carcass of the tire and create a slip angle between the direction the wheel is pointing and the direction the tread is pointing (as shown in Figure 2–17).

The contact patch is deformed and snaps back into place as the twisted tire carcass returns to its original shape. This movement causes a slight delay in the turn-

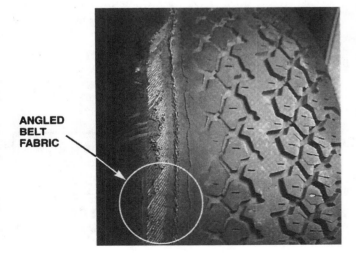

ANGLED BELT FABRIC

Figure 2–16 Notice the angle of the belt material in this worn tire. The angle of the belt fabric can cause a "ply steer" or slight pulling force toward one side of the vehicle.

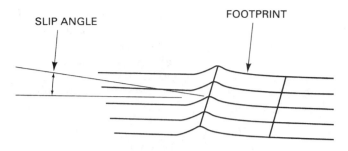

Figure 2–17 Slip angle is the angle between the direction the tire tread is heading and the direction it is pointed.

ing of the vehicle and causes tire wear as the tread rubber moves in relation to the pavement during cornering.

■ RIM WIDTH AND TIRE SIZE

As a general rule, for a given rim width it is best not to change tire width more than 10 mm (either wider or narrower). For a given tire width, it is best not to vary rim width more than 1/2″ in either direction. For example, if the original tire size is 195/70 × 14, then either a 185/70 × 14 (−10 mm) or a 205/70 × 14 (+10 mm) *could* be used on the original rim (wheel) without undue harm *if* the replacement tire has proper clearance with the body and suspension components.

Installing a tire on too narrow a wheel will cause the tire to wear excessively in the center of the tread. Installing a tire on too wide a wheel will cause excessive tire wear on both edges.

See a tire store representative for recommended tire sizes that can be safely installed on your rims.

■ UNIFORM TIRE QUALITY GRADING SYSTEM

The **U.S. Department of Transportation (DOT)** and the **National Highway Traffic Safety Administration (NHTSA)** developed a system of tire grading to help customers better judge the relative performance of tires. The three areas of tire performance are tread wear, traction, and temperature resistance, as shown in Figure 2–18.

> **NOTE:** All tires sold in the United States must have uniform tire quality grading system ratings molded into the sidewall.

Tread Wear

The tread wear grade is a comparison rating based on the wear rate of a standardized tire, tested under carefully controlled conditions, which is assigned a value of 100. A tire rated 200 should have a useful life twice as long as the standard tire's life.

Figure 2–18 Typical "Uniform Tire Quality Grading System" (UTQGS) ratings as imprinted on a tire sidewall.

> **HINT:** The standard tire has a rating for tread wear of 100. This value has generally been accepted to mean a useful life of 20,000 miles of normal driving. Therefore, a tire rated at 200 could be expected to last 40,000 miles, etc.

Tread Wear Rating Number	Approximate Number of mi/km
100	20,000/32,000
150	30,000/48,000
200	40,000/64,000
250	50,000/80,000
300	60,000/96,000
400	80,000/129,000
500	100,000/161,000

The tread wear life of any tire is affected by driving habits (fast stops, starts, and cornering will decrease tread life), tire rotation (or lack of tire rotation), inflation, wheel alignment, road surfaces, and climate conditions.

Traction

Traction performance is rated by the letters *A*, *B*, or *C*, with A the highest.

> **IMPORTANT NOTE:** The traction rating is for **wet braking** distance only! It does not include cornering traction or dry braking performance.

The traction rating is only one of many other factors that affect wet braking traction, including air inflation, tread depth, vehicle speed, and brake performance.

Temperature Resistance

Temperature resistance is rated by the letters *A*, *B*, or *C*, with A the highest rating. Tires generate heat while rotating and flexing during normal driving conditions. A certain amount of heat buildup is desirable because tires produce their highest coefficient of traction at normal operating temperatures. For example, race car drivers frequently swerve their cars left and right during the pace laps, causing increased friction between the tire and the road surface, which warms their tires to operating temperature. However, if temperatures rise too much, a tire can start to come apart—the oils and rubber in the tire start to become a liquid! Grade C is the minimum level that all tires must be able to pass under the current Federal Motor Vehicle Safety Standard No. 109.

■ ALL-SEASON TIRE DESIGNATION

Most all-season tires are rated and labeled as *M & S*, *MS*, or *M + S*, and therefore must adhere to general design features as specified by the Rubber Manufacturers Association (RMA).

Tires labeled M & S are constructed with an aggressive tread design as well as tread compounds and internal construction that are designed for mud and snow. One design feature is that the tire has at least a 25 percent void area. This means that the tread blocks have enough open space around them to allow the blocks to grab and clean themselves of snow and mud. Block angles, dimensional requirements, and minimum cross-sectional width are also a requirement for the M & S designation.

The tread rubber used to make all-season tires is also more flexible at low temperatures. This rubber compound is low-bounce (called **high-hysteresis**) and is more likely to remain in contact with the road surface. The rubber compound is also called *hydrophilic*, meaning that the rubber has an affinity for water (rather than being *hydrophobic* rubber, which repels water).

> **NOTE:** Most vehicle manufacturers recommend that the same *type* of tire be used on all four wheels even though the size of the tire may vary front and rear on some high-performance vehicles. Therefore, if all-season replacement tires are purchased, a complete set of four should be used to be assured of proper handling and uniform traction characteristics. While *tire* manufacturers have been recommending this for years—since the late 1980s—most *vehicle* manufacturers are also recommending that all four tires be of the same construction and tread type to help ensure proper vehicle handling.

Figure 2–19 Typical DOT date code. The last three numbers (227) indicate the week and year the tire was built—22nd week of 1997.

■ DOT TIRE CODE

All tires sold in the United States must be approved by the U.S. Federal Department of Transportation (DOT). DOT tire requirements include resistance to tire damage that could be caused by curbs, chuckholes, and other common occurrences for a tire used on public roads.

> **NOTE:** Most race tires are *not* DOT approved and must never be used on public streets or highways.

Each tire that is DOT approved has a DOT number molded into the sidewall of the tire as shown in Figure 2–19. This number is usually imprinted on only one side of the tire and is usually on the side *opposite the whitewall*. The DOT code includes letters and numbers such as MJP2CBDX264.

The first two letters identify the manufacturer and location. For this example, the first two letters (MJ) mean that the tire was made by the Goodyear Tire and Rubber Company in Topeka, Kansas, USA. The last three numbers are the build date code. The last of these three numbers is the year (1994), and the 26 means that it was built during the 26th week of 1994. The last number is the same for 1984 and 2004, but the style and design of the tire usually change enough after ten years so that it is often easy to correctly identify the decade in which the tire was built.

> **NOTE:** Starting with tires manufactured after January 1, 2000, the tire build date will include four digits rather than three digits. The new code such as "3498" would distinguish the 34th week of 1998 from the 34th week of 2008 ("3408").

Diagnostic Story

Tire Date Code Information Saved Me Money!

This author was looking at a three-year-old vehicle when I noticed that the right rear tire had a build date code newer than the vehicle. I asked the owner, "How badly was this vehicle hit?" The owner stumbled and stuttered a little, then said, "How did you know that an accident occurred?" I told the owner that the right rear tire, while the exact same tire as the others, had a date code indicating that it was only one year old, whereas the original tires were the same age as the vehicle. The last three numbers of the DOT code on the sidewall indicate the week of manufacture (the first two numbers of the three-digit date code) followed by the last number of the year.

The owner immediately admitted that the vehicle slid on ice and hit a curb damaging the right rear tire and wheel. Both the tire and wheel were replaced and the alignment checked. The owner then dropped the price of the vehicle $500! Knowing the date code helps you be sure that you purchase fresh tires and can also help the technician determine if the tires have been replaced. For example, if new tires are found on a vehicle with 20,000 miles, then the technician should check to see if the vehicle may have been involved in an accident or have more miles than indicated on the odometer.

See Appendix 3 in the back of the book for the alphabetical listing of all two-letter manufacturers' DOT tire codes.

■ SPARE TIRES

Most vehicles today come equipped with space-saver spare tires that are smaller than the wheels and tires that are on the vehicle. The reason for the small size is to reduce the size and weight of the entire vehicle and to increase fuel economy by having the entire vehicle weigh less by not carrying a heavy spare tire and wheel around. The style and type of these spare tires have changed a great deal over the last several years, and different makes and types of vehicles use various types of spare tires.

> **CAUTION:** Before using a spare tire, always read the warning label (if so equipped) and understand all use restrictions. For example, some spare tires are not designed to exceed 50 mph (80 km/h) or be driven more than 500 miles (800 km).

Many small space-saving spare tires use a higher than normal air inflation pressure, usually 60 psi (414 kPa). Even though the tire often differs in construction, size, diameter, and width from the vehicle's original tires, it is amazing that the vehicle usually handles the same during normal driving. Obviously, these tires are not constructed with the same durability as a full-size tire and should be removed from service as soon as possible.

> **NOTE:** When was the last time you checked the tire pressure in your spare tire? The spare tire pressure should be checked regularly. See Figure 2–20 for a photo of an extension hose that makes checking and adding air to a spare tire much easier.

Figure 2–20 Here is a clever hint that helps the vehicle owner check the air pressure in the spare tire easily—the extension hose extends from the spare tire through the cover panel in the trunk.

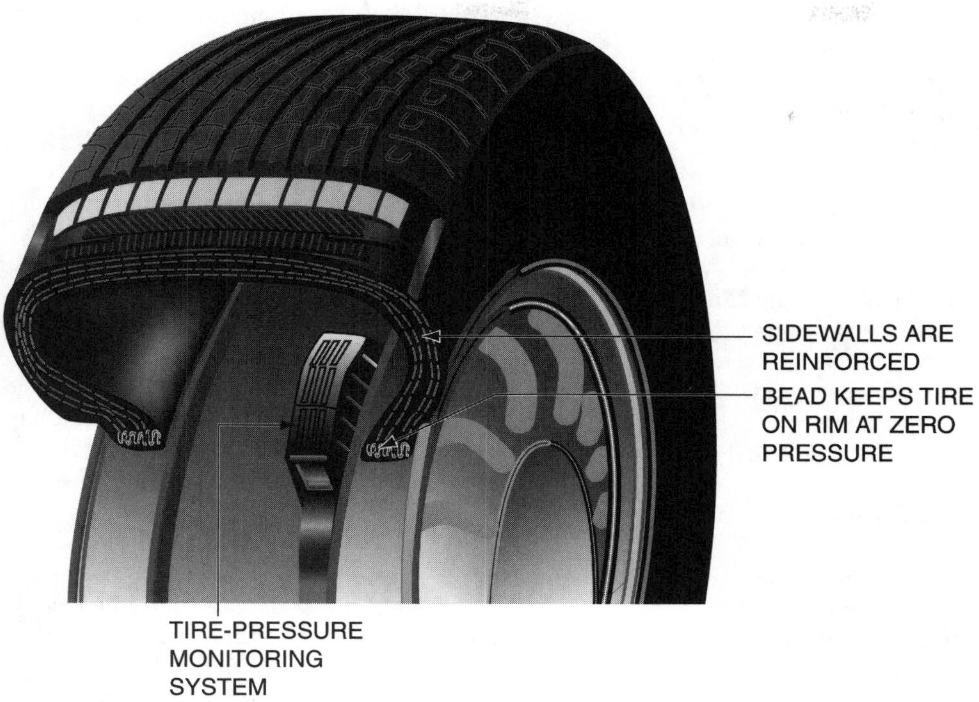

SIDEWALLS ARE REINFORCED

BEAD KEEPS TIRE ON RIM AT ZERO PRESSURE

TIRE-PRESSURE MONITORING SYSTEM

Figure 2–21 Cutaway of a run-flat tire showing the reinforced sidewalls and the required pressure sensor.

■ RUN-FLAT TIRES

Run-flat tires are designed to operate without any air for a limited distance (usually 50 miles at 55 mph). This feature allows vehicle manufacturers to build vehicles without the extra room and weight of a spare tire and jack assembly.

A typical run-flat tire (also called **extended mobility tire [EMT]** or **zero pressure [ZP] tire**) requires the use of an air pressure sensor/transmitter and a dash-mounted receiver to warn the driver that a tire has lost pressure. Because of the reinforced sidewalls, the vehicle handles almost the same with or without air pressure (see Figure 2–21).

> **CAUTION:** Tire engineers warn that rapid cornering should be avoided if a run-flat tire has zero air pressure. The handling during quick maneuvers is often unpredictable and could be dangerous.

■ TIRE SELECTION CONSIDERATIONS

Selecting the proper tire is very important for the proper handling and safety of any vehicle. Do not select a tire by styling or looks alone. For best value and highest satisfaction, follow these guidelines:

Purchasing Suggestions

1. Purchase the same type of tire that came on your vehicle when new.

2. Purchase the same size as the original tire. The width of the tire should be within 10 mm of the width of the original tire. For example, a stock 195/75 × 14 tire acceptable replacement *could be* 185/75 × 14 or 205/75 × 14.

3. Purchase tires with the same speed rating as the original.

Reason Why

1. Chassis and tire engineers spend hundreds of hours testing and evaluating the best tire to use for each vehicle.

2. The size of the tire is critical to the handling of any vehicle. Tire width, size, and aspect ratio affect the following:
   a. Braking effectiveness
   b. Headlight aiming
   c. Vehicle height
   d. Acceleration potential
   e. Speedometer calibration

3. When any vehicle is manufactured, it is optimized for designed use. High-performance tires are generally stiffer and have a speed rating as well. If you purchase non-high-performance tires, the carcass is not as stiff and the suspension is not designed to work with softer tires. Therefore, the cornering and handling, especially fast evasive maneuvers, could be dangerous. The vehicle may be capable of far

4. Purchase four of the same type of tire. Most vehicle manufacturers recommend against installing snow tires or all-season tires on just the drive wheels.

5. Purchase the same *brand* of tire for both front and/or both rear wheels.

more speed than the tires are designed to handle.

4. Every vehicle is designed to function best with four tires of the same size, construction, and tread design unless specifically designed for different sizes of tires front and rear. Different tire styles and tread compounds have different slip angles. It is these different slip angles that can cause a vehicle to handle "funny" or cause the vehicle to get out of control in the event of a sudden maneuver.

5. The sizes of tires are nominal and vary according to exact size (and shape) of the

6. Purchase fresh tires. Tires that are older may have been stored in a hot warehouse where oxygen can attack rubber and cause deterioration.

mold as well as tire design and construction. The same size tire from two different manufacturers is often different in diameter and width. Although the differences should be slight, many vehicles are extremely sensitive to these small differences, and poor vehicle handling, torque steer, and pulling could result if different brands of tires are used.

6. Look at the tire build date code (the last three numbers of the DOT code). Try and purchase four tires with the same or similar date codes.

Frequently Asked Question ???

What Width Rim Should Be Used with What Size Tire?

For best overall vehicle operation and handling, the same size tires should be used that come on the vehicle when new. If a change of tire size is necessary, refer to the tire guide available at most stores that sell or service tires for the recommended alternative size that will fit the width of the rim.

NOTE: Interchangeability does not mean that it is always possible because of differences in rim size, wheel well clearance, and load ratings.

P-75 Series Metric	83 Series Bias	P-70 Series Metric	80 Series Metric	78 Series Alpha Numeric	Rim Width (inch)
P195/75R14	700-14, 735-14	P205/70R14	185-14	E78-14	5–7"
P205/75R14	750-14, 775-14	P215/70R14	195-14	F78-14	5–7.5"
P215/75R14	800-14, 825-14	P225/70R14	205-14	G78-14	5.5–7.5"
P225/75R14	850-14, 855-14	P235/70R14	215-14	H78-14	6.8"
P195/75R15	640-15, 735-15	P205/70R15	175-15	E78-15	5–7"
P205/75R15	670-15, 775-15	P215/70R15	185-15	F78-15	5–7.5"
P215/75R15	710-15, 825-15	P225/70R15	205-15	G78-15	5.5–7.5"
P225/75R15	760-15, 855-15	P235/70R15	215-15	H78-15	6–8"
P235/75R15	820-15, 900-15	P255/70R15	230-15	L78-15	6–8.5"

CAUTION: If changing tire sizes or styles beyond the recommendations as stated here, consult a knowledgeable tire store representative for help in matching wheel and tire combinations to your vehicle.

NOTE: Some tires may be five or more years old when purchased. Always check the date code!

■ WHEELS

The concept of a wheel has not changed in the last 5000 years, but the style and materials used have changed a lot. Early automotive wheels were constructed from wood with a steel band as the tire.

Today's wheels are constructed of steel or aluminum alloy. The center section of the wheel that attaches to the hub is called the **center section** or **spider** because early wheels used wooden spokes that resembled a spider's web. The rubber tire attaches to the rim of the wheel. The rim has two *bead flanges* where the bead of the tire is held against the wheel when the tire is inflated. The shape of this flange is very important and is designated by Tire and Rim Association letters. For example, a wheel designated 14 × 6JJ means that the diameter of the wheel is 14″ and is 6″ wide measured from inside to inside of the flanges. The letters *JJ* indicate the *exact* shape of the flange area (see Figure 2–22). This flange area shape and the angle that the rim drops down from the flange are important because:

- They permit a good seal between the rim and the tire.
- They help retain the tire on the rim in the event of loss of air. This is the reason why modern wheels are called a "safety rim wheel."
- Run-flat tires (tires that are designed to operate without air for a limited distance without damage) often require a specific wheel rim shape.

Wheel Offset

Offset is a very important variable in wheel design. If the center section (spider) is centered on the outer rim, the offset is zero (see Figure 2–23).

Positive Offset

The wheel has a positive offset if the center section is outward from the wheel centerline. Front-wheel-drive vehicles commonly use positive offset wheels to improve the loading on the front wheels and to help provide for a favorable scrub radius.

Negative Offset

The wheel has a negative offset if the center section is inboard (or "dished") from the wheel centerline (see Figure 2–24). Avoid using replacement wheels that differ from the original offset. See Chapter 11 for details on scrub radius and the effect that wheel offset has on steering and suspension geometry.

Back Spacing

Back spacing, also called **rear spacing** or **backside setting,** is the distance between the back rim edge and the wheel center section mounting pad. **This is not the same as offset.** Back spacing can be measured directly with a ruler, as shown in Figure 2–25 on page 43.

Figure 2–22 The size of the wheel is usually cast or stamped into the wheel. This wheel is 5 1/2″ wide. The letters "JJ" refer to the contours of the seat area of the wheel.

TECH TIP

Easy Method to Determine the Bolt Circle of a Five-Lug Wheel

An easy method to determine the approximate bolt circle of a five-lug bolt circle wheel is to use a tape measure and place the end of the tape over the edge of one lug. Then measure the distance to the center of the opposite bolt (see Figure 2–25 on page 43).

Figure 2–23 A cross section of a wheel showing part designations.

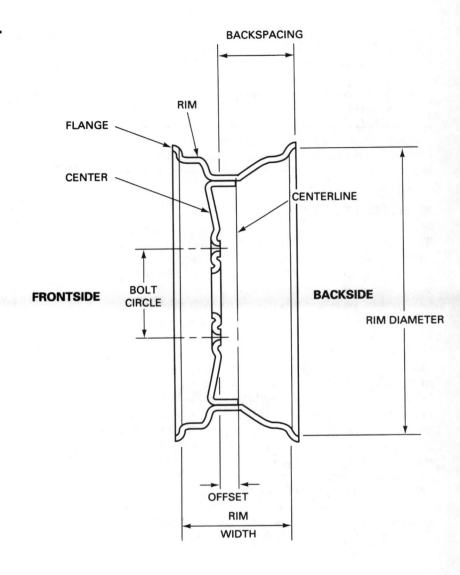

Figure 2–24 Offset is the distance between the centerline of the wheel and the wheel mounting surface.

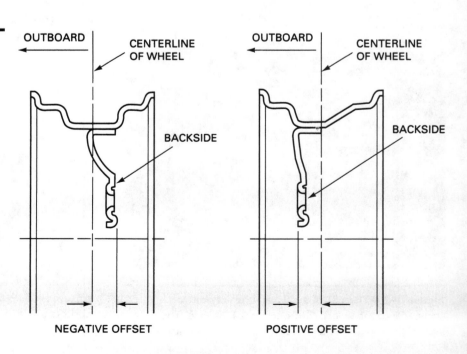

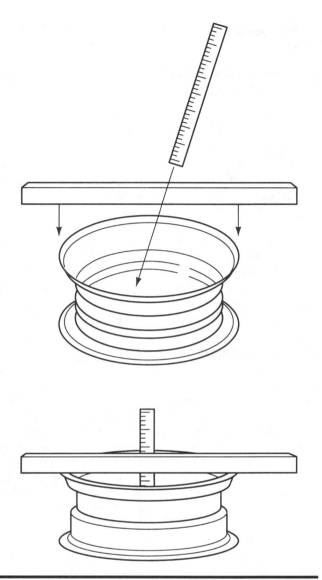

Figure 2–25 Back spacing (rear spacing) is the distance from the mounting pad to the edge of the rim. Most custom wheels use this measurement method to indicate the location of the mounting pad in relation to the rim.

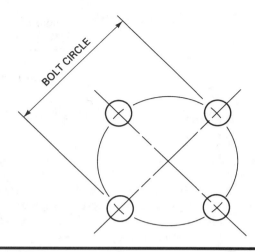

Figure 2–26 Bolt circle is the diameter of a circle that can be drawn through the center of each lug hole or stud.

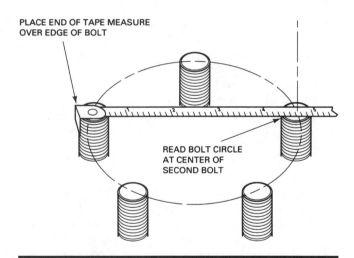

PLACE END OF TAPE MEASURE
OVER EDGE OF BOLT

READ BOLT CIRCLE
AT CENTER OF
SECOND BOLT

Figure 2–27 The easiest method to determine the approximate bolt circle of a five-lug bolt circle wheel.

Determining Bolt Circle

On four-lug axles and wheels, the measurement is simply taken from center to center on opposite studs or holes, as shown in Figure 2–26.

On five-lug axles and wheels, it is a little harder. One method is to measure from the far edge of one bolt hole to the center of the hole two over from the first, as shown in Figure 2–27. Another method for a five-lug wheel is to measure from center to center between two adjacent studs and convert this measurement into bolt circle diameter, as in the chart below (see Figure 2–28).

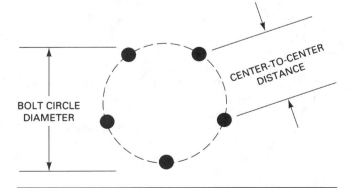

BOLT CIRCLE
DIAMETER

CENTER-TO-CENTER
DISTANCE

Figure 2–28 Measure center-to-center distance and compare the distance with the figures in the chart in the text to determine the diameter for a five-lug bolt circle.

Center-to-Center Distance	Bolt Circle Diameter
2.645″	4 1/2″ bolt circle
2.792″	4 3/4″ bolt circle
2.939″	5″ bolt circle
3.233″	5 1/2″ bolt circle

Steel Wheels

Steel is the traditional wheel material. A steel wheel is very strong due to its designed shape and the fact that it is work hardened during manufacturing. In fact, most of the strength of a steel wheel is due to its work hardening. Painting and baking cycles also increase the strength of a steel wheel. Steel wheels are formed from welded hoops, then flared and joined to stamped spiders.

Aluminum Wheels

Forged and cast aluminum wheels are commonly used on cars and trucks. *Forged* means that the aluminum is hammered or forged under pressure into shape. A forged aluminum wheel is much stronger than a *cast* aluminum wheel.

A cast aluminum wheel is constructed by pouring liquid (molten) aluminum into a mold. After the aluminum has cooled, the cast aluminum wheel is removed from the mold and machined. Aluminum wheels are usually thicker than steel wheels and require special wheel weights when balancing. Coated or covered wheel weights should be used when balancing aluminum wheels to prevent galvanic corrosion damage to the wheel. (See Chapter 3 for details on balancing.) Most aluminum wheels use an alloy of aluminum. Aluminum can be combined (alloyed) with copper, manganese, silicon, or other elements to achieve the physical strength and characteristics for the exact product.

Some racing wheels are made from a lighter weight metal called magnesium. These wheels are called *mag* wheels (an abbreviation for magnesium). True magnesium wheels are not practical for production wheels because their cost and corrosion are excessive compared with steel or aluminum alloy wheels. The term *mag wheel*, however, is still heard when referring to alloy (aluminum) wheels.

HINT: If purchasing replacement aftermarket wheels, check that they are certified by SFI. SFI is the Specialty Equipment Manufacturers Association (SEMA) Foundation, Incorporated. SEMA and SFI are nongovernment agencies that were formed by the manufacturers themselves to establish standards for safety.

Metric Wheels

Several vehicle manufacturers, including Ford Motor Company, equip vehicles with wheels that are 365 mm or 390 mm in diameter rather than the more conventional 14″, 15″, or 16″ wheels.

Metric Code Bead Diameter Tire and Rim

Tire Size	Rim Size
185/65R *365*	*365* × 135TR
195/65R *390*	*390* × 150TR

Note that the rim in this example is 365 mm or 390 mm in diameter and 135 mm or 150 mm wide.

WARNING: Do not attempt to mount an inch code tire on a metric code rim or vice versa. A mismatch of tire size and rim size may result in tire failure and serious or fatal injury.

■ UNSPRUNG WEIGHT

The lighter the wheel and tire assembly, the faster it can react to bumps and dips in the road surface and the better the ride. The chassis and the body of any vehicle are supported by some sort of spring suspension system. It is the purpose of the suspension system to isolate the body of the vehicle from the road surface. Also, for best handling, all four tires must remain in contact with the road. After all, a tire cannot grip the road if it leaves the ground after hitting a bump. The wheel and tire are *unsprung weight* because they are not supported by the vehicle's springs. If heavy wheels or tires are used, every time the vehicle hits a bump, the wheel is forced upward. The heavy mass of the wheel and tire would transmit this force through the spring of the vehicle and eventually to the driver and passengers. Obviously, a much lighter wheel and tire assembly reacts faster to bumps and dips in the road surface. The end result is a smoother riding vehicle with greater control.

An aluminum wheel is *generally* lighter than the same size stamped steel. This is not always the case, however, so before purchasing aluminum wheels, check their weight!

NOTE: Putting oversized tires on an off-road-type vehicle is extremely dangerous. The increased unsprung weight can cause the entire vehicle to leave the ground after hitting a bump in the road. The increased body height necessary to clear the larger tires seriously affects drive shaft angles and wheel alignment angles, making the vehicle very difficult to control.

■ LUG NUTS

Lug nuts are used to hold a wheel to the brake disc, brake drum, or wheel bearing assembly. Most manufacturers use a stud in the brake or bearing assembly with a lug nut to hold the wheel. Some models of VW, Audi, and Mazda use a lug *bolt* that is threaded into a hole in the brake drum or bearing assembly.

> **NOTE:** Some aftermarket manufacturers offer a stud conversion kit to replace the lug bolt with a conventional stud and lug nut.

Typical lug nuts are tapered so that the wheel stud will center the wheel onto the vehicle. Another advantage of the taper of the lug nut and wheel is to provide a suitable surface to prevent the nuts from loosening. The taper, usually 60 degrees, forms a wedge that helps ensure that the lug nut will not loosen. Steel wheels are deformed slightly when the lug nut is torqued down against the wheel mounting flange; be certain that the taper is *toward* the vehicle.

> **NOTE:** Many Chrysler-made products from the late 1950s to the early 1970s had left-handed threads on the wheel studs on the left side of the vehicle and right-handed threads on the right side of the vehicle. More than one technician has broken off wheel studs trying to remove lug nuts from the left side of a Chrysler vehicle. The reason for using left-handed threads on the left side is that the normal direction of rotation tends to tighten, rather than loosen, the lug nuts.

Many alloy wheels use a *shank-nut*-type lug nut that has straight sides without a taper. This style of nut must be used with wheels designed for this nut type. If replacement wheels are used on any vehicle, check with the wheel manufacturer as to the proper type and style lug nut. Figure 2–29 shows several of the many styles of lug nuts that are available.

Size

Lug nuts are sized to the thread size of the stud onto which it screws. The diameter and the number of threads per inch are commonly stated. Since some vehicles use left-hand threads, RH and LH are commonly stated, indicating "right-hand" and "left-hand" threads. A typical size is 7/16-20 RH, where the 7/16 indicates the diameter of the wheel stud and 20 indicates that there are 20 threads per inch. Another common fractional size is 1/2 × 20. Metric sizes such as M12 × 1.5 use a different sizing method.

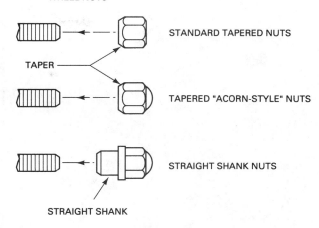

WHEEL NUTS

Figure 2–29 Various styles of lug nuts.

M	metric
12	12-mm diameter of stud
1.5	1.5-mm distance from one thread peak to another

Other commonly used metric lug sizes include M12 × 1.25 and M14 × 1.5. Obviously, metric wheel studs require metric lug nuts.

Tightening Torque

All wheels must be tightened in a star pattern to a specified torque. Always use a torque wrench or torque-absorbing adapters. Both proper tightening sequence and tightness are important to prevent possible damage to the wheel and/or brake rotor or drum. *Never use an air impact wrench to install wheels, except when using torque-absorbing adapters!* See Figure 2–30. Tests performed by skilled technicians using their own air impact wrenches show that tightening torque can vary as much as 20 lb. ft. This uneven torque puts unequal stress on the wheel studs and the wheel mounting surface. (See Chapter 3 for recommended lug nut tightening torque and sequence; see Appendix 2 for the exact wheel lug nut tightening specification for each vehicle.)

TECH TIP

Clip-On Traction Device for Snow

If you live in an area that gets snow, you may want to consider having emergency temporary chains or a device like the one shown in Figure 2–31. After the adapter is attached to the wheel, all that is needed is to clip the plastic traction device onto the adapter and over the tread of the tire.

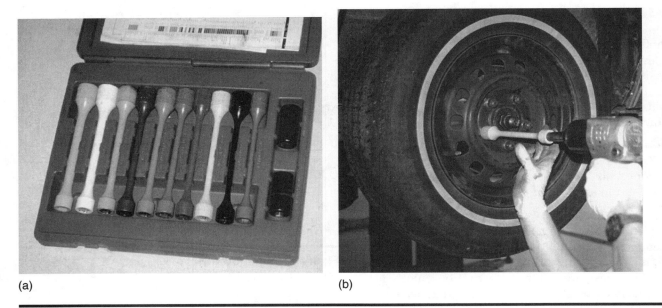

(a)   (b)

Figure 2–30 (a) A typical assortment of torque-limiting adapters to use with an air impact wrench to properly torque wheel nuts. (b) Proper torque is achieved by using an air impact wrench with a torque-limiting socket and keeping the impact on until the lug nut end of the adapter stops rotating. Repeating the torque sequence also helps ensure proper wheel nut torque.

(a)   (b)

Figure 2–31 (a) A clip-on snow traction device as it looks installed on the vehicle. (b) The device can be easily removed by hand from the adapter that is attached to the wheel of the vehicle.

PHOTO SEQUENCE   Checking Tire Pressure and Tread Depth

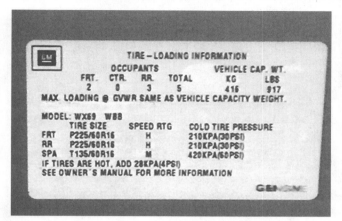

PS2–1 Before checking and adjusting tire pressure, check the tire loading information decal or placard usually located on the driver's door or door jam.

PS2–2 Do not inflate a tire higher than the maximum pressure rating on the tire sidewall.

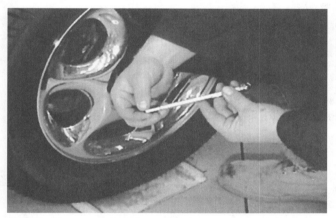

PS2–3 Use a good-quality tire pressure gauge to be assured of accurate tire pressure readings. Be sure the vehicle has not been driven to ensure that the tires are cool when checking tire pressure.

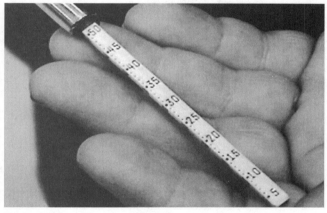

PS2–4 This tire pressure gauge reads from 5 to 50 psi and is a good choice because most passenger car tires fall within the range of this gauge.

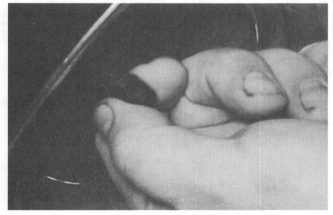

PS2–5 Remove the cap from the tire valve (Schrader valve).

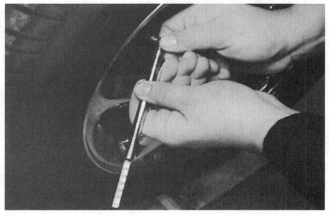

PS2–6 To get accurate tire pressure results, be sure to press the tire gauge straight onto the end of the tire valve in one smooth motion.

Checking Tire Pressure and Tread Depth—continued

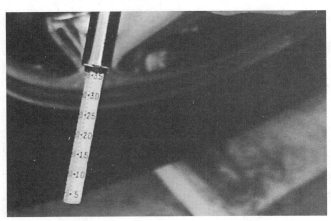

PS2–7 The tire pressure is read at the junction between the housing of the gauge and the movable plunger.

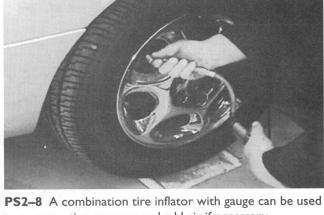

PS2–8 A combination tire inflator with gauge can be used to measure tire pressure and add air if necessary.

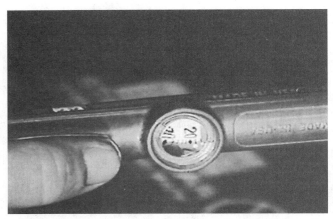

PS2–9 The tire pressure is read through a window on the inflator handle.

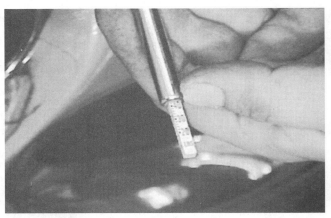

PS2–10 Repeat testing each tire to be sure that all four tires are inflated to the factory recommended inflation. Many vehicle tire information decals specify different tire pressures for front and rear tires.

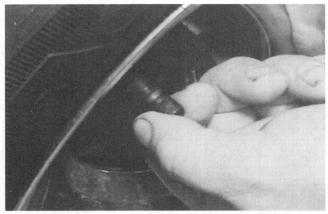

PS2–11 Be sure to check the tire pressure in the spare tire and to reinstall all tire valve caps.

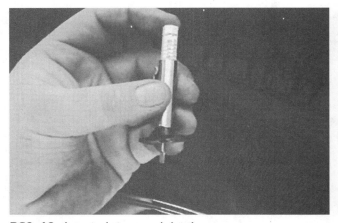

PS2–12 A typical tire tread depth gauge.

Checking Tire Pressure and Tread Depth—continued

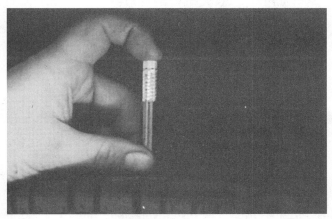

PS2–13 To use a tire tread depth gauge, position the gauge over a groove in the tire and depress the top of the gauge.

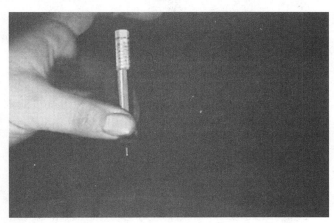

PS2–14 Lift the gauge off the tire to read the depth.

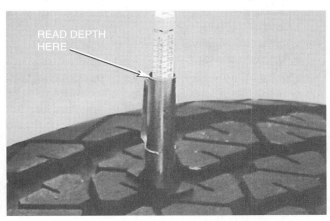

READ DEPTH HERE

PS2–15 Read the tread depth at the junction between the housing and the plunger. The number that is closest to the junction is the depth of the tread.

PS2–16 A penny can also be used to check to see if there is at least 2/32" of tread remaining, an amount often specified by state or local governments as the minimum allowable legal tread depth.

PS2–17 If the top of Lincoln's head is visible, then the tread depth is less than 2/32".

PS2–18 The same tire is measured using a penny, but in another location. In this location, the tread depth is deep enough that the tip of Lincoln's head is covered. All tires should be measured at the minimum tread depth, not at a point of maximum tread depth.

■ SUMMARY

1. New tires have between 9/32″ and 15/32″ tread depth. Wear bars (indicators) show up as a bald strip across the tread of the tire when the tread depth gets down to 2/32″.

2. All tires are assembled by hand from many different materials and chemical compounds. After a green tire is assembled, it is placed into a mold under heat and pressure for about 30 minutes. Tread design and the tire shape are determined by the mold design.

3. A 205/75R × 14 92S tire is 205 mm wide at its widest section and is 75 percent as high as it is wide. The R stands for radial-type construction. The tire is designed for a 14″-diameter rim. The number 92 is the load index of the tire (the higher the number, the more weight the tire can safely support). The S is the speed rating of this tire (S = 112 mph maximum sustained).

4. The uniform tire quality grading system is a rating for tread wear (100, 150, etc.), traction (A, B, C), and temperature resistance (A, B, C).

5. For best overall handling and satisfaction, always select the same size and type of tire that came on the vehicle when new.

6. Replacement wheels should have the same offset as the factory wheels to prevent abnormal tire wear and/or handling problems.

7. All wheels must be secured with the proper size and style of lug nuts. If a wheel stud is broken, it should be replaced immediately to avoid possible wheel damage or loss of vehicle control.

■ REVIEW QUESTIONS

1. List the various parts of a tire and explain how a tire is constructed.

2. Explain the effect that aspect ratio has on ride and handling.

3. List the factors that should be considered when purchasing tires.

4. Explain the three major areas of the uniform tire quality grading system.

5. Explain how to determine proper tire pressure.

■ ASE CERTIFICATION-TYPE QUESTIONS

1. The part of the tire that is under just the tread of a radial tire is called the
   a. Bead
   b. Body (carcass) ply
   c. Belt
   d. Inner liner

2. The aspect ratio of a tire means
   a. Its width to diameter of a wheel ratio
   b. The ratio of height to width
   c. The ratio of width to height
   d. The ratio of rolling resistance

3. A tire is labeled 215/60R × 15 92T; the T indicates
   a. Its speed rating
   b. Its tread wear rating
   c. Its load rating
   d. Its temperature resistance rating

4. The 92 in the tire designation in question 3 refers to the tire's
   a. Speed rating
   b. Tread wear rating
   c. Load rating
   d. Temperature resistance rating

5. Radial tires can cause a vehicle to pull to one side while driving. This is called "radial tire pull" and is often due to the
   a. Angle of the body (carcass) plies
   b. Tire conicity
   c. Tread design
   d. Bead design

6. Tire inflation is very important to the safe and economical operation of any vehicle. Technician A says the maximum pressure should be the pressure imprinted on the sidewall of the tire. Technician B says to inflate the tires to the pressures recommended on the tire information decal or placard on the driver's door. Which technician is correct?
   a. Technician A only
   b. Technician B only
   c. Both A and B
   d. Neither A nor B

7. When purchasing replacement tires, do not change tire width from the stock size by more than
   a. 10 mm
   b. 15 mm
   c. 20 mm
   d. 25 mm

8. What do the letters *JJ* mean in a wheel designation size labeled 14 × 7JJ?
   a. The offset of the rim
   b. The bolt circle code
   c. The back spacing of the rim
   d. The shape of the flange area

9. Front-wheel-drive vehicles usually use what type of wheel?
   a. Negative offset
   b. Positive offset

10. Wheel lug nuts must be tightened
    a. By hand
    b. With a torque wrench
    c. With an air impact wrench
    d. Hand tightened, plus 1/4 turn

Tire and Wheel Service

Objectives: After studying Chapter 3, the reader should be able to:

1. Discuss proper tire mounting procedures.
2. Describe recommended tire rotation methods.
3. Discuss how to properly balance a tire.
4. Describe tire repair procedures.
5. Explain wheel and tire safety precautions.

Proper tire service is extremely important for the safe operation of any vehicle. Premature wear can often be avoided by checking and performing routine service, such as frequent rotation and monthly inflation checks.

■ TIRE MOUNTING RECOMMENDATIONS

1. When removing a wheel from a vehicle for service, mark the location of the wheel and lug stud to ensure that the wheel can be replaced in exactly the same location. This ensures that tire balance will be maintained if the tire/wheel assembly was balanced on the vehicle.
2. Make certain that the wheel has a good, clean metal-to-metal contact with the brake drum or rotor. Grease, oil, or dirt between these two surfaces could cause the wheel lug nuts to loosen while driving.

3. Always check the rim size. For example, simply by looking it is hard to distinguish a 16″ wheel from a 16.5″ wheel used on some trucks (see Figure 3–1). The rim size is marked on the sidewall of the tire, and the rim's diameter and width are stamped somewhere on the wheel.
4. Many tires have been marked with a paint dot or sticker, as shown in Figure 3–2. This mark represents the largest diameter (high point) and/or stiffest portion of the tire. This variation is due to the overlapping of carcass and belt fabric layers, as well as tread and sidewall rubber splices. The tire should be mounted to the rim with this mark lined up with the valve stem. The valve stem hole is typically drilled at the smallest diameter (low point) of the wheel. Mount the tires on the rim with the valve stem matched to (lined up next to) the mark on the tire. This is called **match mounting.**
5. Never use more than 40 psi (275 kPa) to seat a tire bead.
6. Rim flanges must be free of rust, dirt, scale, or loose or flaked rubber buildup prior to mounting the tire (see Figure 3–3).
7. When mounting new tires, do *not* use silicone lubricant on the tire bead. Use special lubricant such as rendered (odorless) animal fat or rubber lubricant to help prevent tire rotation on the rim. This rubber lube is a water-based soap product that is slippery when wet (coefficient of friction less than 0.3) and acts almost as an adhesive when dry (coefficient of friction dry of over 0.5 for natural products and over 1.0 for synthetic products) (see Figure 3–4 on page 53). If the wrong lubricant is used, the rubber in the bead area of the tire can be softened

Figure 3–1 Note the difference in the shape of the rim contour of the 16″ and 16 1/2″ diameter wheels. While it is possible to mount a 16″ tire on a 16 1/2″ rim, it cannot be inflated enough to seat against the rim flange. If an attempt is made to seat the tire bead by overinflating (over 40 psi), the tire bead can break resulting in an explosive force that could cause serious injury or death.

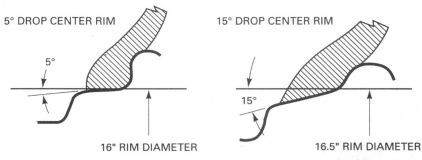

5° DROP CENTER RIM   15° DROP CENTER RIM

16″ RIM DIAMETER   16.5″ RIM DIAMETER

Figure 3–2 For the best possible ride and balance, mount the tire with the mark on the tire lined up with the valve stem. This also allows the tire assembly to be balanced with less weight than if not match mounted.

(a)

(b)

Figure 3–3 (a) Cleaning the bead area of an aluminum (alloy) wheel using a handheld wire brush. The technician is using the tire changer itself to rotate the wheel as the brush is used to remove any remnants of the old tire. (b) Using an electric or air-powered wire brush speeds the process, but care should be exercised to not remove any of the aluminum itself.

HIGH-PERFORMANCE TIP

Fine Tune Handling with Tire Pressure Changes

The handling of a vehicle can be changed by changing tire pressures between the front and rear tires.

Understeer   A term used to describe how a vehicle handles when cornering where additional steering input is needed to maintain the corner or resisting turning into a corner. This is normal handling for most vehicles.

Oversteer   A term used to describe handling where correction while cornering is often necessary because the rear tires lose traction before the front tires.

Tire Pressure	To Decrease Understeer	To Decrease Oversteer
Front tire inflation pressure	Increase	Decrease
Rear tire inflation pressure	Decrease	Increase

CAUTION: Do not exceed the maximum inflation pressure as imprinted on the tire sidewall.

(a)

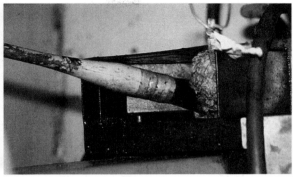

(b)

Figure 3–4 (a) A container of water-based rubber lubricant. This lubricant is mixed with water, and therefore a gallon container will last a long time if properly mixed. (b) Rendered (odorless) animal fat is also used and is recommended by some manufacturers of tire changing equipment.

or weakened. Also, most other lubricants do not increase in friction when they dry like rubber lubricant does. The result can be the rotation of the tire on the rim (wheel), especially during rapid acceleration or braking.

NOTE: Many experts recommend that when new tires are installed the vehicle be driven less than 50 mph (80 km/h) for the first 50 miles (80 km) to allow the tire to adhere to the rim. During this break-in period, the rubber lube used to mount the tire is drying and the tire is becoming fully seated on the rim. By avoiding high speeds, rapid acceleration, and fast braking, the driver is helping to prevent the tire from rotating on the rim.

■ WHEEL MOUNTING TORQUE

Make certain that the wheel studs are clean and dry, and torqued to the manufacturer's specifications.

CAUTION: Most manufacturers warn that the wheel studs should not be oiled or lubricated with grease; this can cause the wheel lug nuts to loosen while driving.

Always tighten lug nuts gradually in the proper sequence (tighten one nut, skip one, and tighten the next nut) (see Figures 3–5 and 3–6). This helps prevent warping the brake drums or rotors, or bending a wheel.

See Appendix 2 for the exact lug nut torque for your vehicle. If the exact torque value is not available, use the chart below as a guide for the usual value based on the size (diameter) of the lug studs.

Stud Diameter	Torque (lb. ft.)
3/8″	35–45
7/16″	55–65
1/2″	75–85
9/16″	95–115
5/8″ (usually only trucks)	125–150
12 mm	70–80
14 mm	85–95

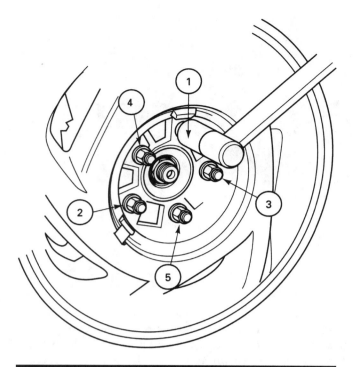

Figure 3–5 Always tighten wheel lug nuts (or studs) in a star pattern to ensure even pressure on the axle flange, brake rotors or drums, and the wheel itself. (Courtesy of Chrysler Corporation)

Figure 3–6 Proper wheel nut torque is essential. Always use a torque wrench or use torque-limiting adapters to tighten wheel nuts. The use of an impact wrench will distort the wheel, hub, and rotor resulting in vibration and brake pedal pulsations.

Many factory-installed and aftermarket wheels use antitheft wheel lug nuts, as shown in Figure 3–7, usually on only one wheel stud. When removing or installing a locked lug nut, be sure the key is held square to the lug nut to prevent damaging either the nut or the key.

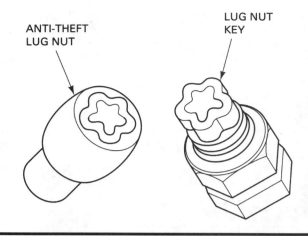

Figure 3–7 Most manufacturers recommend using hand tools rather than an air impact wrench to remove and install lock-type lug nuts to prevent damage. If either the key or the nut is damaged, the nut may be very difficult to remove!

Anytime you install a brand new set of aluminum wheels, retorque the wheels after the first 25 miles. The soft aluminum often compresses slightly, loosening the torque on the wheels.

> **NOTE:** The use of torque-absorbing adapters (torque-limiting shank sockets) on lug nuts with an air impact wrench properly set has proved to give satisfactory results. See Figure 3–8 for a photo of a torque-absorbing adapter.

■ TIRE ROTATION

To ensure long life and even tire wear, it is important to rotate each tire to another location. Some rear-wheel-drive vehicles, for example, may show premature tire wear on

Diagnostic Story

"I Thought the Lug Nuts Were Tight!"

Proper wheel nut torque is critical, as one technician discovered when a customer returned complaining of a lot of noise from the right rear wheel. See Figure 3–9 for a photo of what the technician discovered. The lug (wheel) nuts had loosened and ruined the wheel.

> **CAUTION:** Most vehicle manufacturers also specify that the wheel studs/nuts should not be lubricated with oil or grease. The use of a lubricant on the threads could cause the lug nuts to loosen.

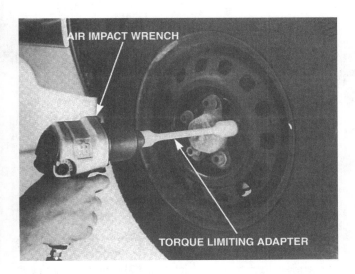

Figure 3–8 A torque-limiting adapter for use with an air impact wrench still requires care to prevent overtightening. The air pressure to the air impact should be limited to 125 psi (860 kPa) in most cases and the proper adapter selected for the vehicle being serviced. The torque adapter absorbs any torque beyond its designed rating. Most adapters are color coded for easy identification as to size of lug nut and torque value.

Figure 3–9 This wheel was damaged because the lug nuts were not properly torqued.

the front tires. The wear usually starts on the outer tread row. This wear usually appears as a front to back (high and low) wear pattern on individual tread blocks. These *blocks of tread* rubber are deformed during cornering, stopping, and turning. This type of tread block wear can cause tire noise and/or tire roughness. Although some shoulder wear on front tires is normal, it can be reduced by proper inflation, alignment, and tire rotation. For best results, tires should be rotated every 6000 miles or six months. See Figure 3–10 for suggested methods of rotation.

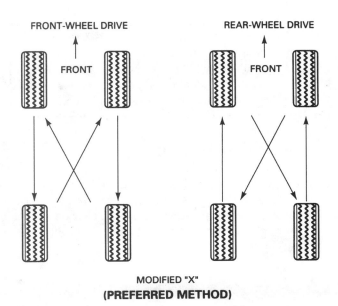

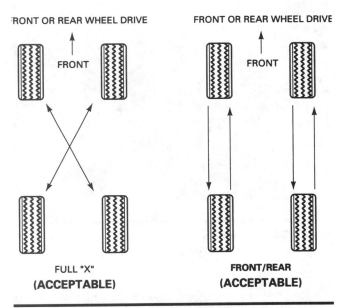

Figure 3–10 The method most often recommended is the modified X method. Using this method, each tire eventually is used at each of the four wheel locations. An easy way to remember the sequence, whether front-wheel drive or rear-wheel drive, is to say to yourself, "Drive wheels straight, cross the nondrive wheels."

> **NOTE:** Radial tires can cause a radial pull due to their construction. If wheel alignment is correct, attempt to correct a pull by rotating the tires front to rear or, if necessary, side to side.

Some tire manufacturers do not recommend rotating the tires on front-wheel-drive vehicles because the front tires often wear three times as fast as the rear tires. They recommend replacing only front tires, because the rear tires often last over 90,000 miles (145,000 km).

HINT: Tire rotation should be done at every *other* oil change. Most manufacturers recommend changing the engine oil every 3000 miles (4800 km) or every three months; tire rotation is recommended every 6000 miles (9600 km), or every six months.

■ TIRE INSPECTION

All tires should be carefully inspected for faults in the tire itself or for signs that something may be wrong with the steering or suspension systems of the vehicle. See Figures 3–11 through 3–16 for examples of common problems.

Figure 3–11 Tire showing excessive shoulder wear resulting from underinflation and/or high-speed cornering.

Figure 3–12 Tire showing excessive wear in the center indicating overinflation or heavy acceleration on a drive wheel.

Figure 3–13 Cuppy-type wear caused by worn suspension components or improper alignment of the rear wheels of a front-wheel-drive vehicle.

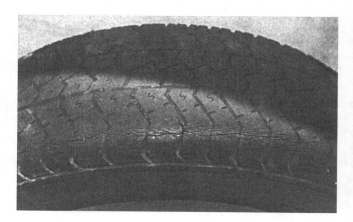

(a)

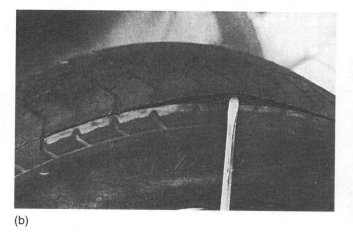

(b)

Figure 3–14 (a) Cracking at the edge of the belt on the tire shoulder can be an early sign of tire failure. (b) Tire failure at belt edge. The owner of the vehicle did not notice the cracking along the shoulder because it was on the inside.

Figure 3–15 Even a gentle bump against a curb can crack an alloy wheel, which can lead to a sudden loss of air pressure and a possible accident.

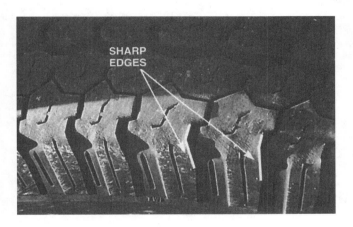

Figure 3–16 Tapered tread block wear caused by high-speed cornering. While an alignment and low tire pressure could also contribute to this wear condition, the best solution is to keep the tires rotated to even out the tire wear.

■ RADIAL RUNOUT

Even though a tire has no visible faults, it can be the cause of vibration. If vibration is felt above 45 mph, regardless of the engine load, the cause is usually an out-of-balance or a defective out-of-round tire. Both of these problems cause a **tramp** or *up-and-down*-type vibration. If the vibration is seen in the hood of the vehicle or felt in the steering wheel, then the problem is usually the *front* tires. If the vibration is felt throughout the entire vehicle or in the seat of your pants, then the rear tires (or drive shaft, in rear-wheel-drive vehicles) are the problem. This can be checked using a runout gauge and checking for **radial runout** (see Figures 3–17 and 3–18). To check radial runout (checking for out of round) and lateral runout (checking for side-to-side movement):

Figure 3–17 Runout gauge (dial gauge) being used to check the radial runout of the tire/wheel assembly.

1. Raise the vehicle so that the tires are off the ground approximately 2″ (5 cm).
2. Place the runout gauge against the tread of the tire in the center of the tread and, while rotating the tire, observe the gauge reading.
3. Note that maximum radial runout should be less than 0.060″ (1.5 mm). Little, if any, tramp will be noticed with less than 0.030″ (0.8 mm) runout. If the reading is over 0.125″ (3.2 mm), replacement of the tire is required.
4. Check all four tires.

Correcting Radial Runout

Excessive radial runout may be corrected by one of several methods:

1. Try relocating the wheel on the mounting studs. Mark one stud and remount the wheel two studs away from its original position. Excessive wheel

---

Frequently Asked Question **???**

I Thought Radial Tires Couldn't Be Rotated!

When radial tires were first introduced by American tire manufacturers in the 1970s, rotating tires side to side was *not* recommended because of concern about belt or tread separation. Since the late 1980s, most tire manufacturers throughout the world, including the United States, have used tire-building equipment specifically designed for radial ply tires. These newer radial tires are constructed so that the tires can now be rotated from one side of the vehicle to the other without fear of causing a separation by the resulting reversal of the direction of rotation.

Figure 3–18 For aggressive tread tires, put masking tape over the tread before measuring runout. The tape allows the dial indicator to roll smoothly over the tread of the tire.

hole and/or stud tolerance may be the cause. If the radial runout is now satisfactory, remark the stud and wheel to prevent a future occurrence of the problem.

2. Remount the tire on the wheel 180 degrees from its original location. This can solve a runout problem, especially if the tire was not match mounted to the wheel originally.

3. If runout is still excessive, remove the tire from the wheel and check the runout of the *wheel*. If the wheel is within 0.035″ (0.9 mm), yet the runout of the tire/wheel assembly is excessive, the problem has to be a defective tire and it should be replaced.

Sometimes a problem within the tire itself can cause a vibration, and yet not show up as being out of round when tested for radial runout. A condition called **radial force variation** can cause a vibration even if correctly balanced. A **tire problem detector (TPD)** can be used to find a defective tire by revolving the tire with normal vehicle weight on the roller and measuring the movement of the spindle with a dial indicator, as shown in Figure 3–19.

NOTE: Some tire balancers are equipped with a roller that is pressed against the tread of the tire to measure radial force variations. Follow the instructions as shown on the balancer display to correct for excessive radial force variations.

■ LATERAL RUNOUT

Another possible problem that tires can cause is a type of vibration called *shimmy*. This *rapid back and forth motion* can be transmitted through the steering linkage to the steering wheel. Excessive runout is usually noticeable by the driver of the vehicle as a side-to-side vibration, especially at low speeds between 5 and 45 mph (8 and 72 km/h). Shimmy can be caused by an internal defect of the tire or a bent wheel. This can be checked using a runout gauge on the side of the tire or wheel to check for *lateral runout*.

Place the runout gauge against the side of the tire and rotate the wheel. Observe the readings. The maxi-

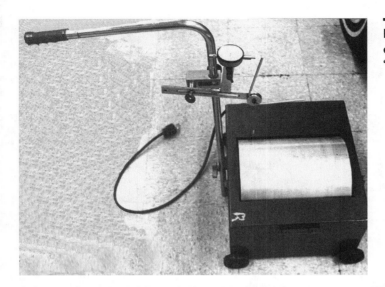

Figure 3–19 Tire problem detector. A similar device equipped with grinding wheels can be used to "true" or "match grind" a tire.

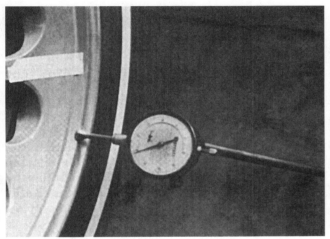

Figure 3–20 Checking the lateral runout of the wheel using a dial gauge (indicator).

mum allowable reading is 0.045″ (1.1 mm). If close to or above 0.045″ (1.1 mm), check on the edge of the wheel to see if the cause of the lateral runout is due to a bent wheel, as shown in Figure 3–20.

Most manufacturers specify a maximum lateral runout of 0.035″ (0.9 mm) for alloy wheels and 0.045″ (1.1 mm) for steel wheels.

Correcting Lateral Runout

Excessive lateral runout may be corrected by one of several methods:

1. Retorque the wheel in the proper star pattern to the specified torque. Unequal or uneven wheel torque can cause excessive lateral runout.
2. Remove the wheel and inspect the wheel mounting flange for corrosion or any other reason that could prevent the wheel from seating flat against the brake rotor or drum surface.

Diagnostic Story

The Greased Wheel Causes a Vibration

Shortly after an oil change and a chassis lubrication, a customer complained of a vibration at highway speed. The tires were checked for excessive radial runout to be certain the cause of the vibration was not due to a defective out-of-round tire. After removing the wheel assembly from the vehicle, excessive grease was found on the inside of the rim. Obviously, the technician who greased the lower ball joints had dropped grease on the rim. After cleaning the wheel, it was checked for proper balance on a dynamic computer balancer and found to be properly balanced. A test drive confirmed that the problem was solved.

3. Check the condition of the wheel or axle bearings. Looseness in the bearings can cause the wheel to wobble.

■ TIRE BALANCING

Proper tire balance is important for tire life, ride comfort, and safety. Tire balancing is needed because of the lack of uniform weight and stiffness (due to splices) and a combination of wheel runout and tire runout. Balancing a tire can compensate for most of these conditions. However, if a tire or wheel is excessively out of round or bent, then replacement of the wheel or tire is required. For more detailed information about vibration causes and correction, see Chapter 13.

Static Balance

The term *static balance* means that the weight mass is evenly distributed around the axis of rotation (see Figure 3–21).

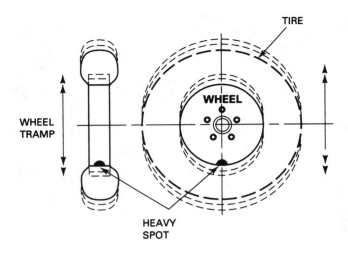

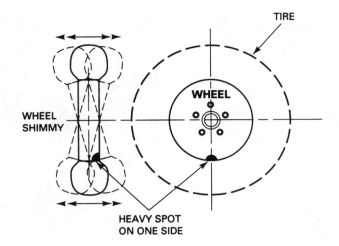

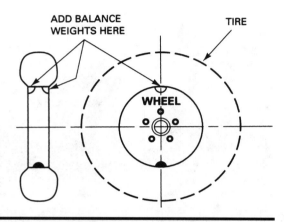

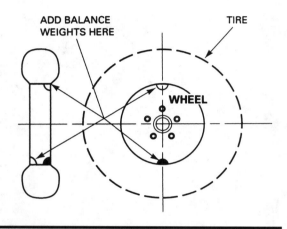

Figure 3–21 Weights are placed opposite the heavy spot of the tire/wheel assembly to balance the tire statically and prevent wheel tramp–type (up-and-down) vibration.

Figure 3–22 Weights are added to correct not only imbalance up and down but also any "wobble" caused by the tire/wheel assembly being out of balance dynamically.

1. For example, if a wheel is spun and stops at different places with each spin, then the tire is statically balanced.
2. **If the static balance is not correct, wheel tramp–type (vertical shake) vibration and uneven tire wear can result.**
3. Static balance can be tested with the tire stationary or while being spun to determine the heavy spot (sometimes called *kinetic balance*).

Dynamic Balance

The term *dynamic balance* means that the centerline of weight mass is in the same plane as the centerline of the wheel (see Figure 3–22).

1. Dynamic balance must be checked while the tire and the wheel are rotated, to determine side-to-side out of balance as well as up-and-down.

2. *Incorrect dynamic balance causes shimmy.* Shimmy-type vibration causes the steering wheel to shake from side to side.

Prebalance Checks

Before attempting to balance any tire, the following should be checked and corrected to ensure good tire balance:

1. Check the wheel bearing adjustment for looseness or wear.
2. Check the radial runout.
3. Check the lateral runout.
4. Remove stones from the tread.
5. Remove grease or dirt buildup on the inside of the wheel.
6. Check for dragging or misadjusted brakes.

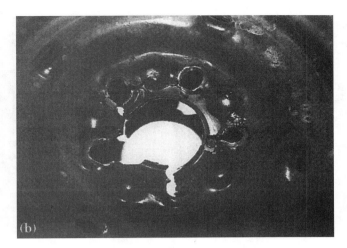

Figure 3–23 (a) A beginning automotive student had just completed a brake job and was ready to install the wheel covers when the instructor noticed the lug nuts had been installed backwards. The tapered end of the nut should go toward the tapered hole in the rim to help center and secure the wheel. Note the gaps around the lug nuts where the hole in the wheel is showing. (b) A wheel that was destroyed because the lug nuts were installed backwards.

7. Check for loose or backward lug nuts (see Figure 3–23).
8. Check for proper tire pressures.
9. Remove all of the old weights.
10. Check for bent or damaged wheel covers.

Diagnostic Story

The Vibrating Ford

A technician was asked to solve a vibration problem on a rear-wheel-drive Ford. During a test drive, the vibration was felt everywhere—the dash, the steering wheel, the front seat, the shoulder belts; everything was vibrating! The technician balanced all four tires on a computer balancer. Even though wheel weights were put on all four wheels and tires, the vibration was even worse than before. The technician re-balanced all four wheels time after time, but the vibration was still present. The shop supervisor then took over the job of solving the mystery of the vibrating Ford. The supervisor balanced one wheel/tire assembly and then tested it again after installing the weights. The balance was way off! The supervisor broke the tire down and found about 1 quart (1 liter) of liquid in the tire! Liquid was found in all four tires. No wonder the tires couldn't be balanced! Every time the tire stopped, the liquid would settle in another location.

The customer later admitted to using a tire stop-leak liquid in all four tires. Besides stop leak, another common source of liquid in tires is water that accumulates in the storage tank of air compressors, which often gets pumped into tires when air is being added. All air compressor storage tanks should be drained of water regularly to prevent this from happening. See Figure 3–24.

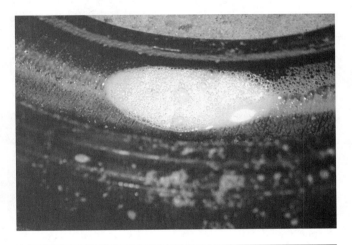

Figure 3–24 Liquid tire stop leak was found in all four tires. This liquid caused the tires to be out-of-balance.

Wheel Weights

Wheel weights are available in a variety of styles and types, including:

1. Clip-on lead weights for standard steel rims.
2. Clip-on weights for Cadillac steel rims. These weights use a longer clip that allows the use of full wheel covers without their hitting the weights near the rim edge.
3. Clip-on weights for alloy (aluminum) wheels.
   a. Uncoated—generally *not* recommended by wheel or vehicle manufacturers because corrosion often occurs where the lead weight contacts the alloy wheel surface.

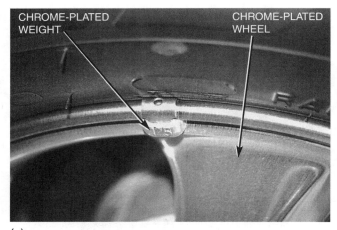

Figure 3–25 (a) A chrome-plated wheel weight used on a new vehicle equipped with chrome-plated aluminum (alloy) wheels. (b) Note the corrosion on this alloy wheel caused by the use of standard lead wheel weights. Using a coated wheel weight would have prevented this damage to an expensive wheel.

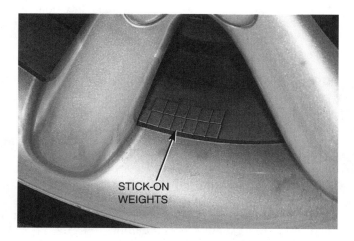

Figure 3–26 Stick-on weights are used from the factory to balance the alloy wheels of this Plymouth Prowler.

b. Coated—lead weights that are painted or coated in a plastic material are usually the *recommended* type of weight to use on alloy wheels (see Figure 3–25).

Weights are usually coated with a nylon or polyester-type material that often matches the color of the aluminum wheels.

4. Stick-on weights come with an adhesive backing that is most often used on alloy wheels as shown in Figure 3–26.

Most wheel weights come in 1/4-ounce (0.25-oz) increments (oz $\times$ 28 = grams).

0.25 oz = 7 gm	1.75 oz = 49 gm
0.50 oz = 14 gm	2.00 oz = 56 gm
0.75 oz = 21 gm	2.25 oz = 63 gm
1.00 oz = 28 gm	2.50 oz = 70 gm
1.25 oz = 35 gm	2.75 oz = 77 gm
1.5 oz = 42 gm	3.00 oz = 84 gm

Bubble Balancer

This type of static balancer is commonly used and is accurate, if calibrated and used correctly. A bubble balancer is portable and can be easily stored away when not in use. It is also easy to use and is relatively inexpensive (see Figure 3–27).

Strobe Balancer

The strobe balancer is being phased out of service by the more expensive computer off-the-vehicle balancers. The strobe balancer is still recommended by many vehicle manufacturers as the method to use to "fine-tune" a difficult-to-balance tire/wheel assembly. It is also used to balance drive shafts. See Chapter 13 for details.

This method spins the tire on the vehicle and uses a bright strobe light to determine the heavy part of the tire. The strobe light is turned on by a magnetic sensor, which

Frequently Asked Question   ???

How Much Is Too Much Weight?

Whenever balancing a tire, it is wise to use as little amount of weight as possible. For most standard-size passenger vehicle tires, most experts recommend that no more than 5.5 oz of weight be added to correct an imbalance condition. If more than 5.5 oz is needed, remove the tire from the wheel (rim) and carefully inspect for damage to the tire or the wheel. If the tire still requires more than 5.5 oz and the wheel is not bent or damaged, replace the tire.

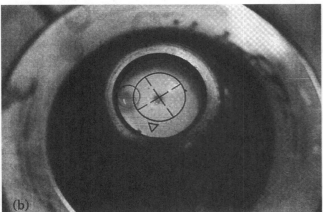

Figure 3–27 (a) Typical portable bubble wheel balancer. (b) When properly balanced, the bubble should be in the center of the cross hairs. The triangle mark is placed on the gauge to be used to line up the valve stem of the tire. This allows the technician to reinstall the tire/wheel assembly back onto the balancer after pounding on the weights in the same location. (c) Most bubble balancers use oil to allow the gauge to float freely. Be careful when moving this type of balance because the oil can easily spill out if tilted.

must be attached to the lower control arm of the vehicle. This sensor triggers the strobe light whenever the heavy part of the tire is straight down at the 6 o'clock position. Depending on the scale which determines the amount of weight needed, the weight is attached at the 12 o'clock position, or exactly opposite the heavy portion as determined by the strobe (see Figure 3–28).

The strobe balancer is usually used to check for static balance (balance up and down); however, the strobe can be used to check dynamic balance by placing the magnetic sensor horizontally against the backing plate of the brakes.

Computer Balancer

Since the mid-1970s, the most popular type of tire balancer is the computer dynamic balancer. Most computer balancers are designed to balance wheels and tires off the vehicle. Computer dynamic balancers spin the tire at a relatively slow speed (approximately 20 mph). Sensors attached to the spindle of the balancer determine the amount and location of weights necessary to balance the tire dynamically. All computer balancers must be programmed with the actual rim size and tire location for the electronic circuits to calibrate

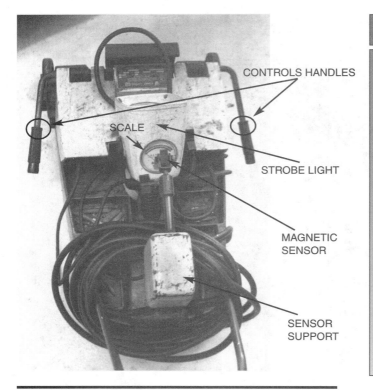

Figure 3–28 Typical strobe balancer used to balance wheels/tires on the vehicle as well as to balance drive shafts.

Figure 3–29 Typical computer balancer.

the required weight locations. Computer balancers are the most expensive type of balancer (see Figure 3–29).

Most computer balancers will be accurate to within 1/4 oz (0.25 oz), while some are accurate to within 1/8 oz (0.125 oz). (Most drivers can feel an out-of-balance of 1 oz or more, but few can feel a vibration caused by just 1/4 oz.) For sensitive drivers or vehicles used for high speeds, such as racing, most computer balancers can be programmed to balance within 1/8 oz (0.125 oz). Refer to the manufacturer's instructions for the exact capabilities and procedures for your computer balancer.

Most vehicle manufacturers specify that no more than 5.5 oz (150 gm) be used to balance any tire, with no more than 3.5 oz (100 gm) used per side of each wheel.

■ REPLACEMENT WHEELS

Whenever a replacement wheel is required, the same offset should be maintained. If wider or larger diameter wheels are to be used, consult a knowledgeable wheel or tire salesperson to determine the correct wheel for your application.

> **CAUTION:** Never remove the weights that are welded to the surface of the brake drum facing the wheel (see Figure 3–30). If replacement wheels do not fit without removing these weights, either replace the brake drum (one without a weight) or select another brand or style of wheel. Removing the weights from a brake drum can cause severe vibration at highway speeds.

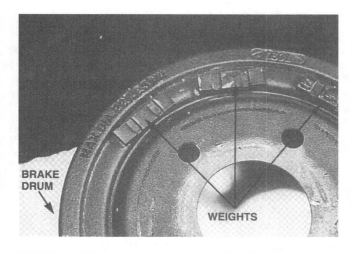

Figure 3–30 Most brake drums do not have this much attached weight.

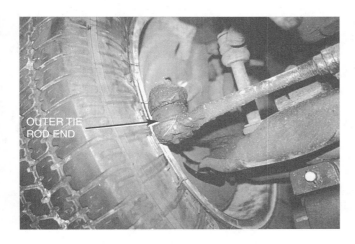

Figure 3–31 Notice that the rim touches the tie rod end. See diagnostic story to find out why.

Diagnostic Story

It Happened to Me—
It Could Happen to You

During routine service, I rotated the tires on a Pontiac TransAm. Everything went well, and I even used a torque wrench to properly torque all of the lug nuts. Then, when I went to drive the car out of the service stall, I heard a horrible grinding sound. When I hoisted the car to investigate, I discovered that the front wheels were hitting the outer tie rod ends (see Figure 3–31). The 16″ wheels had a different back spacing front and rear, and therefore these wheels could not be rotated. Always check replacement or aftermarket wheels for proper fit before driving the vehicle.

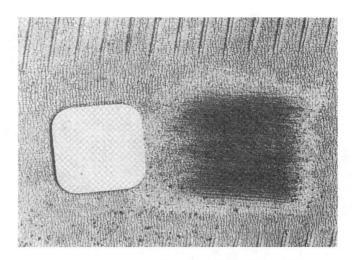

Figure 3–32 The area of the repair should be buffed slightly larger than the patch to be applied.

■ TIRE REPAIR

Tread punctures, nail holes, or cuts up to 1/4″ (2.6 mm) can be repaired. Repairs should be done from the inside of the tire using plugs or patches. The tire should be removed from the rim to make the repair. With the tire off the wheel, inspect the wheel and the tire for hidden damage. The proper steps to follow for a tire repair are:

1. Mark the location of the tire on the wheel.
2. Dismount the tire; inspect and clean the punctured area with a prebuff cleaner. DO NOT USE GASOLINE!
3. Buff the cleaned area with sandpaper or a tire buffing tool until the rubber surface has a smooth, velvet finish (see Figure 3–32).
4. Ream the puncture with a fine reamer from the inside. Cut and remove any loose wire material from the steel belts.
5. Fill the puncture with contour filling material, and cut or buff the material flush with the inner liner of the tire.
6. Apply chemical vulcanizing cement and allow to dry.

NOTE: Most vulcanizing (rubber) cement is highly flammable. Use out of the area of an open flame. Do not smoke when making a tire repair.

7. Apply the patch and use a stitching tool from the center toward the outside of the patch to work any air out from between the patch and the tire (see Figure 3–33). Another excellent tire repair procedure uses a rubber plug. Pull the stem through the hole in the tire as shown in Figure 3–34.
8. Remount the tire on the rim aligning the marks made in Step 1 above. Inflate to recommended pressure and check for air leaks.

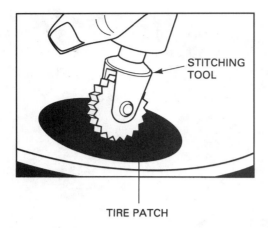

Figure 3–33 A stitching tool being used to force any trapped air out from under the patch.

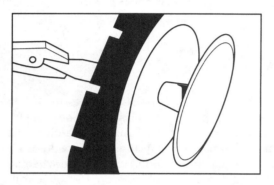

Figure 3–34 A rubber plug being pulled through a hole in the tire. The stem is then cut off flush with the surface of the tire tread.

There are many tire repair products on the market. Always follow the installation and repair procedures exactly per the manufacturer's instructions.

CAUTION: Most experts agree that tire repairs should be done from the inside. Many technicians have been injured and a few killed when the tire they were repairing exploded as a steel reamer tool was inserted into the tire. The reamer can easily create a spark as it is pushed through the steel wires of a steel-belted tire. This spark can ignite a combustible mixture of gases inside the tire caused by using stop leak or inflator cans. Since there is no way a technician can know if a tire has been inflated with a product that uses a combustible gas, always treat a tire as if it could explode.

TECH TIP

Open-End Wrenches Make It Easier

Tire repair is made easier if two open-end wrenches are used to hold the beads of the tire apart. See photo P5–4 on page 73 in the tire repair photo sequence.

Figure 3–35 Directional tires such as this one can be rotated to the opposite side of the vehicle during winter months. When the tire is located on the other side of the vehicle, its rotation is in the opposite direction.

TECH TIP

Dispose of Old Tires Properly

Old tires can no longer be thrown out in the trash. They must be disposed of properly. Tires cannot be buried because they tend to come to the surface. They also trap and hold water, which can be a breeding ground for mosquitoes. Used tires should be sent to a local or regional recycling center where the tires will be ground and used in asphalt paving or other industrial uses. Because there is often a charge to dispose of old tires, it is best to warn the customer of the disposal fee.

HIGH-PERFORMANCE TIP

Wrong Direction May Be Better

High-performance tires are designed for maximum traction when the roads are dry. Many tread designs also allow water to escape from between the tires and the road, thereby improving wet-road traction. A neat trick that is used by tire design and testing engineers on their own vehicles is to mount these directional tires on the opposite side of the vehicle during winter months (see Figure 3–35). The directional tread design works better in the snow when the design is operated in the opposite direction. Although these tires will never come close to providing the traction available from a true snow tire or even an all-season tire design, at least the traction is better than if the tires were left in the "correct" rotation direction. Just be sure to move the tires back to the correct side of the vehicle after the chance of snow is over.

PHOTO SEQUENCE   Mounting a Tire

PS3–1 A typical tire changer of the type that does not touch the wheel and is therefore safe to use with all types of wheels.

PS3–2 The foot pedal controls allow the service technician to break the tire bead, clamp the wheel (rim) to the machine and rotate the tire/wheel assembly, and still have both hands free to help with the changing of the tire.

PS3–3 After removing the valve core and allowing all the air to escape, the tire is placed on the ground and the bead breaker is used on both sides of the wheel to break the bead loose from the rim. All the old wheel weights should also be removed.

PS3–4 After the tire has been broken loose from the rim, the tire/wheel assembly is placed on top of the balancer and the foot control valve depressed, allowing the locking arms to expand to hold the wheel onto the turn plate of the tire changer.

Mounting a Tire—continued

PS3–5 The valve stem is being removed using special pliers.

PS3–6 The top arm of the tire changer is brought down near the top of the wheel. The arm will provide the anchor point for the removal and reinstallation of the tire.

PS3–7 A tire tool (flat bar) is placed between the bead of the tire and the wheel.

PS3–8 The foot pedal is then depressed rotating the table. The tire tool permits the tire to separate from the wheel.

Mounting a Tire—continued

PS3–9 Repeating the procedure on the lower bead allows the tire to be completely removed from the wheel.

PS3–10 The rim area where the tire bead seals is then cleaned. Here the technician is using an air-powered die grinder equipped with a sanding pad to thoroughly clean the area where the new tire will seat.

PS3–11 A new tire valve stem is then pulled through the hole after being lubricated with rubber lube.

PS3–12 Before the new tire is placed on the wheel, both the inside and the outside beads of the tire are coated with rubber lube or tire mounting soap.

Mounting a Tire—continued

PS3–13 The new tire is placed over the rim as far as possible and positioned so that the vertical arm part of the tire changer can be used to guide the tire onto the rim as the wheel is revolved on the turn table.

PS3–14 The service technician is exerting a downward force on the tire as it revolves to keep the bead area down into the drop section of the wheel. This procedure allows the tire to be installed onto the wheel with little effort and without doing harm to either the tire or the wheel.

PS3–15 After the tire has been installed on the wheel, air is blown into the tire through the tire core. The Schrader valve core has not yet been installed. Without the valve, more air is allowed to flow into the tire to make seating the tire on the wheel easier. After the tire is seated, the valve core is installed and inflated to specifications for the vehicle.

PHOTO SEQUENCE   Balancing a Tire

PS3–16 To properly mount a tire and wheel assembly onto a balancer, the proper size cone should be selected. The cone keeps the wheel centered on the shaft of the balancer. The cone fits into the hole in the center of the wheel.

PS3–17 A large retainer and wing nut are then installed and the wing nut tightened by hand.

PS3–18 The computer must know the width of the wheel. Here a measuring caliper is used to determine the width of the wheel.

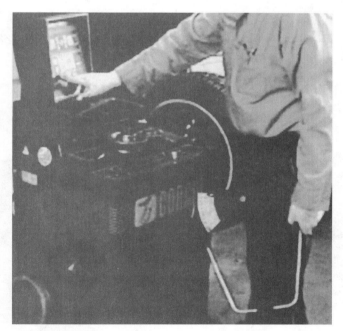

PS3–19 The diameter of the wheel and the distance to the edge of the wheel from the machine are also programmed into the computer balancer.

Balancing a Tire—continued

PS3–20 The safety lid is closed and the balancer is started. The balancer spins the tire/wheel assembly about 20 mph, and sensors inside detect up-and-down as well as side-to-side forces.

PS3–21 The display on the computer balancer indicates the amount of weight to add to each side to achieve proper dynamic balance.

PS3–22 The tire is rotated to the designated location, and the proper size weight is attached to the wheel.

PS3–23 After the weight(s) has been attached to the wheel, the balancer is turned on again to check for proper

PS3–24 When the computer balancer indicates "000", the tire/wheel assembly is properly balanced. The wheel can then be removed from the balancer and reinstalled on the vehicle.

PHOTO SEQUENCE **Repairing a Tire**

PS3–25 The source of the leak was detected by spraying soapy water on the inflated tire. Needlenose pliers are being used to remove the object that caused the leak.

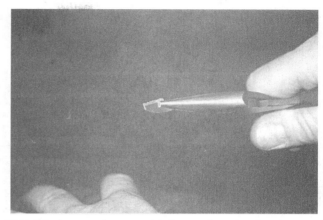

PS3–26 A part of a razor blade was found to be the cause of the flat tire.

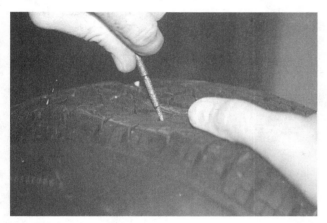

PS3–27 A reamer is being used to clean the puncture hole.

PS3–28 A method used by some technicians is to hold the beads apart with two open-end wrenches. In this case, two line wrenches were used, and this provided more than enough room to gain access to the inside of the tire.

PS3–29 The surrounding area is being buffed using an air-powered die grinder equipped with a special buffing tool specifically designed for this process.

PS3–30 Rubber cement is then applied to the buffed area.

Repairing a Tire—continued

PS3–31 The brush included with the rubber cement makes the job easy. Be sure to cover the entire area around the puncture.

PS3–32 Peal off the paper from the adhesive on the patch and insert the tip of the patch through the puncture from the inside of the tire.

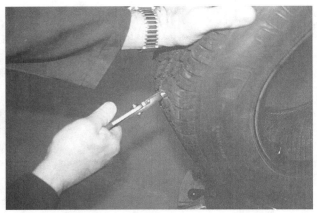

PS3–33 Use a pair of pliers to pull the plug of the patch through the puncture.

PS3–34 A view of the patch on the inside of the tire.

PS3–35 To be assured of an air-tight patch, the adhesive of the patch should be "stitched" to the inside of the tire using a serrated roller called a stitching tool.

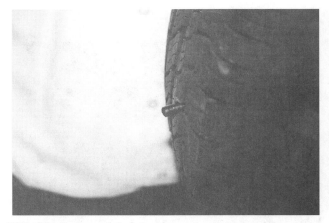

PS3–36 A view of the plug from the outside of the tire after the metal covering used to pierce the puncture is removed from the patch plug. The plug can be trimmed to the level of the tread using side cutters or a knife.

■ SUMMARY

1. For safety and proper vehicle handling, all four tires of the vehicle should be of the same size, construction, and type, except where specified by the manufacturer, such as on some high-performance sports cars.

2. Wheels should always be tightened with a torque wrench to the proper torque in a star pattern.

3. Tires should be rotated every 5000 to 7000 miles (8000 to 11,000 km), or at every other oil change.

4. Wheels should be cleaned around the rim area whenever tires are changed and carefully inspected for cracks or other defects such as excessive lateral or radial runout.

5. Properly balanced tires prolong tire life. Wheel tramp or an up-and-down type of vibration results if the tires are statically out of balance or if the tire is out of round.

6. Dynamic balance is necessary to prevent side-to-side vibration, commonly called shimmy.

7. Only coated or stick-on-type wheel weights should be used on alloy wheels to prevent corrosion damage.

■ REVIEW QUESTIONS

1. List the precautions and recommendations regarding tire selection and maintenance.

2. Determine the proper wheel mounting torque for your vehicle from the guidelines.

3. Determine the bolt circle, wheel diameter, and wheel width.

4. Describe the difference between static and dynamic balance.

■ ASE CERTIFICATION-TYPE QUESTIONS

1. A tire is worn excessively on both edges. The most likely cause of this type of tire wear is
   a. Overinflation
   b. Underinflation
   c. Excessive radial runout
   d. Excessive lateral runout

2. When seating a bead of a tire, never exceed psi
   a. 30
   b. 40
   c. 50
   d. 60

3. For best tire life, most vehicle and tire manufacturers recommend tire rotation every
   a. 3000 miles
   b. 6000 miles
   c. 9000 miles
   d. 12,000 miles

4. What liquid should be used when mounting a tire?
   a. Silicone spray
   b. Grease
   c. Water-based soap
   d. SAE 10W-30 engine oil

5. The most common torque specification range for lug nuts is
   a. 125–150 lb. ft.
   b. 100–120 lb. ft.
   c. 80–100 lb. ft.
   d. 60–80 lb. ft.

6. Which statement is **false?**
   a. Excessive radial runout can cause a tramp-type vibration.
   b. Excessive lateral runout can cause a tramp-type vibration.
   c. A tire out of balance dynamically can cause a shimmy-type vibration.
   d. A tire out of balance statically can cause a shimmy-type vibration.

7. The recommended type of wheel weight to use on aluminum (alloy) wheels is
   a. Lead with plated spring steel clips
   b. Coated (painted) lead weights
   c. Lead weights with longer than normal clips
   d. Aluminum weights

8. Most vehicle and tire manufacturers recommend that no more than _____ ounce balance weight be added to a wheel/tire assembly.
   a. 2.5
   b. 3.5
   c. 4.5
   d. 5.5

9. A vehicle vibrates at highway speed. Technician A says that water in the tire(s) could be the cause. Technician B says that an out-of-round tire could be the cause. Which technician is correct?
   a. Technician A only
   b. Technician B only
   c. Both Technician A and B
   d. Neither Technician A nor B

10. Proper tire inflation pressure is found
    a. On the driver's door or post
    b. In the owner's manual
    c. On the sidewall of the tire
    d. Both a and b

Wheel Bearings and Service

Objectives: After studying Chapter 4, the reader should be able to:

1. Discuss the various types, designs, and parts of automotive antifriction wheel bearings.
2. Describe the symptoms of defective wheel bearings.
3. Explain wheel bearing inspection procedures and causes of spalling and brinelling.
4. List the installation and adjustment procedures for front-wheel bearings.
5. Explain how to inspect, service, and replace rear-wheel bearings and seals.

Bearings allow the wheels of a vehicle to rotate and still support the weight of the entire vehicle.

■ ANTIFRICTION BEARINGS

Antifriction bearings use rolling parts inside the bearing to reduce friction. Four styles of rolling contact bearings include **ball, roller, needle,** and **tapered roller** bearings, as shown in Figure 4–1. All four styles convert sliding friction into rolling motion. All of the weight of a vehicle or load on the bearing is transferred through the rolling part.

In a ball bearing, all of the load is concentrated into small spots where the ball contacts the inner and outer race (rings). See Figure 4–2. While ball bearings cannot support the same weight as roller bearings, there is less friction in ball bearings and they generally operate at higher speeds.

A roller bearing, having a greater (longer) contact area, can support heavier loads than a ball bearing (see Figure 4–3).

A needle bearing is a type of roller bearing that uses smaller rollers called **needle rollers.**

The clearance between the diameter of the ball or straight roller is manufactured into the bearing to provide the proper **radial clearance,** and is *not adjustable.*

Tapered Roller Bearings

The most commonly used automotive wheel bearing is the tapered roller bearing. Not only is the bearing itself tapered but the rollers are tapered as well. By design, this type of bearing can withstand **radial** (up and down) as well as **axial** (thrust) loads in one direction (see Figure 4–4).

Most non-drive-wheel bearings are tapered rollers rotating between races and assembled on an angle. The taper allows more weight to be handled by the friction-reducing bearings because the weight is directed over the entire length of each roller rather than concentrated on a small spot, as with ball bearings. The rollers are held in place by a **cage** between the **inner race** (also called the **inner ring,** or **cone**) and the **outer race** (also called the **outer ring,** or **cup**). Tapered roller bearings must be loose in the cage to allow for heat expansion. Tapered roller bearings should always be adjusted with a certain amount of free play to allow for heat expansion. On non-drive-axle vehicle wheels, the cup is tightly fitted to the wheel hub and the cone is loosely fitted to the wheel spindle. New bearings come packaged with rollers, cage, and inner race assembled together, wrapped with moisture-resistant paper (see Figure 4–5).

Figure 4–1 Rolling contact bearings include (left to right) ball, roller, needle, and tapered roller.

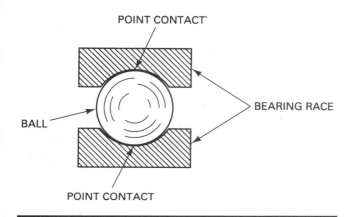

Figure 4–2 Ball bearing point contact.

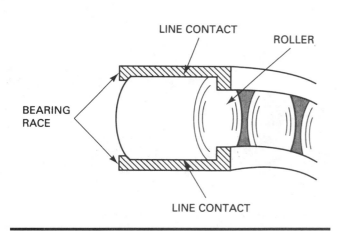

Figure 4–3 Roller bearing line contact.

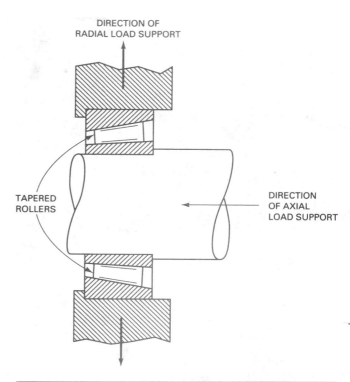

Figure 4–4 A tapered roller bearing will support a radial load and an axial load in only one direction.

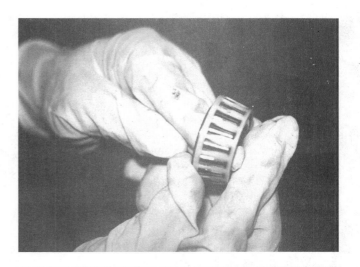

Figure 4–5 Many tapered roller bearings use a plastic cage to retain the rollers.

■ INNER AND OUTER WHEEL BEARINGS

Most rear-wheel-drive vehicles use an inner and an outer wheel bearing on the front wheels. The inner wheel bearing is always the larger because it is designed to carry most of the vehicle weight and transmit this weight to the suspension through to the spindle (see Figure 4–6). Between the inner wheel bearing and the

Figure 4–6 Non-drive-wheel hub with inner and outer tapered roller bearings. By angling the inner and outer bearings in opposite directions, axial (thrust) loads are supported in both directions.

spindle, there is a grease seal, which prevents grease from getting onto the braking surface.

■ STANDARD BEARING SIZES

Bearings use standard dimensions for inside diameter, widths, and outside diameters. The standardization of bearing sizes helps interchangeability. The dimensions that are standardized include bearing bore size (inside diameter), bearing series (light to heavy usage), and external dimensions. When replacing a wheel bearing, note the original bearing brand name and number. Replacement bearing catalogs usually have crossover charts from one brand to another. The bearing number is usually the same because of the interchangeability and standardization within the wheel bearing industry.

■ SEALED FRONT-WHEEL-DRIVE BEARINGS

Most front-wheel-drive vehicles use a sealed nonadjustable front-wheel bearing. This type of bearing can include either two preloaded tapered roller bearings or a double-row ball bearing. This type of sealed bearing is also used on the rear of many front-wheel-drive vehicles (see Figure 4–7).

Double-row ball bearings are often used because of their reduced friction and greater seize resistance. Figures 4–8 and 4–9 show sealed bearings.

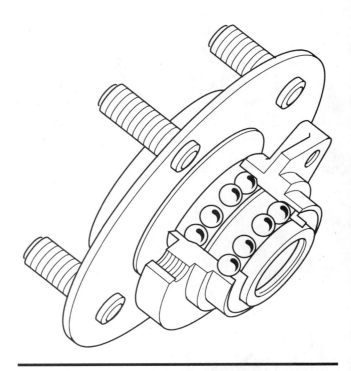

Figure 4–7 This front-wheel bearing assembly is a double-row ball bearing design. It is a prelubricated sealed bearing. The bearing assembly is a loose fit in the steering knuckle. The drive axle shaft is a splined fit through the bearing.

Figure 4–8 Hub and bearing assembly pulled from the knuckle and drive axle shaft.

■ BEARING GREASES

Vehicle manufacturers specify the type and consistency of grease for each application. The service technician should know what these specifications mean. Grease is an oil with a thickening agent to allow it to be installed in places where a liquid lubricant would not stay. Greases

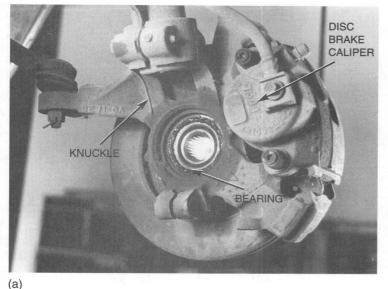

(a)   (b)

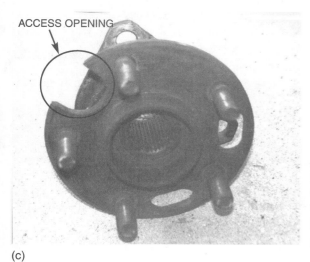

Figure 4–9 (a) Front-wheel-drive bearing shown after removal of
the drive axle shaft. (b) Bearing assembly as removed from a front-
wheel-drive General Motors vehicle. The entire assembly as shown
is replaced as a unit. (c) The wheel mounting flange is notched to
provide access to the bearing-to-knuckle bolts.

(c)

are named for their thickening agents such as aluminum,
barium, calcium, lithium, or sodium.

The **American Society for Testing Materials
(ASTM)** specifies the consistency using a penetration test.

The **National Lubricating Grease Institute
(NLGI)** specifies grease by designation as to its use:

"GC" designation is acceptable for wheel bearings.

"LB" designation is acceptable for chassis lubrication.

Many greases are labeled with both GC and LB and
are therefore acceptable for both wheel bearings and
chassis use, such as in lubricating ball joints and tie
rods. NLGI also uses the penetration test as a guide to
assign the grease a number. Low numbers are very
fluid, and higher numbers are more firm. For example,
a typical grease used for wheel bearings is labeled
NLGI #2 6C. See the following chart.

National Lubricating Grease Institute (NLGI) Numbers

NLGI Number	Relative Consistency
000	Very fluid
00	Fluid
0	Semifluid
1	Very soft
2	Soft (typically used for wheel bearings)
3	Semifirm
4	Firm
5	Very firm
6	Hard

Timken OK Load is from a test that determines the
maximum load the lubricant will carry. The *OK Load* is
the maximum weight that can be applied without scor-
ing the test block.

*More rolling bearings are destroyed by overlubrica-
tion than by underlubrication, because the heat generated in the bearings cannot be transferred easily to the*

Figure 4–10 Typical lip seal with a garter spring.

air through the excessive grease. Bearings should never be filled beyond one-third to one-half of their grease capacity by volume. Molybdenum disulfide is added to grease in amounts up to 10 percent for use as a multipurpose lubrication on automotive equipment parts such as chassis joints, steering joints, U-joints, and king pins.

■ SEALS

Seals are used in all vehicles to keep lubricant, such as grease, from leaking out and to prevent dirt, dust, or water from getting into the bearing or lubricant. Two general applications of seals include static and dynamic. **Static seals** are used between two surfaces that do not move. **Dynamic seals** are used to seal between two surfaces that move. Wheel bearing seals are dynamic seals that must seal between rotating axle hubs and the stationary spindles or axle housing. Most dynamic seals use a synthetic rubber lip seal encased in metal. The lip is often held in contact with the moving part with the aid of a **garter spring,** as seen in Figure 4–10.

The sealing lip should be installed toward the grease or fluid being contained (see Figure 4–11).

■ SYMPTOMS AND DIAGNOSIS OF DEFECTIVE BEARINGS

Wheel bearings control the positioning and reduce the rolling resistance of vehicle wheels. Whenever a bearing fails, the wheel may not be kept in position and noise is usually heard. Symptoms of defective wheel bearings include:

1. A hum, rumbling, or growling noise, which increases with vehicle speed.

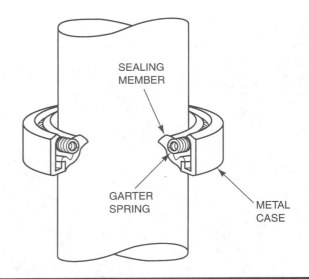

Figure 4–11 A garter spring helps hold the sharp lip edge of the seal tight against the shaft. (Courtesy of Dana Corporation)

2. Roughness felt in the steering wheel, which changes with vehicle speed or cornering.
3. Looseness or excessive play in the steering wheel, especially while driving over rough road surfaces.
4. A loud grinding noise produced in severe cases by a defective front-wheel bearing.
5. Pulling to one side during braking.
6. Bearing roughness—with the vehicle off the ground, rotate the wheel by hand, listening and feeling carefully.
7. Bearing looseness—grasp the wheel at the top and bottom and wiggle it back and forth.

> **NOTE:** Excessive looseness in the wheel bearings can cause a low brake pedal.

If any of these symptoms are present, carefully clean and inspect the bearings.

■ NON-DRIVE-WHEEL BEARING INSPECTION AND SERVICE

1. Hoist the vehicle safely.
2. Remove the wheel.
3. Remove the brake caliper assembly and support it with a coat hanger or other suitable hook to avoid allowing the caliper to hang by the brake hose.
4. Remove the grease cap (dust cap). See Figure 4–12.
5. Remove the old cotter key and discard.

> **NOTE:** The term *cotter* as in cotter key or cotter pin is derived from the old English verb meaning "to close or fasten."

TECH TIP ✔

Watch Out for Bearing Overload

It is not uncommon for vehicles to be overloaded. This is particularly common with pickup trucks and vans. Whenever there is a heavy load, the axle bearings must support the entire weight of the vehicle, including its cargo. If a bump is hit while driving with a heavy load, the balls of a ball bearing or the rollers of a roller bearing can dent the race of the bearing. **This dent or imprint is called brinelling,** after Johann A. Brinell, a Swedish engineer who developed a process of testing for surface hardness by pressing a hard ball with a standard force into a sample material to be tested.

Once this imprint is made, the bearing will make noise whenever the roller or ball rolls over the indentation. Continued use causes wear to occur on all of the balls or rollers, with eventual failure. While this may take months, the *cause* of the bearing failure is often overloading of the vehicle. Avoid shock loads and overloading for safety and for longer vehicle life.

Figure 4–13 After wiggling the brake rotor slightly, the washer and outer bearing can be easily lifted out of the wheel hub.

Figure 4–14 Some technicians remove the inner wheel bearing and the grease seal at the same time by jerking the rotor off the spindle after reinstalling the spindle nut. Although this is a quick and easy method, sometimes the bearing is damaged (deformed) from being jerked out of the hub using this procedure.

Figure 4–12 Removing the grease cap with grease cap pliers.

6. Remove the spindle nut (castle nut).
7. Remove the washer and the outer wheel bearing (see Figure 4–13).
8. Remove the bearing hub from the spindle. The inner bearing will remain in the hub and may be removed (simply lifted out) after the grease seal is pried out (see Figure 4–14).
9. Most vehicle and bearing manufacturers recommend cleaning the bearing thoroughly in solvent or acetone. If there is no acetone, clean the solvent off the bearings with denatured alcohol to make certain that the thin solvent layer is

completely washed off and dry. *All solvent must be removed or allowed to dry from the bearing because the new grease will not stick to a layer of solvent.*

10. Carefully inspect the bearings and the races for the following:
    - The outer race for lines, scratches, or pits
    - The cage should be round; if the cage has straight sections, this is an indication of an overtightened adjustment or the cage has been dropped.

If either of these is observed, then the bearing must be replaced, including the outer race. Failure to

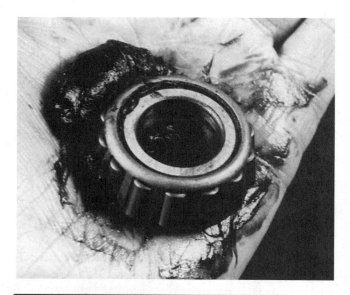

Figure 4–15 When packing grease into a cleaned bearing, force grease around each roller as shown.

replace the outer race (which is included when you purchase a bearing) could lead to rapid failure of the new bearing.

11. Pack the cleaned or new bearing thoroughly with clean, new, approved wheel bearing grease using hand packing or a wheel bearing packer. Always clean out all of the old grease before applying the recommended type of new grease. *Because of compatibility problems, it is not recommended that greases be mixed* (see Figure 4–15).

> **NOTE:** Some vehicle manufacturers do *not* recommend that stringy-type wheel bearing grease be used. Centrifugal force can cause the grease to be thrown outward from the bearing. Because of the stringy texture, the grease may not flow back into the bearing after it has been thrown outward. The result is lack of lubrication and eventual bearing failure.

12. Place a thin layer of grease on the outer race.
13. Apply a thin layer of grease to the spindle, being sure to cover the outer bearing seat, inner bearing seat, and shoulder at the grease seal seat.
14. Install a new **grease seal** (also called a **grease retainer**) flush with the hub (see Figure 4–16).
15. Place approximately 3 tablespoons of grease into the grease cavity of the wheel hub. Excessive grease could cause the inner grease seal to fail with the possibility of grease getting on the brakes. Place the rotor with the inner bearing and seal in place over the spindle until the grease seal rests on the grease seal shoulder.
16. Install the outer bearing and the bearing washer.

Figure 4–16 Installing a grease seal with a special tool after installing the inner bearing.

17. Install the spindle nut and, while rotating the tire assembly, tighten to about 12–30 lb. ft. with a wrench to seat the bearing correctly in the race (cup) and on the spindle (see Figure 4–17).
18. While still rotating the tire assembly, loosen the nut approximately 1/2 turn and then *hand tighten only* (about 5 in. lb.).

> **NOTE:** If the wheel bearing is properly adjusted, the wheel will still have about 0.001″ to 0.005″ (0.03 to 0.13 mm) end play. This looseness is necessary to allow the tapered roller bearing to expand when hot and not bind or cause the wheel to lock up.

19. Install a new cotter key. (An old cotter key could break a part off where it was bent and lodge in the bearing, causing major damage.)

> **HINT:** Most vehicles use a cotter key that is 1/8″ in diameter by 1 1/2″ long.

20. If the cotter key does not line up with the hole in the spindle, loosen slightly (no more than 1/16″ of

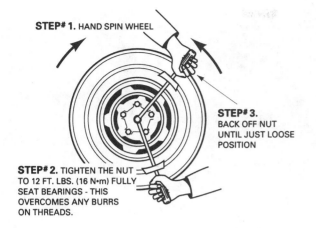

STEP# 1. HAND SPIN WHEEL

STEP# 3. BACK OFF NUT UNTIL JUST LOOSE POSITION

STEP# 2. TIGHTEN THE NUT TO 12 FT. LBS. (16 N•m) FULLY SEAT BEARINGS - THIS OVERCOMES ANY BURRS ON THREADS.

STEP# 5. LOOSEN NUT UNTIL EITHER HOLE IN THE SPINDLE LINES UP WITH A SLOT IN THE NUT –THEN INSERT COTTER PIN.

NOTICE: BEND ENDS OF COTTER PIN AGAINST NUT. CUT OFF EXTRA LENGTH TO PREVENT INTERFERENCE WITH DUST CAP.

STEP# 4. HAND "SNUG-UP" THE NUT

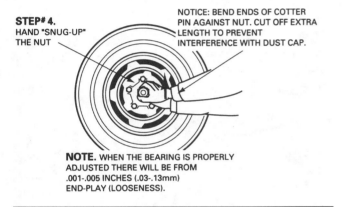

NOTE. WHEN THE BEARING IS PROPERLY ADJUSTED THERE WILL BE FROM .001-.005 INCHES (.03-.13mm) END-PLAY (LOOSENESS).

Figure 4–17 The wheel bearing adjustment procedure as specified for rear-wheel-drive General Motors vehicles. (Courtesy of Oldsmobile Division)

a turn) until the hole lines up. *Never tighten more than hand tight.*

21. Bend the cotter key ends up and around the nut, not over the end of the spindle where the end of the cotter key could rub on the grease cap, causing noise (see Figure 4–18).
22. Install the grease cap (dust cap) and the wheel cover.

CAUTION: Clean grease off disc brake rotors or drums after servicing the wheel bearings. Use a brake cleaner and a rag. Even a slight amount of grease on the friction surfaces of the brakes can harm the friction lining and/or cause brake noise.

■ FRONT-WHEEL-DRIVE SEALED BEARING REPLACEMENT

Most front-wheel-drive vehicles use a sealed bearing assembly that is bolted to the steering knuckle and sup-

Figure 4–18 Properly installed cotter key.

TECH TIP

Wheel Bearing Looseness Test

Looseness in a front-wheel bearing can allow the rotor to move whenever the front wheel hits a bump, forcing the caliper piston in and the brake pedal to kick back, causing the feeling that the brakes are locking up.

Loose wheel bearings are easily diagnosed by removing the cover of the master cylinder reservoir and watching the brake fluid as the front wheels are turned left and right with the steering wheel. If the brake fluid moves while the front wheels are being turned, caliper piston(s) are moving in and out because of loose wheel bearing(s). If everything is okay the brake fluid should *not* move.

ports the drive axle. Many designs incorporate the splined drive hub that transfers power from the drive axle to the wheels that are bolted to the hub (see Figures 4–19, 4–20, and 4–21).

Many front-wheel-drive vehicles use a bearing that must be pressed off the steering knuckle. Special aftermarket tools are also available to remove many of the bearings without removing the knuckle from the vehicle. Check the factory service manual and tool manufacturers for exact procedures for the vehicle being serviced.

Diagnosing a defective front bearing on a front-wheel-drive vehicle is sometimes confusing. A defective wheel bearing is usually noisy while driving straight.

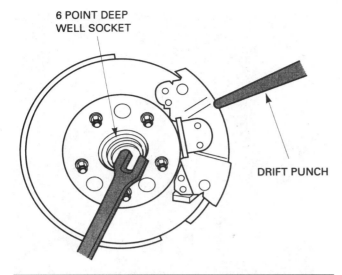

Figure 4–19 Removing the drive axle shaft hub nut. This nut is usually very tight, and the drift (tapered) punch wedged into the cooling fins of the brake rotor keeps the hub from revolving when the nut is loosened. (Courtesy of Oldsmobile Division)

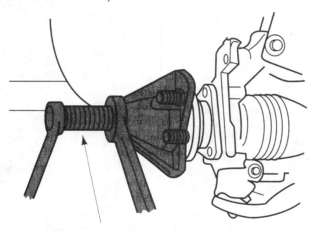

Figure 4–20 A special puller makes the job of removing the hub bearing from the knuckle easy without damaging any component. (Courtesy of Oldsmobile)

The noise increases with vehicle speed (wheel speed). A drive axle shaft U-joint (CV joint) can also be the cause of noise on a front-wheel-drive vehicle, but it usually makes *more noise* while turning and accelerating. See Chapter 6 for further CV joint analysis.

■ REAR AXLE BEARING AND SEAL REPLACEMENT

The rear bearings used on rear-wheel-drive vehicles are constructed and serviced differently from other types of wheel bearings. Rear axle bearings are either

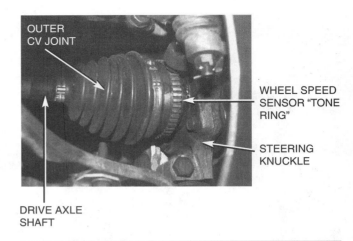

Figure 4–21 Be careful not to nick or damage the wheel speed sensor used for input information for traction control and antilock braking functions.

Figure 4–22 Retainer-plate-type rear axles can be removed by first removing the brake drum. The fasteners can be reached through a hole in the axle flange.

sealed or lubricated by the rear-end lubricant. The rear axle must be removed from the vehicle to replace the rear axle bearing. Two basic types of axle retaining methods are the **retainer-plate-type** and the **C-lock** methods.

The retainer-plate-type rear axle uses four fasteners that retain the axle to the axle housing. To remove the drive shaft and the rear axle bearing and seal, the retainer bolts or nuts must be removed.

HINT: If the axle flange has an access hole, then a retainer-plate-type axle is used (see Figure 4–22).

Figure 4–23 (a) To remove the axle from this vehicle equipped with a retainer-plate-type rear axle, the brake drum was placed back onto the axle studs backwards so that the drum itself can be used as a slide hammer to pull the axle out of the axle housing. (b) A couple of pulls and the rear axle is pulled out of the axle housing.

T E C H   T I P   ✔

The Brake Drum Slide-Hammer Trick

To remove the axle from a vehicle equipped with a retainer-plate-type rear axle, simply use the brake drum as a slide hammer to remove the axle from the axle housing (see Figure 4–23). If the brake drum does not provide enough force, a slide hammer can also be used to remove the axle shaft.

Figure 4–24 A slide-hammer-type axle puller being used to remove a rear axle.

The hole or holes in the wheel flange permit a socket wrench access to the fasteners. After the fasteners have been removed, the axle shaft must be removed from the rear axle housing. With the retainer-plate-type rear axle, the bearing is press fit onto the axle and the bearing cup (outer race) is also press fit into the axle housing tube. (See Figures 4–24, 4–25, and 4–26 for ways to remove the axle shaft.)

Vehicles that use C-clips use a straight roller bearing supporting a semifloating axle shaft inside the axle housing. The straight rollers do not have an inner race; the rollers ride on the axle itself. If a bearing fails, both the axle and the bearing usually need to be replaced. The

Figure 4–25 The brake backing plate came off with the rear axle and bearing on this vehicle. The backing plate made it more difficult to press off the old bearing.

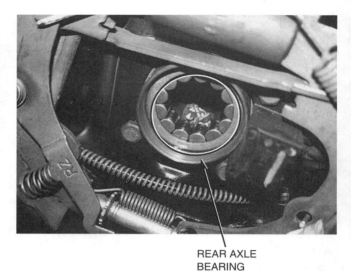

REAR AXLE
BEARING

Figure 4–27 Rear axle bearing after the axle shaft has been removed.

Figure 4–26 The ball bearings fell out onto the ground when this axle was pulled out of the axle housing. Diagnosing the cause of the noise and vibration was easy on this vehicle.

"C" washer

Figure 4–28 The C-clip (C washer) can be seen after removing the differential cover plate. The C-clip fits into a groove in the axle.

outer bearing race holding the rollers is pressed into the rear axle housing. The axle bearing is usually lubricated by the rear-end lubricant, and a grease seal is located on the outside of the bearing, as shown in Figure 4–27.

NOTE: Some replacement bearings are available that are designed to ride on a fresh, unworn section of the old axle. These bearings allow the use of the original axle, saving the cost of a replacement axle.

The C-clip-type rear axle retaining method requires that the differential cover plate be removed (see Figure 4–28).

After removal of the cover, the differential pinion shaft has to be removed before the C-clip that retains the axle can be removed (see Figure 4–29).

NOTE: When removing the differential cover, rear axle lubricant will flow from between the housing and the cover. Be sure to dispose of the old rear axle lubricant in the environmentally approved way, and refill with the proper type and viscosity (thickness) of rear-end lubricant. Check the vehicle specifications for the recommended grade.

(a)

(b)

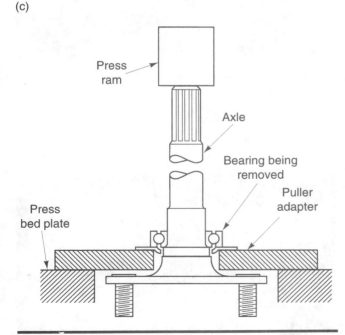
(c)

Figure 4–29 (a) Removing the pinion shaft lock bolt. (b) After the lock bolt has been removed, the pinion shaft can be removed. (c) The axle can be pushed inward slightly to allow the C-clip to be removed. After the C-clip has been removed, the axle can be easily pulled out of the axle housing.

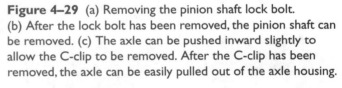

Figure 4–30 Using a hydraulic press to press an axle bearing from the axle. When pressing a new bearing back onto the axle, pressure should only be on the inner bearing race to prevent damaging the bearing.

Once the C-clip has been removed, the axle simply is pulled out of the axle tube. Axle bearings with inner races are pressed onto the axle shaft and must be pressed off using a hydraulic press. A bearing retaining collar should be chiseled or drilled into to expand the collar to allow it to be removed (see Figure 4–30).

Always follow the manufacturer's recommended bearing removal and replacement procedures. Always replace the rear axle seal whenever replacing a rear axle bearing. See Figure 4–31 for an example of seal removal.

See Figure 4–32 for an example of a rear axle bearing with a broken outer race.

When refilling the differential, check for a tag or lettering as to the correct lubricant, as shown in Figure 4–33.

Always check the differential vent to make sure it is clear (see Figure 4–34).

A clogged vent can cause excessive pressure to build up inside the differential and cause the rear axle seals to leak. If rear-end lubricant gets on the brake linings, the brakes will not have the proper friction and the linings themselves will be ruined and must be replaced.

Figure 4–31 Removing an axle seal using the axle shaft as the tool.

Figure 4–32 This axle bearing came from a high-mileage vehicle. Noise was first noticed when turning because weight transfer increased the load on the bearing. Later, a rumbling sound occurred all the time, increasing in noise level as the vehicle speed increased.

■ BEARING FAILURE ANALYSIS

Whenever a bearing is replaced, the old bearing must be inspected and the cause of the failure eliminated. See Figures 4–35 through 4–41 on pages 88–90 for examples of normal and abnormal bearing wear.

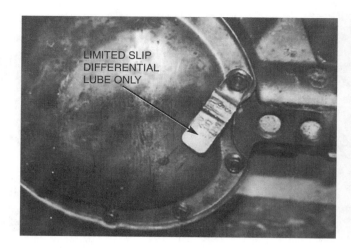

Figure 4–33 Most limited-slip differentials have an identifying tag indicating the proper lubricant to use in the unit.

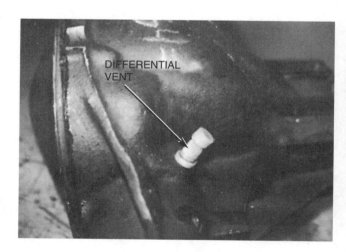

Figure 4–34 All differentials are equipped with a vent. If the vent becomes clogged, pressure can increase inside the differential causing a leak.

Figure 4–35 A normally worn bearing. If it doesn't have too much play, it can be reused. (Courtesy of SKF USA Inc.)

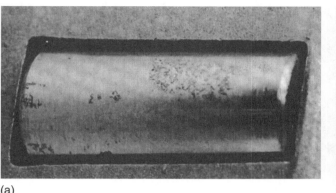
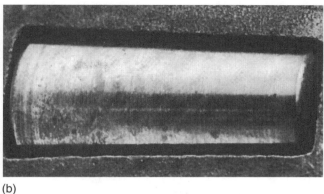

(a)   (b)

Figure 4–36 (a) When corrosion etches into the surface of a roller or race, the bearing should be discarded. (b) If light corrosion stains can be removed with an oil-soaked cloth, the bearing can be reused. (Courtesy of SKF USA Inc.)

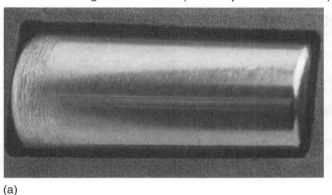

(a)   (b)

Figure 4–37 (a) When just the end of a roller is scored, it is from excessive preload. Discard the bearing. (b) A more advanced case of pitting. Under load, it will rapidly lead to "spalling." (Courtesy of SKF USA Inc.)

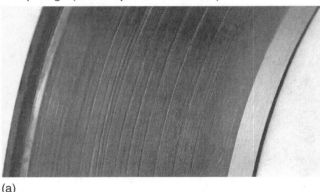

(a)   (b)

Figure 4–38 (a) Always check for faint grooves in the race. This bearing should not be reused. (b) Grooves like this are often matched by grooves in the race (above). Discard the bearing. (Courtesy of SKF USA Inc.)

A wheel bearing may fail for several reasons, including:

Metal Fatigue

Long vehicle usage, even under normal driving conditions, causes metal to fatigue. Cracks often appear. Eventually these cracks expand downward into the metal from the surface. The metal between the cracks can break out into small chips, slabs, or scales of metal. This process of breaking up is called **spalling** (see Figure 4–42 on page 91).

Shock Loading

Shock loading causes dents to be formed in the race of a bearing, which eventually leads to bearing failure. See the Tech Tip entitled "Watch Out for Bearing Overload" and Figure 4–43 on page 91.

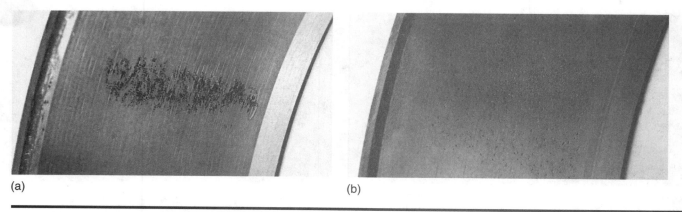

(a) (b)

Figure 4–39 (a) Regular patterns of etching in the race are from corrosion. This bearing should be replaced. (b) Light pitting comes from contaminants being pressed into the race. Discard the bearing. (Courtesy of SKF USA Inc.)

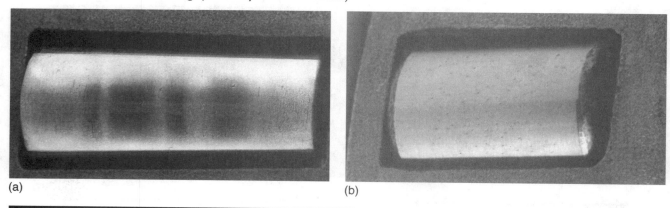

(a) (b)

Figure 4–40 (a) This bearing is worn unevenly. Notice the stripes. It shouldn't be reused. (b) Any damage that causes low spots in the metal renders the bearing useless. (Courtesy of SKF USA Inc.)

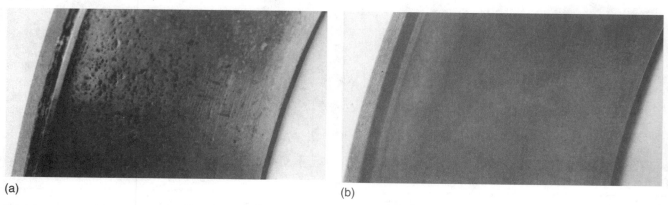

(a) (b)

Figure 4–41 (a) In this more advanced case of pitting, you can see how the race has been damaged. (b) Discoloration is a result of overheating. Even a lightly burned bearing should be replaced. (Courtesy of SKF USA Inc.)

Electrical Arcing

Electrical current flowing through a bearing can cause arcing and damage to the bearing, as shown in Figure 4–44.

Electrical current can result from electrical welding on the vehicle without a proper ground connection.

Without a proper return path, the electrical flow often travels throughout the vehicle, attempting to find a ground path. Always place the ground cable as close as possible to the area being welded. Another very common cause of bearing failure due to electrical arcing is a poor body ground wire connection between the body

(a)

(b)

Figure 4–42 (a) Pitting eventually leads to "spalling," a condition where the metal falls away in large chunks. (b) In this "spalled" roller, the metal has actually begun to flake away from the surface. (Courtesy of SKF USA Inc.)

Figure 4–43 These dents result from the rollers "hammering" against the race. It is called brinelling. (Courtesy of SKF USA Inc.)

Figure 4–44 This condition results from an improperly grounded arc welder. Replace the bearing. (Courtesy of SKF USA Inc.)

of the vehicle and the engine. All electrical current for accessories, lights, sound systems, etc., must return to the negative (−) terminal of the battery. If this ground wire connection becomes loose or corroded, the electrical current takes alternative paths to ground. The engine and entire drivetrain are electrically insulated by rubber mounts. Even the exhaust system is electrically insulated from the body or from the vehicle with rubber insulating **hangers.** Suspension and wheel bearings are also insulated electrically by rubber control arm and shock absorber **bushings.** However, dirt is a conductor of electricity, especially when wet. Dirt acts as an electrical conductor and allows electrical current to flow through the suspension system *through the bearing* to the engine block where the main starter ground cable connects to the battery negative (−) terminal.

Therefore, whenever any bearing is replaced in the chassis or drive line systems such as wheel bearings, U-joints, or drive axle shaft bearings, always check to see that the ground wires between the body of the vehicle and the engine block or negative (−) terminal of the battery are okay and the connections at both ends are clean

and tight. If there is any doubt as to whether the body ground wires are okay, additional wires can always be added between the body and the engine without causing any harm.

TECH TIP

What's That Sound?

Defective wheel bearings usually make noise. The noise most defective wheel bearings make sounds like noisy snow tires. Wheel bearing noise will remain constant while driving over different types of road surfaces, while tire tread noise usually changes with different road surfaces. In fact, many defective bearings have been ignored by vehicle owners and technicians because it was thought that the source of the noise was the aggressive tread design of the mud and snow tires. Always suspect defective wheel bearings whenever you hear what seems to be extreme or unusually loud tire noise.

PHOTO SEQUENCE Wheel Bearing Service

PS4–1 After safely hoisting the vehicle, remove the wheel(s) and then the dust (grease) cap.

PS4–2 Remove the old cotter key and discard it.

PS4–3 Remove the spindle nut and the washer.

PS4–4 Gently wiggle the disc brake rotor slightly to help free the outer wheel bearing.

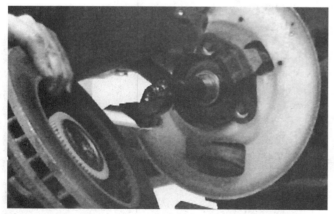

PS4–5 Grasp the rotor and slide the rotor off the spindle.

PS4–6 Use a seal puller to remove the old seal. After the grease seal has been removed, remove the inner wheel bearing.

Wheel Bearing Service—continued

PS4–7 Thoroughly clean the wheel bearings. Cleaning solvent often leaves a residue on the surface of the bearings that should be removed using denatured alcohol. Brake cleaner can also be used. Thoroughly inspect the bearings for damage. Discard any bearing if a fault is discovered.

PS4–8 After the bearing has been thoroughly cleaned, it can be repacked using a bearing packer.

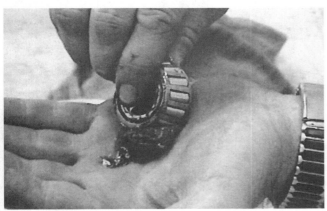

PS4–9 Wheel bearings can also be packed by hand. Place wheel bearing grease into your palm and force the grease between the inner and outer race of the bearing.

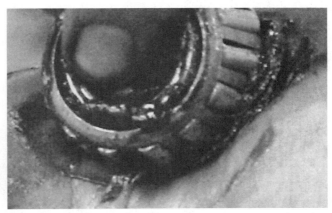

PS4–10 The grease should squirt out from between the roller bearing cage as shown.

PS4–11 Add some grease to the outer surfaces of the bearing.

PS4–12 Place the inner bearing back into the rotor after applying a thin layer of grease to the outer race. Install a new grease seal using a hammer on a grease seal installation tool to properly seat the outside rim of the seal.

Wheel Bearing Service—continued

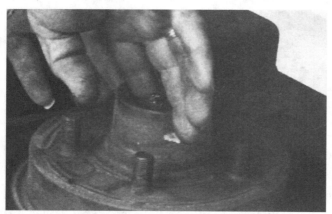

PS4–13 Apply a thin layer of grease to the outer bearing race.

PS4–14 Apply a thin layer of grease to the spindle.

PS4–15 Install the rotor onto the spindle followed by the outer bearing, washer, and spindle nut.

PS4–16 Tighten the spindle nut snugly while rotating the disc brake rotor to seat the bearings. Loosen the nut, and while still rotating the rotor, tighten the nut finger tight.

PS4–17 Install a new cotter key and cut off any excess.

PS4–18 Carefully install the dust (grease) cap. Avoid denting the cap.

■ SUMMARY

1. Wheel bearings support the entire weight of a vehicle and are used to reduce rolling friction. Ball and straight roller-type bearings are nonadjustable, whereas **tapered** roller-type bearings must be adjusted for proper clearance.

2. Most wheel bearings are standardized sizes.

3. Most front-wheel-drive vehicles use sealed bearings, either two preloaded tapered roller bearings or double-row ball bearings.

4. Bearing grease is an oil with a thickener. The higher the NLGI number of the grease, the thicker or harder its consistency.

5. Defective wheel bearings usually make more noise while turning because more weight is applied to the bearing as the vehicle turns.

6. A defective bearing can be caused by metal fatigue that leads to **spalling** or shock loads that cause **brinelling,** bearing damage from electrical arcing due to poor body ground wires, or improper electrical welding on the vehicle.

7. Tapered wheel bearings must be adjusted by hand tightening the spindle nut after properly seating the bearings. A new cotter key must always be used.

8. All bearings must be serviced, replaced, and/or adjusted using the vehicle manufacturer's recommended procedures as stated in the service manual.

■ REVIEW QUESTIONS

1. List three common types of automotive antifriction bearings.

2. Explain the adjustment procedure for a typical tapered roller wheel bearing.

3. List four symptoms of a defective wheel bearing.

4. Describe how the rear axle is removed from a C-clip-type axle.

■ ASE CERTIFICATION-TYPE QUESTIONS

1. Which type of automotive bearing can withstand radial and thrust loads, yet must be adjusted for proper clearance?
   a. Roller bearing
   b. Tapered roller bearing
   c. Ball bearings
   d. Needle roller bearing

2. Most sealed bearings used on the front wheels of front-wheel-drive vehicles are usually which type?
   a. Roller bearing
   b. Single tapered roller bearing
   c. Double-row ball bearing
   d. Needle roller bearing

3. On a bearing that has been shock loaded, the race (cup) of the bearing can be dented. This type of bearing failure is called
   a. Spalling
   b. Arcing
   c. Brinelling
   d. Fluting

4. The bearing grease most often specified is rated NLGI
   a. #00
   b. #0
   c. #1
   d. #2

5. A non-drive-wheel bearing adjustment procedure includes a final spindle nut tightening torque of
   a. Finger tight
   b. 5 in. lb.
   c. 12–30 lb. ft.
   d. 10–15 lb. ft. plus 1/16 turn

6. After a non-drive-wheel bearing has been properly adjusted, the wheel should have how much end play?
   a. Zero
   b. 0.001 to 0.005″
   c. 0.10 to 0.30″
   d. 1/16″ to 3/32″

7. The differential cover must be removed before removing the rear axle on which type of axle?
   a. Retainer plate
   b. C-clip
   c. Press fit
   d. Welded tube

8. What part *must* be replaced when servicing a wheel bearing on a non-drive wheel?
   a. The bearing cup
   b. The grease seal
   c. The cotter key
   d. The retainer washer

9. A rear axle uses a C-clip-type axle retainer. The rear axle bearing is noisy and requires replacement. Technician A says that the axle itself may also require replacement. Technician B says that the rear axle seal should also be replaced. Which technician is correct?
   a. Technician A only
   b. Technician B only
   c. Both Technician A and B
   d. Neither Technician A nor B

10. A defective wheel bearing usually sounds like
    a. Marbles in a tin can
    b. A noisy tire
    c. A baby rattle
    d. A clicking sound—like a ballpoint pen

Drive Axle Shafts and CV Joints

A drive axle shaft transmits engine torque from the transmission or transaxle (if front-wheel drive) to the rear axle assembly or drive wheels (see Figures 5–1 and 5–2).

Driveshaft is the term used by the Society of Automotive Engineers (SAE) to describe the shaft between the transmission and the rear axle assembly on a rear-wheel-drive vehicle. General Motors, Chrysler, and some other manufacturers use the term **propeller shaft** or **prop shaft** to describe this same part. The SAE term will be used throughout this textbook.

A typical driveshaft is a hollow steel tube. A splined end yoke is welded onto one end that slips over the splines of the output shaft of the transmission (see Figure 5–3). An end yoke is welded onto the other end of the driveshaft. Some driveshafts use a center support bearing.

■ DRIVESHAFT DESIGN

Most driveshafts are constructed of hollow steel tubing. *The forces are transmitted through the surface of the driveshaft tubing.* The surface is, therefore, in tension, and cracks can develop on the outside surface of the driveshaft due to metal fatigue. Driveshaft tubing can bend and, if dented, can collapse (see Figure 5–4 on page 98).

Most rear-wheel-drive cars and light trucks use a one- or two-piece driveshaft. A steel tube driveshaft has a maximum length of about 65" (165 cm). Beyond this critical length, a **center support bearing** must be used, as shown in Figure 5–5 on page 98. A center support bearing is also called a **steady bearing** or **hanger bearing.**

Some vehicle manufacturers use aluminum driveshafts; these can be as long as 90" (230 cm) with no problem. Many extended cab pickup trucks and certain vans use aluminum driveshafts to eliminate the need (and expense) of a center support bearing. Composite material driveshafts are also used in some vehicles. These carbon-fiber-plastic driveshafts are very strong yet lightweight and can be made in extended lengths without the need for a center support bearing.

To dampen driveshaft noise, it is common to line the inside of the hollow driveshaft with cardboard. This helps eliminate the tinny sound whenever shifting between drive and reverse in a vehicle equipped with an automatic transmission (see Figure 5–6 on page 99).

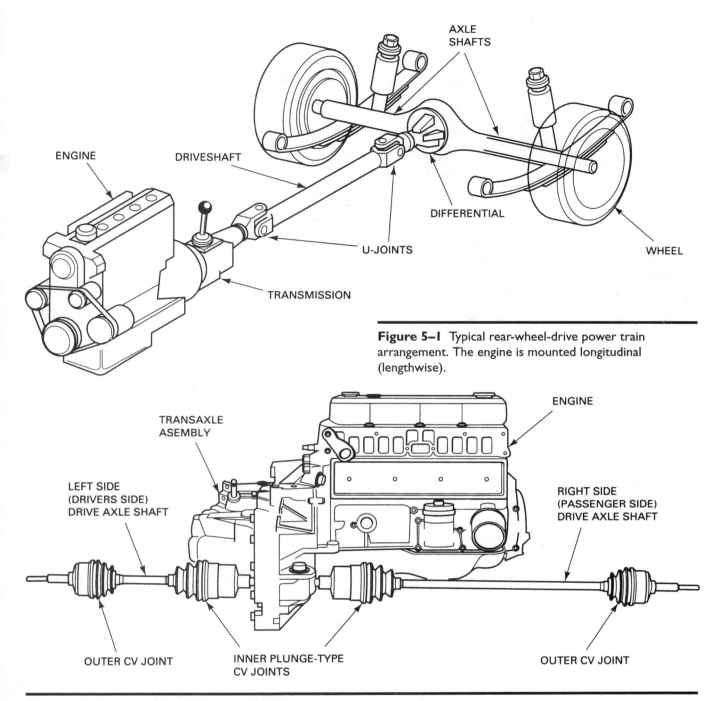

AXLE
SHAFTS

ENGINE

DRIVESHAFT

DIFFERENTIAL

U-JOINTS

WHEEL

TRANSMISSION

Figure 5–1 Typical rear-wheel-drive power train arrangement. The engine is mounted longitudinal (lengthwise).

ENGINE

TRANSAXLE
ASEMBLY

LEFT SIDE
(DRIVERS SIDE)
DRIVE AXLE SHAFT

RIGHT SIDE
(PASSENGER SIDE)
DRIVE AXLE SHAFT

OUTER CV JOINT

INNER PLUNGE-TYPE
CV JOINTS

OUTER CV JOINT

Figure 5–2 Typical front-wheel-drive power train arrangement. The engine is usually mounted transversely (sideways).

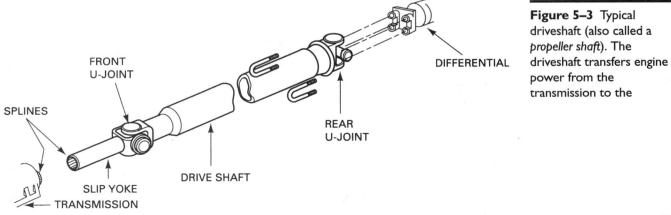

FRONT
U-JOINT

DIFFERENTIAL

SPLINES

REAR
U-JOINT

SLIP YOKE

DRIVE SHAFT

TRANSMISSION

Figure 5–3 Typical driveshaft (also called a *propeller shaft*). The driveshaft transfers engine power from the transmission to the

Figure 5–4 This driveshaft failed because it had a slight dent caused by a rock. When engine torque was applied, the driveshaft collapsed, twisted, and then broke.

■ DRIVESHAFT BALANCE

All driveshafts are balanced. Generally, any driveshaft whose rotational speed is greater than 1000 rpm must be balanced. Driveshaft balance should be within 1/2 of 1 percent of the driveshaft weight. (This is one of the biggest reasons why aluminum or composite driveshafts can be longer because of their light weight.) (See Chapter 13 for driveshaft balancing procedures.)

Driveshafts are often not available by make, model, and year of vehicle. There are too many variations at the factory, such as transmission type, differential, or U-joint type. To get a replacement driveshaft, it is usually necessary to know the series of U-joints (type or style of U-joint) and the center-to-center distance between the U-joints.

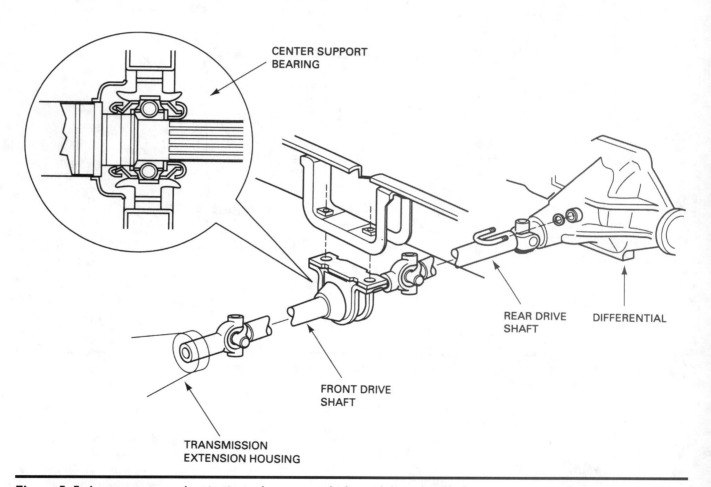

Figure 5–5 A center support bearing is used on many vehicles with long driveshafts.

Figure 5–6 Some driveshafts use rubber between an inner and outer housing to absorb vibrations and shocks to the driveline.

■ U-JOINT DESIGN AND OPERATION

Universal joints (U-joints) are used at both ends of a driveshaft. U-joints allow the wheels and the rear axle to move up and down and remain flexible and still transfer power to the drive wheels. A simple universal joint can be made from two Y-shaped yokes connected by a crossmember called a **cross** or **spider.** The four arms of the cross are called **trunnions.** See Figure 5–7 for a line drawing of a simple U-joint with all part names identi-

fied. A similar design is the common U-joint used with a socket wrench set.

Most U-joints are called **cross yoke** or **Cardan universal joints.** Cardan is named for a 16th-century Italian mathematician who worked with objects which moved freely in any direction. Torque from the engine is transferred through the U-joint. The engine drives the U-joint at a constant speed, but the output speed of the U-joint changes because of the angle of the joint. The speed changes twice per revolution. *The greater the angle, the greater the change in speed (velocity)* (see Figure 5–8).

If only one U-joint were used in a driveline, this change in speed of the driven side (output end) would generate vibrations in the driveline. To help reduce vibration, another U-joint is used at the other end of the driveshaft. If the angles of both joints are nearly equal, the acceleration and deceleration of one joint is offset by the alternate deceleration and acceleration of the second joint. *It is very important that both U-joints operate at about the same angle to prevent excessive driveline vibration* (see Figure 5–9).

Acceptable Working Angles

Universal joints used in a typical driveshaft should have a *working angle* of 1/2 degree to 3 degrees (see Figure 5–10). The working angle is the angle between the driving end and the driven end of the joint. If the driveshaft is perfectly straight (0 degree working angle), then the needle bearings inside the bearing cap are not revolving because there is no force (no difference in angles) to cause the rotation of the needle bearings. If the needle bearings do not rotate, they can exert a constant pressure in one place and damage the bearing journal. If a two-piece drive shaft is used, one U-joint (usually the front) runs at a small working angle of about 1/2 degree, just enough to keep the needle bearings rotating. The

Figure 5–7 A simple universal joint (U-joint).

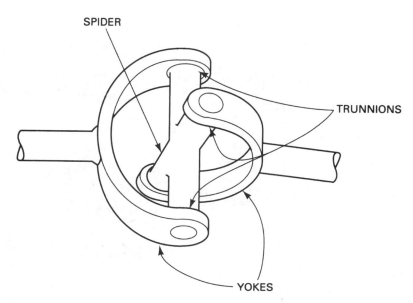

SPIDER

TRUNNIONS

YOKES

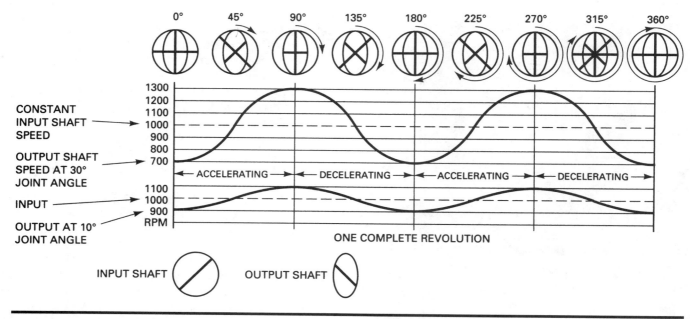

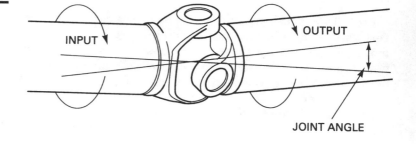

Figure 5–8 How the speed difference on the output of a typical U-joint varies with the speed and the angle of the U-joint. At the bottom of the chart, the input speed is a constant 1000 rpm, while the output speed varies from 900 rpm to 1100 rpm when the angle difference in the joint is only 10 degrees. At the top part of the chart, the input speed is a constant 1000 rpm, yet the output speed varies from 700 to 1200 rpm when the angle difference in the joint is changed to 30 degrees. (Courtesy of Dana Corporation)

Figure 5–9 The joint angle is the difference between the angles of the joint. (Courtesy of Dana Corporation)

other two U-joints (from the center support bearing and rear U-joint at the differential) operate at typical working angles of a single-piece driveshaft.

If the U-joint working angles differ by more than 1/2 degree, a vibration is usually produced that is *torque sensitive*. As the vehicle is first accelerated from a stop, engine torque can create unequal driveshaft angles by causing the differential to rotate on its suspension support arms. This vibration is most noticeable when the vehicle is heavily loaded and being accelerated at lower speeds. The vibration usually diminishes at higher speeds due to decrease in the torque being transmitted. If the driveshaft angles are excessive (over 3 degrees), a vibration is usually produced that increases as the speed of the vehicle (and driveshaft) increases. (See Chapter 13 for additional information on vibration diagnosis and correction.)

Figure 5–10 The angle of this rear U-joint is noticeable.

Figure 5–11 A double Cardan U-joint.

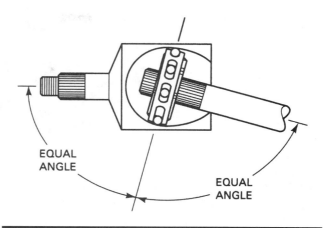

EQUAL
ANGLE

EQUAL
ANGLE

Figure 5–12 A constant velocity (CV) joint can operate at high angles without a change in velocity (speed) because the joint design results in equal angles between input and output. (Courtesy of TRW Inc.)

■ CONSTANT VELOCITY JOINTS

Constant velocity joints (commonly called CV joints) are designed to rotate without changing speed. Regular U-joints are usually designed to work up to 12 degrees of angularity. If two Cardan-style U-joints are joined together, the angle at which this **double Cardan** joint can function is about 18 degrees to 20 degrees (see Figure 5–11).

Double Cardan U-joints were first used on large rear-wheel-drive vehicles to help reduce driveline-induced vibrations, especially when the rear of the vehicle was fully loaded and driveshaft angles were at their greatest. As long as a U-joint (either single or double Cardan) operates in a straight line, the driven shaft will rotate at the same constant speed (velocity) as the driving shaft. As the angle increases, the driven shaft speed or velocity varies during each revolution. This produces pulsations and a noticeable vibration or surge. The higher the shaft speed and the greater the angle of the joint, the greater the pulsations.

NOTE: Many four-wheel-drive light trucks use standard Cardan-style U-joints in the front drive axles. If the front wheels are turned sharply and then accelerated, the entire truck often shakes due to the pulsations created by the speed variations through the U-joints. This vibration is normal and cannot be corrected. It is characteristic of this type of design and is usually not noticeable in normal driving.

The first constant velocity joint was designed by Alfred H. Rzeppa (pronounced shep'pa) in the mid 1920s. The **Rzeppa joint** transfers power through six round balls that are held in position midway between the two shafts. This design causes the angle between the shafts to be equally split regardless of the angle (see Figure 5–12).

Because the angle is always split equally, torque is transferred equally without the change in speed (velocity) that occurs in Cardan-style U-joints. This style joint results in a constant velocity between driving and driven shafts. It can also function at angles greater than simple U-joints can, up to 40 degrees.

NOTE: CV joints are also called **LÖBRO joints,** the brand name of an original equipment manufacturer.

While commonly used today in all front-wheel-drive vehicles and many four-wheel-drive vehicles, its first use was on the front-wheel-drive 1929 Cord. Built in Auburn, Indiana, the Cord was the first front-wheel-drive car to use a CV-type drive axle joint.

Outer CV Joints

The Rzeppa-type CV joint is most commonly used as an outer joint on most front-wheel-drive vehicles (see Figure 5–13). The outer joint must:

1. Allow up to 40 degrees or more of movement to allow the front wheels to turn
2. Allow the front wheels to move up and down through normal suspension travel in order to provide a smooth ride over rough surfaces
3. Be able to transmit engine torque to drive the front wheels

Outer CV joints are called **fixed joints.** The outer joints are also attached to the front wheels. They are more likely to suffer from road hazards that often can cut through the protective outer flexible boot (see Figure 5–14). Once this boot has been split open, the special high-quality grease is thrown out and contaminants

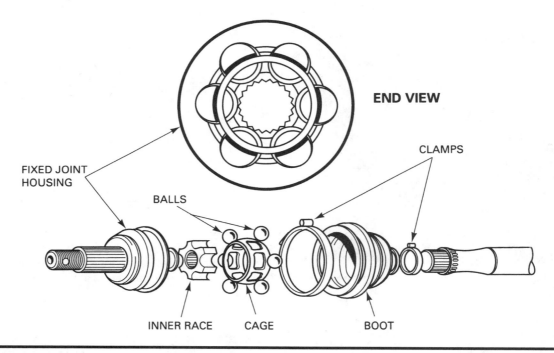

END VIEW

CLAMPS

FIXED JOINT
HOUSING

BALLS

INNER RACE   CAGE   BOOT

Figure 5–13 A Rzeppa fixed joint. This type of CV joint is commonly used at the wheel side of the drive axle shaft This joint can operate at high angles to compensate for suspension travel and steering angle changes. (Courtesy of Dana Corporation)

Figure 5–14 A drive axle shaft being removed from a vehicle because of the torn boot. This joint was ruined and was making a very loud clicking noise especially when turning.

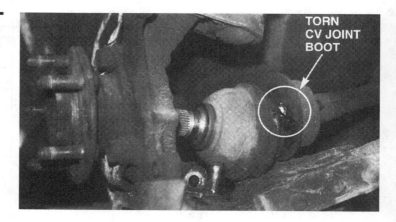

TORN
CV JOINT
BOOT

such as dirt and water can enter. Some joints cannot be replaced individually if worn (see Figure 5–15).

NOTE: Research has shown that in as few as eight hours of driving time, a CV joint can be destroyed by dirt, moisture, and a lack of lubrication if the boot is torn. The technician should warn the owner as to the possible cost involved in replacing the CV joint itself whenever a torn CV boot is found.

Inner CV Joints

Inner CV joints attach the output of the transaxle to the drive axle shaft. Inner CV joints are therefore inboard, or toward the center of the vehicle (see

Figure 5–16). Inner CV joints have to be able to perform two very important movements:

1. Allow the drive axle shaft to move up and down as the wheels travel over bumps
2. Allow the drive axle shaft to change length as required during vehicle suspension travel movements (lengthening and shortening as the vehicle moves up and down; same as the slip yoke on a conventional RWD driveshaft). Inner CV joints are also called **plunge joints.**

Drive Axle Shafts

Unequal length drive axle shafts (also called **half shafts**) result in unequal drive axle shaft angles to the

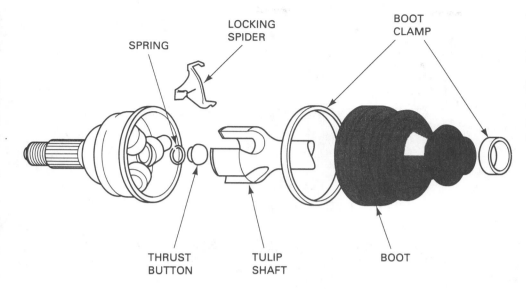

SPRING

LOCKING
SPIDER

BOOT
CLAMP

THRUST
BUTTON

TULIP
SHAFT

BOOT

Figure 5–15 A tripod fixed joint. This type of joint is found on some Renault and Japanese vehicles. If the joint wears out, it is to be replaced with an entire drive axle shaft assembly. (Courtesy of Dana Corporation)

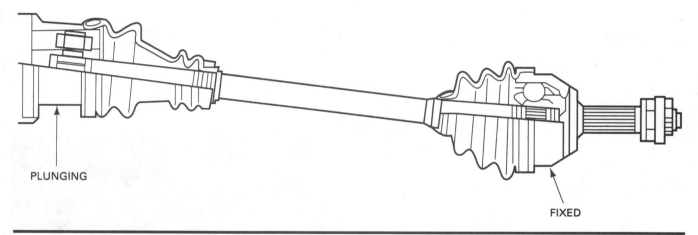

PLUNGING

FIXED

Figure 5–16 The fixed outer joint is required to move in all directions because the wheels must turn for steering as well as move up and down during suspension movement. The inner joint has to be able to not only move up and down but also plunge in and out as the suspension moves up and down. (Courtesy of Dana Corporation)

front drive wheels (see Figure 5–17). This unequal angle often results in a pull on the steering wheel during acceleration. This pulling to one side during acceleration due to unequal engine torque being applied to the front drive wheels is called **torque steer**. To help reduce the effect of torque steer, some vehicles are manufactured with an intermediate shaft that results in equal drive axle shaft angles. Both designs use fixed outer CV joints with plunge-type inner joints.

Typical types of inner CV joints that are designed to move axially, or *plunge*, include:

1. Double offset
2. Tripod
3. Cross groove

Frequently Asked Question **???**

What Is That Weight for on the Drive Axle Shaft?

Some drive axle shafts are equipped with what looks like a balance weight (see Figure 5–18 on page 110). It is actually a dampener weight used to dampen out certain driveline vibrations. The weight is not used on all vehicles and may or may not appear on the same vehicle depending on engine, transmission, and other options. The service technician should always try to replace a defective or worn drive axle shaft with the exact replacement, so when replacing an entire drive axle shaft, the technician should always follow the manufacturers' instructions regarding either transferring or not transferring the weight to the new shaft.

UNEQUAL LENGTH DRIVESHAFT

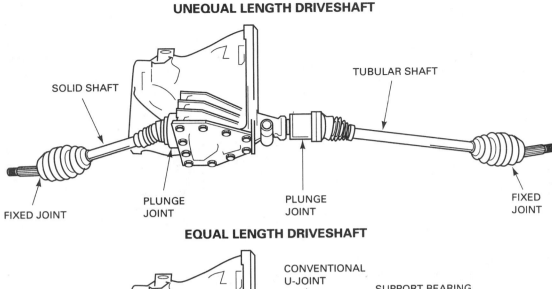

Figure 5–17
Unequal length driveshafts result in unequal drive axle shaft angles to the front drive wheels. This unequal angle side-to-side often results in a steering of the vehicle during acceleration called *torque steer*. By using an intermediate shaft, both drive axles are the same angle and torque steer effect is reduced. (Courtesy of Dana Corporation)

EQUAL LENGTH DRIVESHAFT

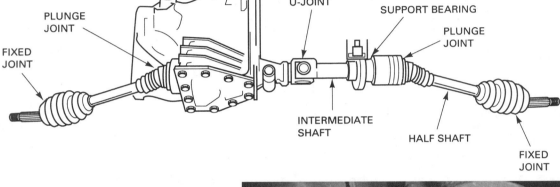

See Figures 5–19, 5–20, and 5–21 on pages 105–106 for examples of each type.

CV joints are also used in rear-wheel-drive vehicles and in many four-wheel-drive vehicles, such as the rear of the FWD van shown in Figure 5–22 on page 106.

CV Joint Boot Materials

The pliable boot surrounding the CV joint must be able to remain flexible under all weather conditions and still be strong enough to avoid being punctured by road debris. There are four basic types of boot materials used over CV joints:

1. **Natural rubber** (black) uses a bridge-type stainless steel clamp to retain.
2. **Silicone rubber** (gray) is a high-temperature-resistant material that is usually only used in places that need heat protection, such as the inner CV joint of a front-wheel-drive vehicle.
3. **Hard thermoplastic** (black) is a hard plastic material requiring heavy-duty clamps and a lot of torque to tighten (about 100 lb. ft.).

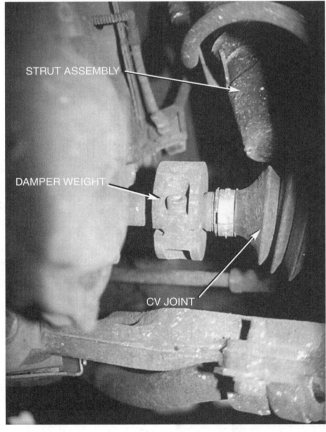

Figure 5–18 A typical drive axle shaft with dampener weight.

TRIPOD TYPE PLUNGE JOINT

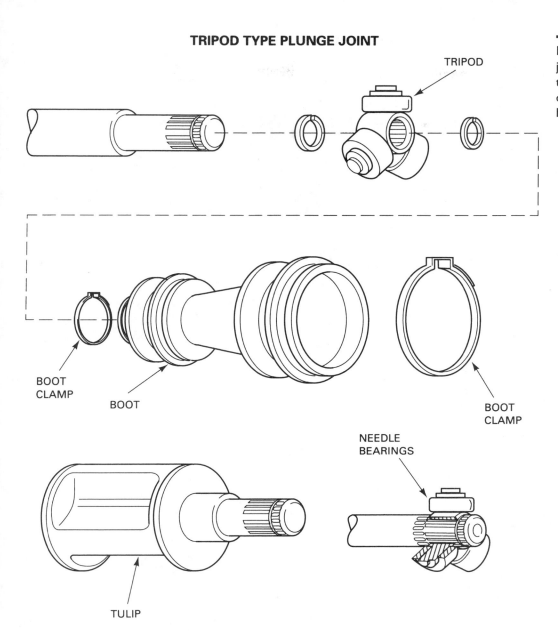

TRIPOD

BOOT
CLAMP

BOOT

BOOT
CLAMP

NEEDLE
BEARINGS

TULIP

NOTE: *CARE MUST BE TAKEN OR TRIPOD ROLLERS MAY COME OFF TRIPOD.*

4. **Urethane** (usually blue) is a type of boot material usually found in an aftermarket part. See Figure 5–23 on page 107 for examples of various types of CV joint boots as used on Chrysler vehicles.

NOTE: Some aftermarket companies offer a split-style replacement CV joint boot. Being split means that the boot can be replaced without having to remove the drive axle shaft. Vehicle manufacturers usually do *not* recommend this type of replacement boot because the joint cannot be disassembled and properly cleaned with the drive axle still in the vehicle. The split boots must also be kept perfectly clean (a hard job to do with all the grease in the joint) in order to properly seal the seam on the split boot.

It is important that boot seals be inspected regularly and replaced if damaged. The inboard (plunging joint) can often pump water into the joint around the seals or through small holes in the boot material itself because the joint moves in and out. Seal retainers are used to provide a leak-proof connection between the boot seal and the housing or axle shaft.

CV Joint Grease

CV joints require special greases. Grease is an oil with thickening agents. Greases are named for the thickening agents used.

Most CV joint grease is molybdenum disulfide–type grease, commonly referred to as *moly* grease. The exact composition of grease can vary depending on the CV joint manufacturer. *The grease supplied with a*

Figure 5–20 A cross-groove plunge joint is used on many German front-wheel-drive vehicles and as both inner and outer joints on the rear of vehicles that use an independent-type rear suspension. (Courtesy of Dana Corporation)

CROSS GROOVE PLUNGE JOINT

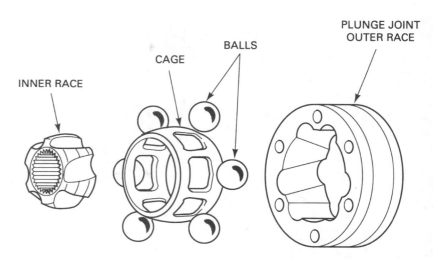

DOUBLE OFFSET BALL TYPE PLUNGE JOINT

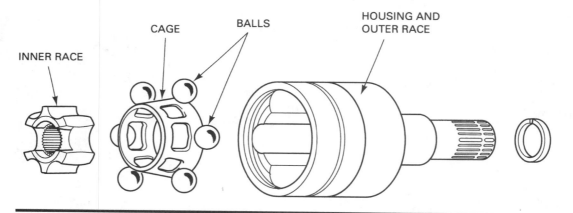

Figure 5–21 Double offset ball–type plunge joint. (Courtesy of Dana Corporation)

Figure 5–22 An inner plunge–type CV joint on the rear of a four-wheel-drive van. The rear differential gets its power from the front-wheel-drive engine and transaxle through a viscous coupling and driveshaft.

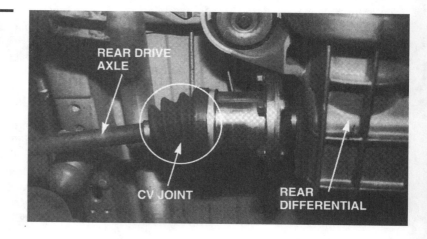

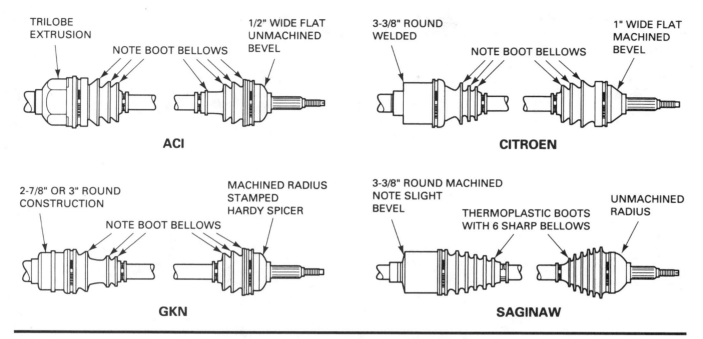

Figure 5–23 Getting the correct boot kit or parts from the parts store is more difficult on many Chrysler front-wheel-drive vehicles because Chrysler has used four different manufacturers for their axle shaft assemblies. (Courtesy of Dana Corporation)

replacement CV joint or boot kit should be the only grease used.

The exact mix of chemicals, viscosity (thickness), wear, and corrosion-resistant properties varies from one CV joint application to another. Some technicians mistakenly think that the *color* of the grease determines in which CV joint it is used. The color—such as black, blue, red, or tan—is used to identify the grease during manufacturing and packaging as well as to give the grease a consistent, even color (due to blending of various ingredients in the grease).

The exact grease to use depends on many factors, including:

1. The type (style) of CV joint. For example, outer (fixed) and inner (plunging) joints have different lubricating needs.
2. The location of the joint on the vehicle. For example, inner CV joints are usually exposed to the greatest amount of heat.
3. The type of boot. The grease has to be compatible with the boot material.

■ DIFFERENTIALS

A differential is used on the drive wheels of vehicles to allow for the transmission of torque to the drive wheels and to permit the drive wheels to rotate at different speeds. A differential is also called a **rear end** or, ab-

breviated, simply **diff.** Whenever any vehicle makes a turn, the outside wheel must travel a greater distance than the inside wheel (see Figure 5–24).

The driveshaft applies torque to the **pinion gear** that meshes below the centerline of a **ring gear.** This type of gear set is called a **hypoid gear set** and requires gear lubrication specifically designed for this type of service. The ring gear is attached to a **differential case** that also contains small beveled **spider gears** or **pinion gears.** A **pinion shaft** passes through the two pinion gears in the case. In mesh with the pinion gears are two **side gears** that are splined to the inner ends of the axles (see Figure 5–25).

Differential Operation

When traveling straight, both rear wheels are turning at the same speed. The ring gear and the case with the pinion shaft rotate the differential pinion gears. The teeth on the differential pinion gears mesh with and apply torque to the axle side gears and axles (see Figure 5–26).

When turning a corner, the outside wheel is turning faster than the inside wheel because the outer wheel has to travel a greater distance.

The outer axle and side gear rotate faster than the inside axle and its side gear. This difference in speed between the side gears causes the differential pinion gears to rotate on the differential pinion shaft. Thrust washers are used to allow the pinion gears and side gears to move without wearing the internal surfaces of the differential case.

Figure 5–24 The difference between the travel distance of the drive wheels is controlled by the differential. (Courtesy of Oldsmobile Division, GMC)

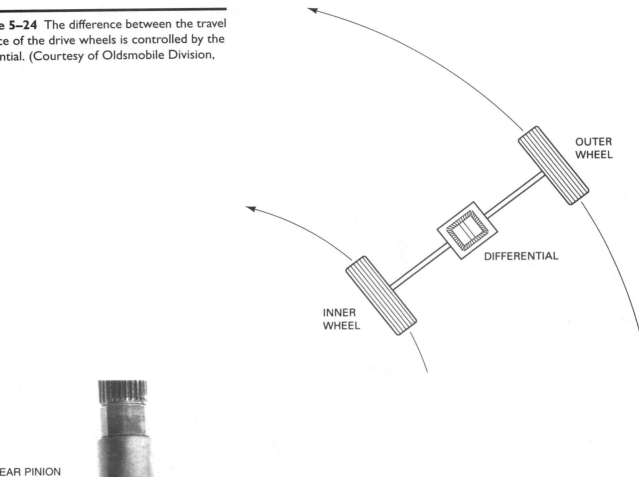

Figure 5–25 Pinion gears and side gears mounted in the case. The ring gear is bolted to the case and meshes with the pinion gear. This entire assembly is placed in the differential *housing*.

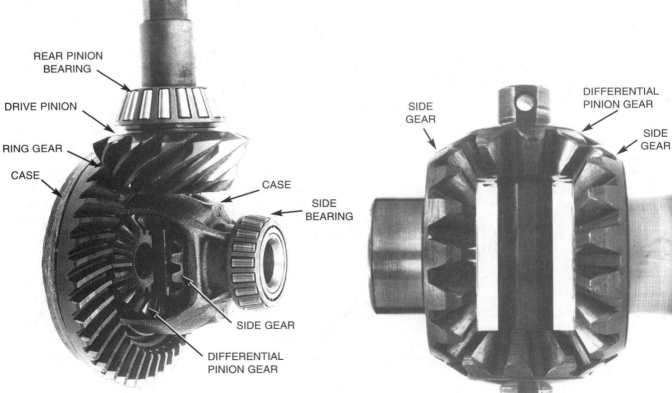

Figure 5–26 Differential pinion gears assembled with two side gears.

NOTE: These thrust washers can easily become worn if the drive wheels are allowed to spin on a slippery surface such as ice or snow. The wheel with the least traction receives the engine torque and spins at *twice* the speed indicated by the speedometer. These thrust washers can also wear if unequal size tires are used on the drive wheels.

Because a conventional differential allows for the drive wheels to rotate at different speeds, when one wheel has little or no traction, the engine torque is applied to the wheel with the *least* traction and causes one wheel to spin on ice or other slippery surfaces.

Limited Slip Differentials

Limited slip differentials use clutches and other parts to limit the amount of slippage that can occur from one wheel to another. Different vehicle and axle manufacturers use different names for their design of limited slip differentials. Some differentials, such as the Eaton locking differential, will lock, sending torque to both rear wheels without any slippage. Other designs allow a limited amount of torque to be transferred to the wheel with the lower traction, hence, the term *limited slip* (see Figure 5–27).

Some transaxles (transmission and differentials in one unit) are equipped with limited slip differentials. Figure 5–28 is a photo of a silicone viscous coupling final drive from a front-wheel-drive vehicle. This type of coupling permits both drive wheels to receive the same amount of engine torque.

If there is very little difference in speeds between the two drive wheels, the silicone fluid is thinner and allows for small changes in speed to allow for road bumps, dips, and turns. If there is a big difference in speed, the viscous silicone fluid increases in viscosity and tends to act as a solid, thereby holding both drive wheels together at the same speed.

Differential Lubricant

Because all differentials use hypoid gear sets, a special lubricant is necessary because the gears both roll and slide between their meshed teeth. Gear lubes are specified by the **American Petroleum Institute (API).**

GL-1	Straight mineral oil
GL-2	Worm-type gear lubricant
GL-3	Mild-type EP lubricant (will *not* protect hypoid gears)
GL-4	Mild-type EP lubricant suitable for manual transmissions/transaxles
GL-5	EP-type; OK for hypoid gears (meets military Mil-L2150B type requirements)

EP additives are sulfurized esters and organic sulfur-phosphorous compounds. They are useful as lubricants between steel and steel. They are not effective between steel and soft metal. In some cases, they accelerate corrosive wear of the soft metal. Lubricants containing EP additives should *not* be used with babbitt-, aluminum-, copper-, or bronze-bearing materials. Most differentials require:

1. SAE 80W-90 GL-5 (see Figure 5–29 on page 111)
2. SAE 75W-90 GL-5
3. SAE 80W GL-5

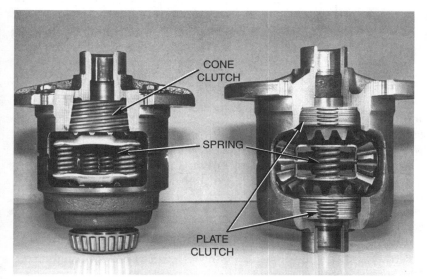

Figure 5–27 Cutaway sections of limited-slip-type differential cases showing a cone-type clutch (left) and a plate-type clutch (right).

(a)

(b)

(c)

Figure 5–28 (a) A manual transaxle assembly showing the final drive unit (differential) and viscous coupling that acts as a limited-slip differential on this front-wheel-drive vehicle. (b) The inside of the viscous coupling consists of thin metal discs. (c) Viscous silicone fluid is used between the metal discs.

NOTE: Always check the *exact* specification before adding or replacing rear axle lubricant. For example, manual transmissions or transaxles may require any one of four possible lubricants, including:

SAE 80W-90 GL-5

STF (synchromesh transmission fluid) similar to ATF, but with friction characteristics designed for manual transmissions

ATF (automatic transmission fluid)

Engine oil (usually SAE 5W-30)

Limited slip differentials (often abbreviated LSD) usually use an additive that modifies the friction characteristics of the rear axle lubricant to prevent chattering while cornering. Figure 5–30 shows typical differential fluid inspection and fill plug.

■ FOUR-WHEEL-DRIVE SYSTEMS

Two-wheel-drive vehicles use engine torque to turn either the front or the rear wheels. A differential is re-

quired to allow the drive wheels to travel different distances and speeds while cornering or driving over bumps or dips in the road. A four-wheel-drive vehicle, therefore, requires two differentials: one for the front wheels and one for the rear wheels.

NOTE: The term 4×4 means a four-wheeled vehicle that has engine torque applied to all four wheels (four-wheel drive). A 4×2 means a four-wheeled vehicle that has engine torque applied to only two wheels (two-wheel drive).

Four-wheel-drive vehicles require more than just two differentials. The front and the rear wheels of a four-wheel-drive vehicle also travel different distances and speeds whenever cornering or running over dips or rises in the road. There are three different methods used to allow for front-to-rear driveline speed variation.

Method 1: Locking Hubs

Engine power from the transmission is applied directly to the rear differential through the *transfer case* (see

Figure 5–29 A container of GL-5 SAE 80W-90 gear lubricant.

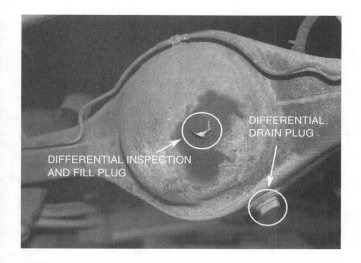

Figure 5–30 Some differentials come equipped with a drain plug as well as an inspection and fill plug making it easier to change the differential lubricant.

Figure 5–31). The transfer case permits the driver to select a low-speed, high-power gear ratio inside the transfer case while in four-wheel drive. These positions and their meanings include:

4H   Four-wheel drive with no gear reduction in the transfer case.

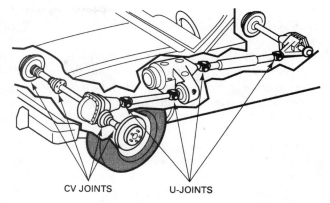

CV JOINTS   U-JOINTS

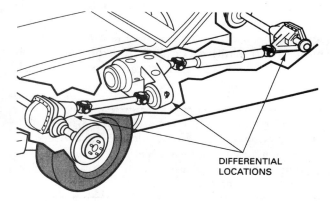

DIFFERENTIAL LOCATIONS

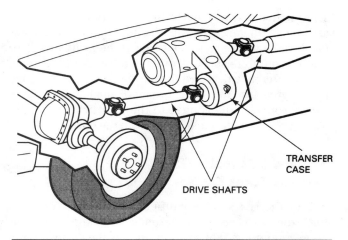

TRANSFER CASE

DRIVE SHAFTS

Figure 5–31 Many light trucks and sport utility vehicles use a transfer case to provide engine power to all four wheels and to allow a gear reduction for maximum power to get through mud or snow. (Courtesy of Dana Corporation)

4L   Four-wheel drive with gear reduction. Use of this position is usually restricted to low speeds on slippery surfaces.

2H   Two-wheel drive (rear wheels only) in high range, meaning no gear reduction in the transfer case.

> **CAUTION:** Check the owner's manual or service manual for the recommended procedure to follow when changing from one position to another in the transfer case. Some manuals require that the vehicle be stopped before selecting between two- and four-wheel drive, and between high and low range.

The transfer case also applies torque to the front differential. Torque is then applied to the front wheels through the drive axles to the **locking hubs.** In normal 4H driving on hard surfaces, the front hubs *must* be in the unlocked position. On loose road surfaces that can absorb and allow for tire slippage due to the different tire speeds front to back, the front hubs are locked. Some four-wheel-drive vehicles, such as the GM S-10 pickup truck, use a clutch or disconnect device that disengages the front drive axle shaft from the front wheels. This clutch is built into the front differential housing.

> **CAUTION:** Failure to unlock the front wheel hubs while driving on a hard road surface can cause serious driveline vibrations and damage to driveshafts, U-joints, and bearings, as well as to the transfer case, transmission, and even the engine.

Method 2: Auto-Locking Hubs

This method uses a clutch arrangement built into the hub assembly. Whenever driving on smooth, hard road surfaces, the hubs *free wheel* and allow the front wheels to rotate at different speeds from the rear wheels. When the speed difference between the wheels and the front drive axle is great, the hubs will automatically lock and allow engine torque to be applied to the front wheels. Figure 5–32 shows an auto-locking hub with the cover removed.

Method 3: Full-Time Four-Wheel Drive

This method uses a center differential to allow front and rear wheels to travel at different speeds under all operating conditions. While this method is the easiest to operate both on and off the road, the center differential

Figure 5–32 Automatic-locking hub with the cover removed.

can cause the vehicle to get stuck in mud or snow even though it is a "four-wheel-drive vehicle." All open-style differentials allow for speed differences. Torque is applied equally to both drive wheels. If one wheel has little traction, the other wheel will not receive equal torque to propel the vehicle. This is why many vehicles are spinning just one wheel when stuck on ice or snow.

If a center differential is used, and one rear wheel starts to spin, all of the engine power is applied to the spinning wheel and not to the front wheels, where it is most needed. The most common solution to this problem is to lock the center differential to prevent this from happening. A **viscous coupling** is commonly used on many four-wheel-drive vehicles to provide an automatic lockup of the center differential. A viscous coupling is a

Figure 5–33 A viscous coupling used between the front transaxle and the rear differential of this all-wheel-drive mini van.

type of fluid clutch. When the speed difference between the front and rear wheels is high enough, the silicone fluid inside the coupling stiffens to reduce the speed difference between the front and rear driveshafts. This center differential lock, combined with a limited-slip-type rear differential, greatly helps the vehicle maintain traction under all road and weather conditions (see Figure 5–33).

■ ALL-WHEEL DRIVE

Some cars and light trucks are equipped with an all-wheel-drive system that uses a transfer case with a center differential and only one speed (high). Low-range gear reduction is not used. A viscous coupling is usually incorporated into the center differential to provide superior all-weather traction. Combined with a limited slip differential in the rear, and sometimes also in the front, an all-wheel drive system can provide ideal road traction under all driving conditions without any action by the driver.

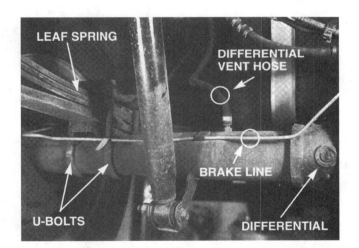

Figure 5–34 The differential vent on this truck uses a hose which is routed upward and ends above the level of the tires. Placing the vent this high prevents water from getting into the differential if the vehicle were driven off road or in deep water.

PHOTO SEQUENCE   Removing and Installing a Driveshaft

PS5–1 Start the removal of a driveshaft (propeller shaft) by safely hoisting the vehicle and removing the rear U-joint attaching bolts.

PS5–2 This truck is equipped with a center support bearing assembly that needs to be unbolted from the crossmember.

PS5–3 This Chevrolet C-30 truck also uses bolts to retain the front U-joint to the output shaft of the transmission.

PS5–4 Removing the driveshaft from underneath the vehicle.

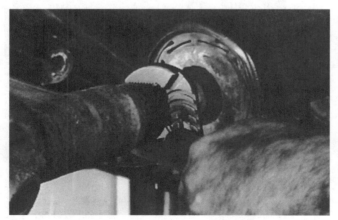

PS5–5 Loosening the shaft seal retainer to allow easy separation of the two-piece driveshaft.

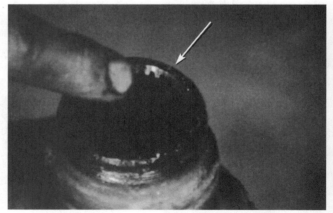

PS5–6 Note the wider spline on the yoke of the driveshaft. This one wider tooth prevents the driveshaft from being installed out of phase.

Removing and Installing a Driveshaft—continued

PS5–7 The corresponding mating spline on the shaft.

PS5–8 After thoroughly cleaning and lubricating the spline, reassemble the two-piece driveshaft.

PS5–9 Carefully check the condition of the U-joints by moving the joints by hand and feeling for any binding or roughness. Before reinstalling the driveshaft, double-check that the U-joints at both ends are parallel to each other and therefore are in phase.

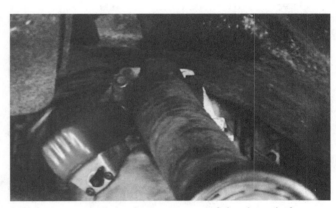

PS5–10 Reinstall the front portion of the driveshaft.

PS5–11 Reinstall the center support bearing and torque to factory specifications.

PS5–12 Reinstall the rear U-joint to the differential and torque the retaining bolts to factory specifications.

■ SUMMARY

1. The driveshaft of a rear-wheel-drive vehicle transmits engine power from the transmission to the differential.

2. driveshaft length is usually limited to about 65" due to balancing considerations unless a two-piece or a composite material shaft is used.

3. Universal joints (U-joints) allow the driveshaft to transmit engine power while the suspension and the rear axle assembly are moving up and down during normal driving conditions.

4. Acceptable working angles for a Cardan-type U-joint are 1/2 to 3 degrees. Some angle is necessary to cause the roller bearings to rotate; a working angle of greater than 3 degrees can lead to driveline vibrations.

5. Constant velocity (CV) joints are used on all front-wheel-drive vehicles and many four-wheel-drive vehicles to provide a smooth transmission of torque to the drive wheels regardless of angularity of the wheel or joint.

6. Outer or fixed CV joints commonly use a Rzeppa design, while inner CV joints are the plunging or tripot type.

7. In a typical differential in a rear-wheel-drive vehicle, the driveshaft applies engine torque to the pinion gear that meshes with and turns a ring gear attached to the differential case. The axles are splined to side gears that rotate with the differential case and mesh with the pinion gears. When a corner is turned, the inside wheel slows down, causing the pinion gear to rotate on the pinion shaft.

8. Limited-slip differentials limit the amount of slippage that can occur between the two drive wheels of the drive axle. Special hypoid gear lubricant is required in all differentials, and limited-slip units usually require a special friction additive.

9. Four-wheel-drive vehicles use two differentials, one for the front wheels and one for the rear wheels. Locking hubs with or without a center differential and transfer case are used to transmit engine torque to all four wheels.

■ REVIEW QUESTIONS

1. Explain why Cardan-type U-joints on a driveshaft must be within 1/2 degree working angles.

2. Describe how the differential allows the outside drive wheel to travel faster than the inside drive wheel while the vehicle is turning a corner.

3. Explain why many four-wheel-drive vehicles should not be driven on a smooth, hard surface with the front hubs locked.

4. What makes a constant velocity joint able to transmit engine torque through an angle at a constant velocity?

5. What type of grease must be used in CV joints?

■ ASE CERTIFICATION-TYPE QUESTIONS

1. The name most often used to describe the universal joints on a conventional rear-wheel-drive vehicle driveshaft is
   a. Trunnion   c. CV
   b. Cardan   d. Spider

2. A rear-wheel-drive vehicle shudders or vibrates when first accelerating from a stop. The vibration is less noticeable at higher speeds. The most likely cause is
   a. driveshaft unbalance
   b. Excessive U-joint working angles
   c. Unequal U-joint working angles
   d. Brinelling of the U-joint

3. All driveshafts are balanced.
   a. True   b. False

4. The maximum difference between the front and rear working angle of a driveshaft is
   a. 1/4 degree   c. 1 degree
   b. 1/2 degree   d. 3 degrees

5. An all-wheel-drive vehicle has how many differentials?
   a. One   c. Three
   b. Two   d. Four

6. The owner of a full-size four-wheel-drive pickup truck complains that whenever turning sharply and accelerating rapidly, a severe vibration is created. Technician A says that the transfer case may be defective. Technician B says that this is normal if conventional Cardan U-joints are used to drive the front wheels. Which technician is correct?
   a. Technician A only
   b. Technician B only
   c. Both Technician A and B
   d. Neither Technician A nor B

7. The outer CV joints used on front-wheel-drive vehicles are
   a. Fixed type
   b. Plunge type

8. The proper grease to use with a CV joint is
   a. Black chassis grease
   b. Dark blue EP grease
   c. Red moly grease
   d. The grease that is supplied with the boot kit

9. The pinion gears in a differential spin on the pinion shaft whenever the vehicle is driven
   a. On a straight road
   b. Around a curve or corner

10. Which component is *not* used on a vehicle equipped with all-wheel drive?
    a. Front differential
    b. Rear differential
    c. Center differential
    d. Locking hubs

Drive Axle Shaft and CV Joint Service

The driveshaft of a typical rear-wheel-drive (RWD) vehicle rotates about three times faster than the wheels. This is due to the gear reduction that occurs in the differential. The differential not only provides gear reduction but also allows for a difference in the speed of the rear wheels that is necessary whenever turning a corner.

The driveshaft rotates the same speed as the engine if the transmission ratio is 1 to 1 (1:1). The engine speed, in revolutions per minute (rpm), is transmitted through the transmission at the same speed. In lower gears, the engine speed is many times faster than the output of the transmission. Most transmissions today, both manual and automatic, have an overdrive gear. This means that at highway speeds, the driveshaft is rotating faster than the engine (the engine speed is de-

creased or overdriven to help reduce engine speed and improve fuel economy).

The driveshaft must travel up and down as the vehicle moves over bumps and dips in the road while rotating and transmitting engine torque to the drive wheels. The driveshaft and universal joints should be carefully inspected whenever any of the following problems or symptoms occur:

1. Vibration or harshness at highway speed
2. A clicking sound whenever the vehicle is moving either forward or in reverse

NOTE: A click-click-click sound while moving in reverse is usually the first indication of a defective U-joint. This clicking occurs in reverse because the needle bearings are being forced to rotate in a direction opposite the usual.

3. A clunking sound whenever changing gears, such as moving from drive to reverse

■ DRIVESHAFT AND U-JOINT INSPECTION

The driveshaft should be inspected for the following:

1. Any dents or creases caused by incorrect hoisting of the vehicle or by road debris.

CAUTION: A dented or creased driveshaft can collapse, especially when the vehicle is under load. This collapse of the driveshaft can cause severe damage to the vehicle and may cause an accident.

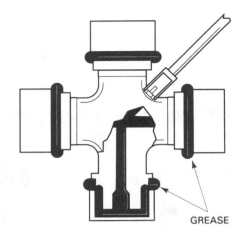

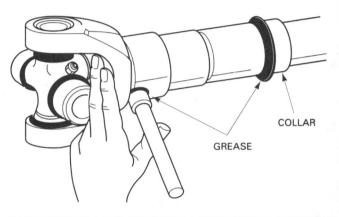

Figure 6–2 All U-joints and spline collars equipped with a grease fitting should be greased four times a year as part of a regular lubrication service. (Courtesy of Dana Corporation)

Figure 6–1 Notice how the needle bearings have worn grooves into the bearing surface of the U-joint.

2. Undercoating, grease, or dirt buildup on the driveshaft can cause a vibration.
3. Undercoating should be removed using a suitable solvent and a rag. Always dispose of used rags properly.

The U-joints should be inspected every time the vehicle chassis is lubricated, or four times a year. Most original equipment (OE) U-joints are permanently lubricated and have no provision for greasing. If there is a grease fitting, the U-joint should be lubricated by applying grease with a grease gun (see Figures 6–2 and 6–3).

Besides periodic lubrication, the driveshaft should be grabbed and moved to see if there is any movement of the U-joints. If *any* movement is noticed when the driveshaft is moved, the U-joint is worn and must be replaced.

> **NOTE:** U-joints are not serviceable items and cannot be repaired. If worn or defective, they must be replaced.

U-joints can be defective and still not show noticeable free movement. *A proper U-joint inspection can be performed only by removing the driveshaft from the vehicle.*

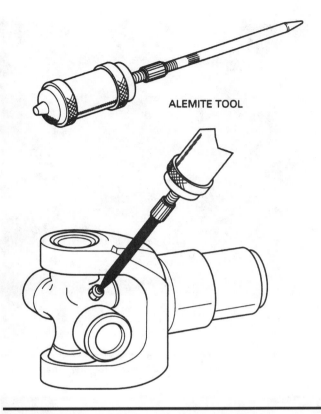

ALEMITE TOOL

Figure 6–3 Many U-joints require a special grease gun tool to reach the grease fittings. (Courtesy of Dana Corporation)

Figure 6–4 Always mark the original location of U-joints before disassembly. (Courtesy of Dana Corporation)

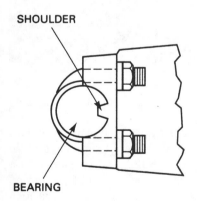

SHOULDER

BEARING

U-BOLT TYPE

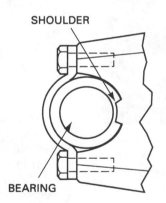

SHOULDER

BEARING

STRAP TYPE

Figure 6–5 Two types of retaining methods that are commonly used at the rear U-joint at the differential. (Courtesy of Dana Corporation)

Before removing the driveshaft, always mark the position of all mating parts to ensure proper reassembly. White correction fluid, also known as "White Out" or "Liquid Paper," is an easy and fast-drying marking material (see Figure 6–4).

To remove the driveshaft from a rear-wheel-drive vehicle, remove the four fasteners at the rear U-joint at the differential (see Figure 6–5).

Push the driveshaft forward toward the transmission and then down and toward the rear of the vehicle. The driveshaft should slip out of the transmission spline and can be removed from underneath the vehicle.

HINT: With the driveshaft removed, transmission lubricant can leak out of the rear extension housing. To prevent a mess, use an old spline the same size as the one being removed or place a plastic bag over the extension housing to hold any escaping lubricant. A rubber band can be used to hold the bag onto the extension housing.

To inspect U-joints, move each joint through its full travel, making sure it can move (articulate) freely and equally in all directions (see Figure 6–6).

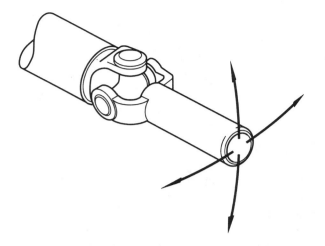

Figure 6–6 The best way to check any U-joint is to remove the driveshaft from the vehicle and move each joint in all directions. A good U-joint should be free to move without binding. (Courtesy of Dana Corporation)

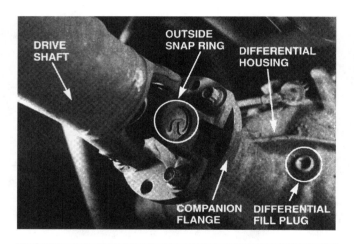

Figure 6–7 Typical U-joint that uses an outside snap ring. This style joint bolts directly to the companion flange that is attached to the pinion gear in the differential.

■ U-JOINT REPLACEMENT

All movement in a U-joint should occur between the trunnions and the needle bearings in the end caps. The end caps are press-fit to the yokes, which are welded to the driveshaft. There are three types of retainers used to keep the bearing caps on the U-joints: the outside snap ring (see Figure 6–7), the inside retaining ring (see Figure 6–8), and injected synthetic (usually nylon).

After removing the retainer, use a press to separate the U-joint from the yoke, as in Figure 6–9.

U-joints that use synthetic retainers must be separated using a press and a special tool to press onto both

Figure 6–8 Removing an inside retaining ring (snap ring).

sides of the joint in order to shear the plastic retainer, as shown in Figure 6–11.

Replacement U-joints use spring clips instead of injected plastic. Remove the old U-joint from the yoke, as shown in Figure 6–12 on page 122, and replace with a new U-joint. Replacement U-joints should be *forged* (never cast) and use up to thirty-two needle bearings (also called **pin bushings**) instead of just twenty-four needle bearings used in lower-quality U-joints. Replacement U-joints usually have a grease fitting so that the new replacement U-joint can be properly lubricated (see Figure 6–13 on page 122).

After removing any dirt or burrs from the yoke, press in a new U-joint. Rotate the new joint after installation to make sure it moves freely, without binding or stiffness. If a U-joint is stiff, it can cause a vibration.

HINT: If a U-joint is slightly stiff after being installed, strike the U-joint using a brass punch and a light hammer. This often frees a stiff joint and is often called "relieving the joint." The shock aligns the needle bearings in the end caps.

Figure 6–9 Pushing the lower bushing cap out of a U-joint by pushing the upper bushing cap inward using a press.

Figure 6–10 A new U-joint being pressed onto the yoke using a special fixture. Note that tape has been placed over the other end caps to keep them from falling off.

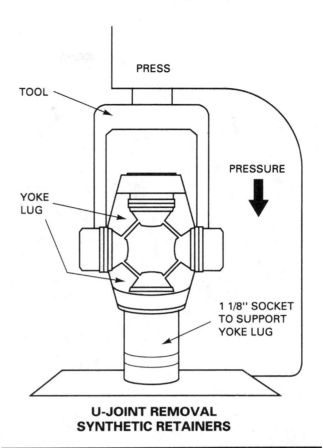

U-JOINT REMOVAL SYNTHETIC RETAINERS

Figure 6–11 A special tool being used to press apart a U-joint that is retained by injected plastic. Heat from a propane torch may be necessary to soften the plastic to avoid exerting too much force on the U-joint.

U-Joint Working Angles

Unequal or incorrect U-joint working angles can cause severe vibrations. Drive shaft and U-joint angles may change from the original factory setting due to one or more of the following:

1. Defective or collapsed engine or transmission mounts
2. Defective or sagging springs, especially the rear springs due to overloading or other causes
3. Accident damage or other changes to the chassis of the vehicle
4. Vehicle modification that raises or lowers the ride height

Replace any engine or transmission mount that is cracked or collapsed. When a mount collapses, the engine drops from its original location. Now the driveshaft angles are changed and a vibration may be felt.

Rear springs often sag after many years of service or after being overloaded. This is especially true of pickup trucks. Many people carry as much as the cargo bed can hold, often exceeding the factory-recommended carry capacity or gross vehicle weight (GVW) of the vehicle.

Figure 6–12 Removing the worn cross from the yoke.

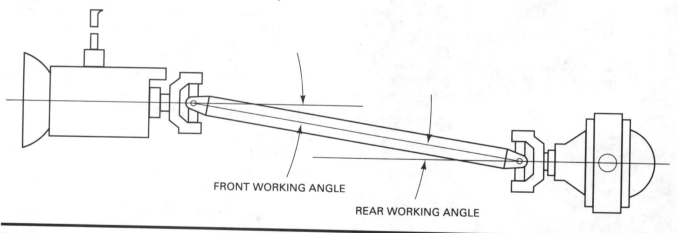

Figure 6–13 When installing a new U-joint, position the grease fitting on the inboard side (toward the driveshaft tube) and in alignment with the grease fitting of the U-joint at the other end. (Courtesy of Dana Corporation)

FRONT WORKING ANGLE

REAR WORKING ANGLE

Figure 6–14 The working angle of most U-joints should be at least 1/2 degree (to permit the needle bearing to rotate in the U-joints) and should not exceed 3 degrees or a vibration can occur in the driveshaft especially at higher speeds. The difference between the front and rear working angles should be within 1/2 degree of each other. (Courtesy of General Motors Corporation, Service Technology Group)

To measure U-joint and driveshaft angles, the vehicle must be hoisted using an axle contact or drive-on-type lift so as to maintain the same driveshaft angles as the vehicle has while being driven. (See Chapter 13 for a detailed explanation of vibration correction, including driveshaft balancing.)

The working angles of the two U-joints on a driveshaft should be within 0.5 degree of each other in order to cancel out speed changes (see Figure 6–14). To measure the working angle of a U-joint, follow these steps:

Step 1 Place an **inclinometer** (a tool used to measure angles) on the rear U-joint bearing cap. Level the bubble and read the angle. (See Figure 6–15. The reading is 19.5 degrees.)

Step 2 Rotate the driveshaft 90 degrees and read the angle of the rear yoke. For example, this reading is 17 degrees.

Step 3 Subtract the smaller reading from the larger reading to obtain the working angle of the joint. In this example it is 2.5 degrees ($19.5° - 17° = 2.5°$).

Figure 6–15 Inclinometer reads 19.5 degrees at this rear U-joint.

Repeat the same procedure for the front U-joint. The front and rear working angles should be within 0.5 degree. If the two working angles are not within 0.5 degree, shims can be added to bring the two angles closer together. The angle of the rear joint is changed by installing a tapered

---

CAUTION: Use caution whenever using wedges between the differential and the rear leaf spring to restore the correct U-joint working angle. Even though wedges are made to raise the front of the differential, the tilt often prevents rear-end lubricant from reaching the pinion bearing, resulting in pinion bearing noise and eventual failure.

**TECH TIP**

Quick and Easy Backlash Test

Whenever a driveline clunk is being diagnosed, one possible cause is excessive backlash (clearance) between the ring gear teeth and differential pinion teeth in the differential. Another common cause of excessive differential backlash is too much clearance between differential carrier pinion teeth and side gear teeth. A quick test to check backlash involves three easy steps:

Step 1   Hoist the vehicle on a frame contact lift, allowing the drive wheels to be rotated.

Step 2   Have an assistant hold one drive wheel and the driveshaft to keep them from turning.

Step 3   Move the other drive wheel, observing how far the tire can rotate. This is the amount of backlash in the differential; it should be less than 1" (25 mm) of movement measured *at the tire*.

If the tire can move more than 1" (25 mm), then the differential should be inspected for wear and parts replaced as necessary. If the tire moves *less* than 1" (25 mm), then the backlash between the ring gear and pinion is probably *not* the cause of the noise.

---

shim between the leaf spring and the axle, as shown in Figure 6–16.

The angle of the front joint is changed by adding or removing shims from the mount under the transmission (see Figure 6–17).

■ CV JOINT DIAGNOSIS

When a CV joint wears or fails, the most common symptom is noise while driving. An outer fixed CV joint will

---

Figure 6–16 Placing a tapered metal wedge between the rear leaf spring and the rear axle pedestal to correct rear U-joint working angles.

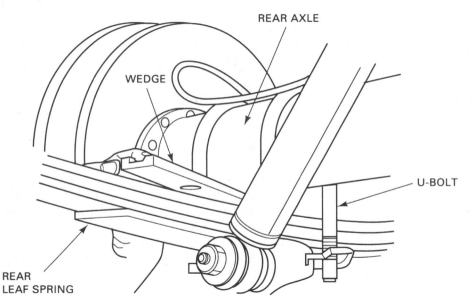

REAR AXLE

WEDGE

U-BOLT

REAR LEAF SPRING

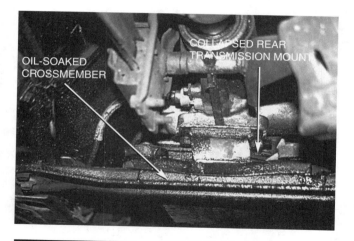

Figure 6–17 A transmission oil pan gasket leak allowed automatic transmission fluid (ATF) to saturate the rear transmission mount rubber, causing it to collapse. After replacing the defective mount, proper driveshaft angles were restored and the driveline vibration was corrected.

Figure 6–18 Typical drive axle shaft assembly complete with hub and bearing, inner and outer CV joints.

most likely be heard when turning sharply and accelerating at the same time. This noise is usually a clicking sound. While inner joint failure is less common, a defective inner CV joint often creates a loud clunk while accelerating from rest. To help verify a defective joint, drive the vehicle in reverse while turning and accelerating. This almost always will reveal a defective outer joint.

■ REPLACEMENT SHAFT ASSEMBLIES

Front-wheel-drive vehicles were widely used in Europe and Japan long before they became popular in North America. The standard repair procedure used in these countries is the replacement of the entire drive assembly if there is a CV joint failure. Replacement boot kits are rarely seen in Europe because it is felt that even a

slight amount of dirt or water inside a CV joint is unacceptable. Vehicle owners simply wait until the joint wear causes severe noise, and then the entire assembly is replaced.

Entire drive axle shaft assemblies can easily be replaced and the defective unit sent to a company for remanufacturing. Even though cost to the customer is higher, the parts and repair shop does not have to inventory every type, size, and style of boot kit and CV joint. Service procedures and practices, therefore, vary according to location and the availability of parts. For example, some service technicians use replacement drive axle assemblies from salvage yards with good success (see Figure 6–18).

> **NOTE:** Some drive axle shafts have a weight attached between the inner and outer CV joint. This is a damper weight. It is not a balance weight, and it need not be transferred to the replacement drive axle shaft (half shaft) unless instructed to do so in the directions that accompany the replacement shaft assembly.

■ CV JOINT SERVICE

The hub nut must be removed whenever servicing a CV joint or shaft assembly on a front-wheel-drive vehicle. Since these nuts are usually torqued to almost 200 lb. ft. (260 N-m), keep the vehicle on the ground until the hub nut is loosened and then follow these steps (see Figure 6–19):

Step 1 Remove the front wheel and hub nut.

> **NOTE:** Most manufacturers warn against using an air impact wrench to remove the hub nut. The impacting force can damage the hub bearing.

Step 2 To allow the knuckle room to move outward enough to remove the drive axle shaft, some or all of the following will have to be disconnected:
    a. Lower ball joint or **pinch bolt** (see Figure 6–20)
    b. Tie rod end (see Figure 6–21 on page 126)
    c. Stabilizer bar link
    d. Front disc brake caliper

Step 3 Remove the splined end of the axle from the hub bearing. Sometimes a special puller may be necessary, but in most cases the shaft can be tapped inward through the hub bearing with a light hammer after installing the hub nut temporarily to protect the threads (see Figures 6–22 and 6–23 on page 126).

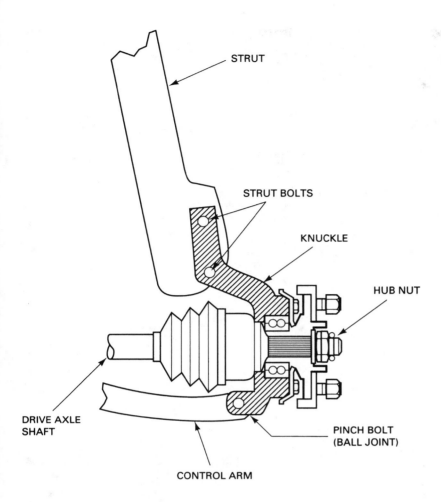

STRUT

STRUT BOLTS

KNUCKLE

HUB NUT

DRIVE AXLE
SHAFT

PINCH BOLT
(BALL JOINT)

CONTROL ARM

Figure 6–19 The hub nut must be removed before the hub bearing assembly or drive axle shaft can be removed from the vehicle.

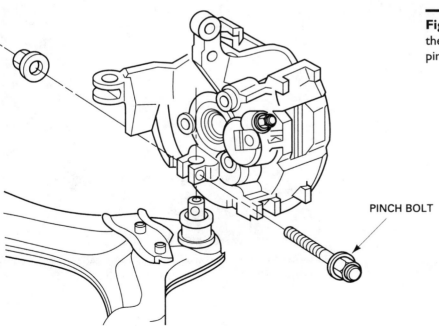

PINCH BOLT

Figure 6–20 Many knuckles are attached to the ball joint on the lower control arm by a pinch bolt. (Courtesy of General Motors)

Figure 6–21 A tie rod end puller can be used to separate the tapered stud of the tie rod from the tapered hole on the steering knuckle after the retaining nut is removed. Without a puller, a tie rod end can be separated from a knuckle by prying upward on the tie rod end and striking the knuckle with a heavy hammer. The shock "breaks the taper" releasing the tie rod end stud. (Courtesy of General Motors)

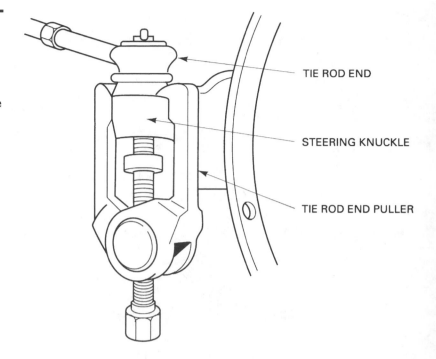

TIE ROD END

STEERING KNUCKLE

TIE ROD END PULLER

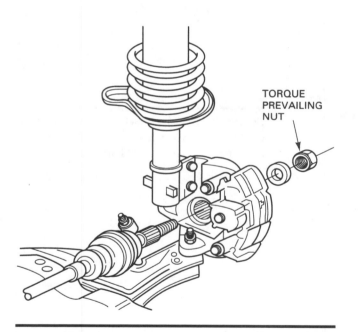

TORQUE PREVAILING NUT

Figure 6–22 Many drive axles are retained by torque prevailing nuts that must not be reused. Torque prevailing nuts are slightly deformed or contain a plastic insert that holds the nut tight (retains the torque) to the shaft without loosening. (Courtesy of General Motors)

Figure 6–23 The drive axle shaft has just been removed from this vehicle. Note that the disc brake caliper and rotor are still attached to the knuckle.

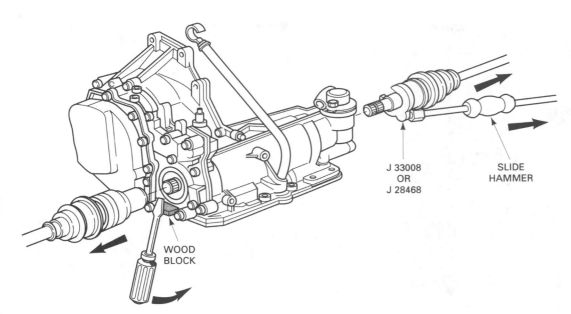

Figure 6–24 Most inner CV joints can be separated from the transaxle with a pry bar. (Courtesy of General Motors)

J 33008
OR
J 28468

SLIDE
HAMMER

WOOD
BLOCK

Step 4 Use a pry bar or special tool with a slide hammer as shown in Figure 6–24 and remove the inner joint from the transaxle. Some German axles are retained by fasteners that have to be removed rather than by clips that can be pulled out.

Step 5 Disassemble, clean and inspect all components (see Figures 6–25 through 6–34 on pages 128–130).

Step 6 Replace the entire joint if there are *any* worn parts. Pack *all* the grease that is supplied into the assembly or joint (see Figure 6–35 on page 131). Assemble the joint and position the boot in the same location as marked. Before clamping the last seal on the boot, be sure to release trapped air to prevent the boot from expanding when heated and collapsing when cold, as in Figure 6–36 on page 131. This is sometimes called *burping the boot.* Clamp the boot according to the manufacturer's specifications.

Step 7 Reinstall the drive axle shaft in the reverse order of removal, and torque the drive axle nut to factory specifications (see Figure 6–37 on page 131).

TECH TIP

Spline Bind Cure

Driveline clunk often occurs in rear-wheel-drive vehicles when shifting between drive and reverse or when accelerating from a stop. Often the cause of this noise is excessive clearance between the teeth of the ring and pinion in the differential. Another cause is called *spline bind,* where the changing rear pinion angle creates a binding in the spline when the rear springs change in height. For example, when a pickup truck stops, the weight transfers toward the front and unloads the rear springs. The front of the differential noses downward and forward as the rear springs unload. When the driver accelerates forward, the rear of the truck squats downward, causing the driveshaft to be pulled rearward when the front of the differential rotates upward. This upward movement on the spline often causes the spline to bind and make a loud clunk when the bind is finally released.

The method recommended by vehicle manufacturers to eliminate this noise is to follow these steps:

1. Remove the driveshaft.
2. Clean the splines on both the driveshaft yoke and the transmission output shaft.
3. Remove any burrs on the splines with a small metal file (remove all filings).
4. Apply a high-temperature grease to the spline teeth of the yoke. Apply grease to each spline, but do not fill the splines. Synthetic chassis grease is preferred because of its high temperature resistance.
5. Reinstall the driveshaft.

Figure 6–25 When removing a drive axle shaft assembly, use care to avoid pulling the plunge joint apart.

Figure 6–26 If other service work requires that just one end of the drive axle shaft be disconnected from the vehicle, be sure that the free end is supported to prevent damage to the protective boots or allowing the joint to separate. The method shown using a shop cloth to tie up one end may not be very pretty, but at least the technician took precautions to support the end of the shaft while removing the transaxle to replace a clutch.

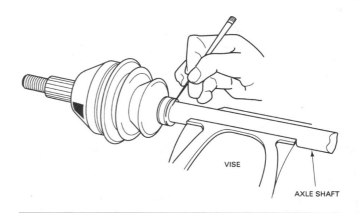

Figure 6–27 With a scribe, mark the location of the boots before removal. The replacement boots must be in the same location.

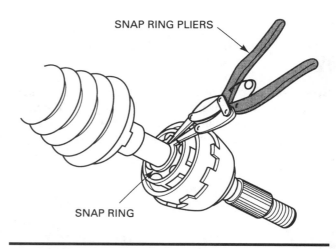

Figure 6–28 Most CV joints use a snap ring to retain the joint to the drive axle shaft

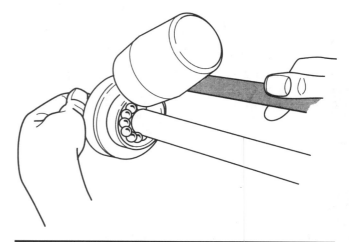

Figure 6–29 After releasing the snap ring, most CV joints can be tapped off the shaft using a brass or shot-filled plastic (dead-blow) hammer.

Figure 6–30 Typical outer CV joint after removing the boot and the joint from the drive axle shaft. This joint was removed from the vehicle because a torn boot was found. After disassembly and cleaning, this joint was found to be okay and was put back into service. Even though the grease looks terrible, there was enough grease in the joint to provide enough lubrication to prevent any wear from occurring.

Figure 6–32 The cage can be removed from the joint housing after removing all of the balls by aligning the ball openings with the lands between the outer ball races.

Figure 6–31 The cage of this Rzeppa-type CV joint is rotated so that one ball at a time can be removed. Some joints require that the technician use a brass punch and a hammer to move the cage.

Figure 6–33 Tripod (tripot) spider with rollers pulled from the outer race.

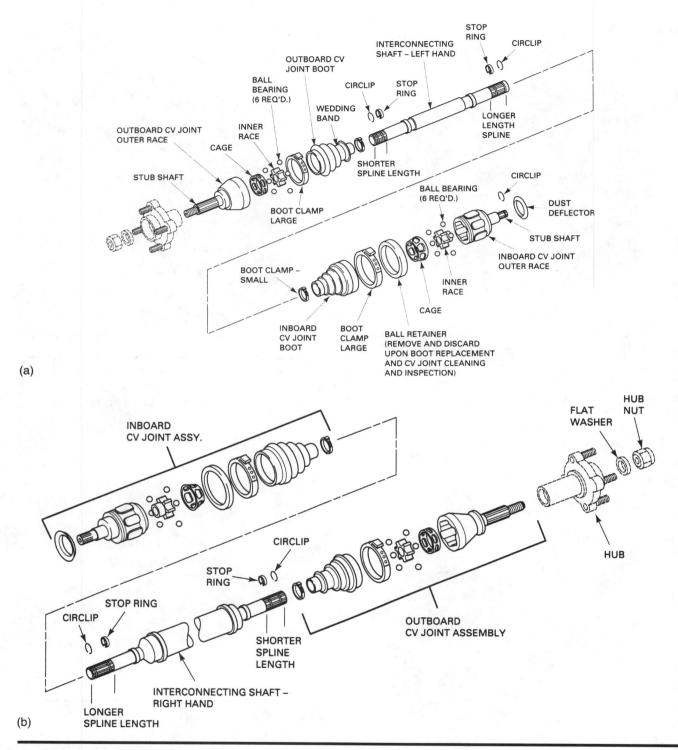

(a)

(b)

Figure 6–34 Typical left (a) and right (b) drive axle shaft assemblies showing all parts including the location of stop rings and circlips. (Courtesy of Ford Motor Company)

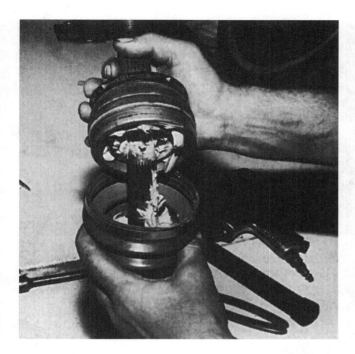

Figure 6–35 Be sure to use *all* of the grease supplied with the replacement joint or boot kit. Use only the grease supplied and do not use substitute grease.

Figure 6–36 Before you finish clamping the CV joint boot, burp the trapped air from the boot.

Figure 6–37 A screwdriver placed into the cooling slots of the rotor and wedged against the caliper is being used to keep the drive axle from rotating as the drive axle spindle nut is being torqued to specifications using a torque wrench. An alternative method would involve installing the spindle nut snugly, installing the wheel and tire assembly, and then lowering the vehicle to the ground before torquing the spindle nut to specifications. Most vehicle manufacturers also recommend using a *new* replacement spindle nut.

(a)

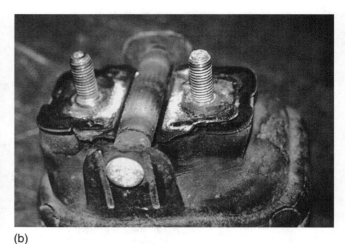

(b)

(c)

Figure 6–38 (a) Before the engine mount could be replaced, the vehicle was hoisted and a tall safety stand was placed under the engine. The stand was adjustable, thereby allowing the technician to raise the engine slightly to get the old mount out. (b) The old engine mount was torn and shorter than the replacement. (c) The engine had to be raised higher to get the new (noncollapsed) engine mount installed.

Diagnostic Story 📖

The Vibrating Buick

The owner of a front-wheel-drive Buick complained that it vibrated during acceleration only. The vehicle would also pull toward one side during acceleration. An inspection discovered a worn (cracked) engine mount. After replacing the mount, the CV joint angles were restored and both the vibration and the pulling to one side during acceleration were solved. See Figure 6–38.

PHOTO SEQUENCE   Drive Axle Shaft Service

PS6–1 Start the servicing of a drive axle shaft by positioning the hoist correctly under the vehicle.

PS6–2 Remove the hub cap or lug nut cover.

PS6–3 Loosen the drive axle shaft nut using a breaker bar while the vehicle is still resting on the ground. This axle shaft nut is often tightened to 200 lb. ft. of torque, which makes it difficult to remove, especially after the vehicle is raised on a hoist.

PS6–4 After loosening the drive axle shaft retaining nut, hoist the vehicle to a good working height.

PS6–5 Remove the front wheel(s).

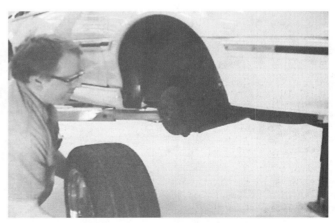

PS6–6 Place the wheels out of the work area for safety.

Drive Axle Shaft Service—continued

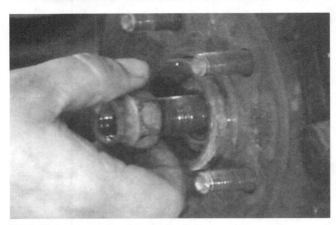

PS6–7 Remove the drive axle shaft nut that was loosened when the vehicle was still on the ground. Discard the nut.

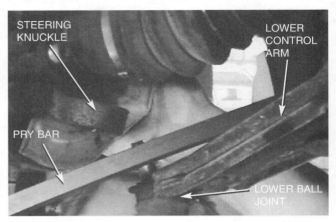

PS6–8 Separate the lower arm from the steering knuckle by disconnecting the lower ball joint. After removing the retaining nut, use a pry bar and exert a downward force on the lower control arm as an assistant strikes the knuckle around the ball joint with a hammer to separate.

PS6–9 To allow room to remove the drive axle shaft on this vehicle, it is necessary to disconnect the outer tie rod end from the strut assembly. Start the separation process by removing (and discarding) the cotter key from the outer tie rod retaining nut.

PS6–10 After removing the retaining nut, tap the joint with a hammer to "break the taper" of the joint.

PS6–11 A special tool is often necessary to push the drive axle shaft through the splines in the bearing hub.

PS6–12 The hub assembly including the disc brake caliper and rotor can be pulled outward off the splines of the drive axle shaft.

Drive Axle Shaft Service—continued

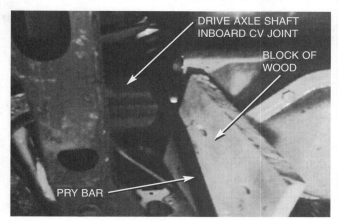

PS6–13 Use a pry bar to separate the inboard CV joint from the transaxle. Use a block of wood against the transaxle to prevent damage to the case of the transaxle. A special slide hammer tool is recommended by some vehicle manufacturers to separate the drive axle shaft from the transaxle.

PS6–14 After the circlip at the end of the drive axle shaft has been released from the transaxle, the entire drive axle shaft assembly can be removed from the vehicle.

PS6–15 After repairing or replacing the boot on the drive axle shaft, it can be replaced in the vehicle.

PS6–16 Carefully insert the splines of the drive axle shaft into the splines of the side gear inside the transaxle.

PS6–17 Guide the front hub over the outboard end of the drive axle shaft. In this case the strut is still attached to the body and is supporting the weight of the caliper, rotor, and hub assembly, so this procedure is relatively easy.

PS6–18 After the drive axle shaft has been installed through the bearing hub, use a pry bar to lower the lower control arm to help reinstall the lower ball joint into the steering knuckle.

Drive Axle Shaft Service—continued

PS6–19 A tap with a hammer is often necessary to seat the tapered ball joint stud into the tapered hole in the steering knuckle. If this step is not performed, the entire ball joint stud can turn instead when the retaining nut is being installed.

PS6–20 Reinstall the outer tie rod end to the strut assembly.

PS6–21 Install a new cotter key everywhere a cotter key is used.

PS6–22 Install a new drive axle shaft retaining nut, but do not torque it to its final reading until the vehicle is lowered to the ground.

PS6–23 Install the front wheel(s) and torque the lug nuts to specifications, or use the correct torque-absorbing adapter and an air impact as shown here.

PS6–24 After the vehicle is lowered to the ground, use a torque wrench and torque the drive axle shaft retaining nut to factory specifications. Reattach the lug nut cover and test-drive the vehicle to check for proper operation.

■ SUMMARY

1. A defective U-joint often makes a *clicking* sound when the vehicle is driven in reverse. Severely defective U-joints can cause driveline vibrations or a *clunk* sound when the transmission is shifted from reverse to drive or from drive to reverse.

2. Incorrect driveshaft working angles can result from collapsed engine or transmission mounts.

3. Driveline clunk noise can often be corrected by applying high-temperature chassis grease to the splines of the front yoke on the driveshaft.

4. CV joints require careful cleaning, inspection, and lubrication, using specific CV joint grease.

■ REVIEW QUESTIONS

1. List two items that should be checked when inspecting a driveshaft.

2. List the steps necessary to measure driveshaft U-joint working angles.

3. Describe how to replace a Cardan-type U-joint.

4. Explain the proper steps to perform when replacing a CV joint.

■ ASE CERTIFICATION-TYPE QUESTIONS

1. Two technicians are discussing U-joints. Technician A says that a defective U-joint could cause a loud clunk when the transmission is shifted between drive and reverse. Technician B says a worn U-joint can cause a clicking sound only when driving the vehicle in reverse. Which technician is correct?
   a. Technician A only
   b. Technician B only
   c. Both Technician A and B
   d. Neither Technician A nor B

2. Incorrect or unequal U-joint working angles are most likely to be caused by
   a. A bent driveshaft
   b. A collapsed engine or transmission mount
   c. A dry output shaft spline
   d. Defective or damaged U-joints

3. A defective outer CV joint will usually make a
   a. Rumbling noise
   b. Growling noise
   c. Clicking noise
   d. Clunking noise

4. The last step after installing a replacement CV boot is to
   a. "Burp the boot"
   b. Lubricate the CV joint with chassis grease
   c. Mark the location of the boot on the drive axle shaft
   d. Separate the CV joint before installation

5. Cardan-type U-joints should be removed from a driveshaft yoke using
   a. A special tool
   b. A torch
   c. A press
   d. A torque wrench

6. Two technicians are discussing CV joints. Technician A says that the entire front suspension has to be disassembled to remove most CV joints. Technician B says that most CV joints are bolted onto the drive axle shaft. Which technician is correct?
   a. Technician A only
   b. Technician B only
   c. Both Technician A and B
   d. Neither Technician A nor B

7. The splines of the driveshaft yoke should be lubricated to prevent
   a. A vibration
   b. Spline bind
   c. Rust
   d. Transmission fluid leaking from the extension housing

8. It is recommended by many experts that an air impact wrench not be used to remove or install the drive axle shaft nut because the impacting force can damage the hub bearing.
   a. True
   b. False

9. Front and rear driveshaft U-joint working angles should be within _____ degrees of each other.
   a. 0.5
   b. 1.0
   c. 3.0
   d. 4.0

10. A defective (collapsed) engine mount on a front-wheel-drive vehicle can cause a vibration.
   a. True
   b. False

Steering System Components and Operation

Objectives: After studying Chapter 7, the reader should be able to:

1. Identify steering system components.
2. Describe how the movement of the steering wheel causes the front wheels to turn.
3. Discuss the components and operation of power steering pumps.
4. List the components of a typical power recirculating ball nut steering gear system.
5. Describe the operation of a power rack and pinion steering system.
6. Explain the operation of electric and four-wheel steering.

The steering system consists of those parts required to turn the front wheels with the steering wheel.

■ STEERING SYSTEM COMPONENTS

Steering Column

The typical vehicle requires about three complete revolutions (turns) of the steering wheel to rotate the front wheels from full left to full right. The front wheels rotate up to 45 degrees while turning. The steering wheel is bolted to a splined shaft in the steering column (see Figure 7–1). Since the late 1960s, the steering column and shaft have been sectional and are designed to collapse in the event of an accident as shown in Figure 7–2 on page 140.

Intermediate Shaft

Steering forces are transferred from the steering column to an **intermediate shaft** and a **flexible coupling** (see Figure 7–3 on page 140). The flexible coupling is also called a **rag joint** or **steering coupling disc** and is used to insulate noise, vibration, and harshness from being transferred from the steering components up through the steering column to the driver (see Figure 7–4 on page 140).

Many steering systems use one or more U-joints between the steering column and the steering gear, as shown in Figure 7–5 on page 141.

Conventional Steering Gear

The rotation of the steering wheel is transferred to the front wheels through a steering gear and linkage. The intermediate shaft is splined to a **worm gear** inside a conventional steering gear. Around the worm gear is a nut with gear teeth that meshes with the teeth on a section of a gear called a **sector gear.** The sector gear is part of a **pitman shaft**, also known as a **sector shaft**, as shown in Figure 7–6 on page 141.

Ball, roller, or needle bearings support the sector shaft and the worm gear shaft, depending on the make and model of gear assembly.

As the steering wheel is turned, the movement is transmitted through the steering gear to an arm attached to the bottom end of the pitman shaft. This arm is called the **pitman arm.** Whenever the steering wheel is turned, the pitman arm moves. Many designs of steering gears have been used in the past. Since the late 1950s and early 1960s, most manufacturers have used a recirculating ball nut–type design. The term *recirculating ball* comes

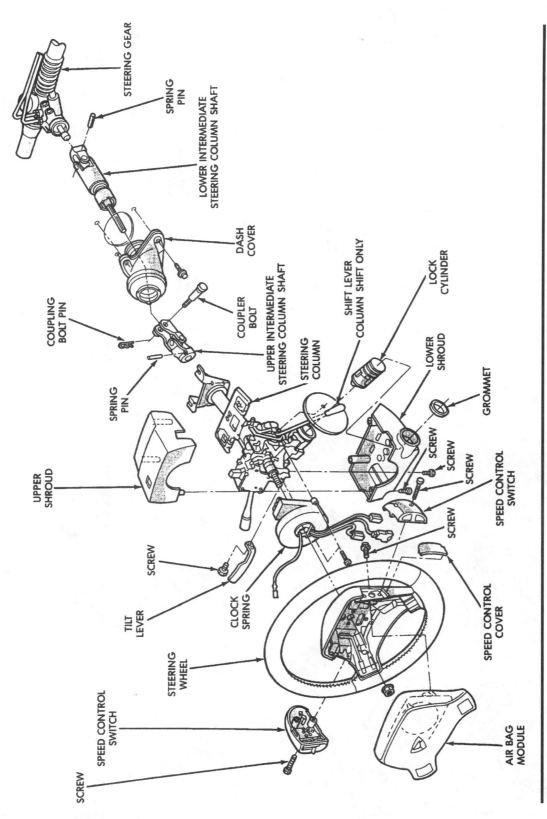

Figure 7-1 Typical steering column and related components. (Courtesy of Chrysler Corporation)

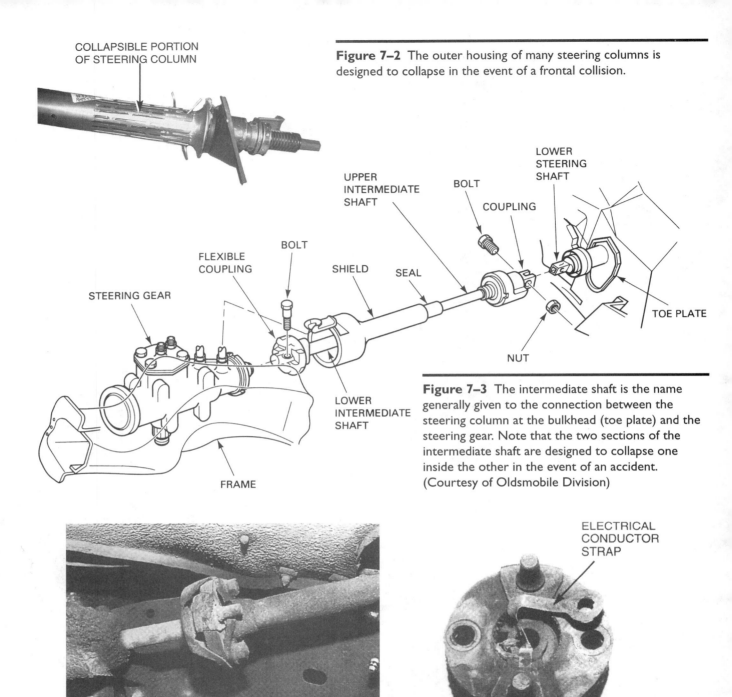

COLLAPSIBLE PORTION OF STEERING COLUMN

Figure 7–2 The outer housing of many steering columns is designed to collapse in the event of a frontal collision.

UPPER INTERMEDIATE SHAFT

BOLT

COUPLING

LOWER STEERING SHAFT

FLEXIBLE COUPLING

BOLT

SHIELD

SEAL

STEERING GEAR

TOE PLATE

NUT

LOWER INTERMEDIATE SHAFT

FRAME

Figure 7–3 The intermediate shaft is the name generally given to the connection between the steering column at the bulkhead (toe plate) and the steering gear. Note that the two sections of the intermediate shaft are designed to collapse one inside the other in the event of an accident. (Courtesy of Oldsmobile Division)

(a)

ELECTRICAL CONDUCTOR STRAP

(b)

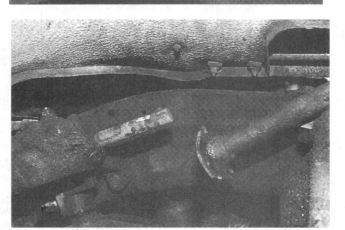

(c)

Figure 7–4 (a) Typical flexible coupling on an old pickup truck. (b) After removal of the joint, the electrical conductor strap provides the electrical ground for the horn ring. Because the flexible coupling isolates noise and vibration, the fabric also insulates the steering column shaft from the rest of the vehicle body. (c) Note the splines on the stub shaft of the steering gear. This flexible coupling was replaced because the splines in the coupling itself had rusted away resulting in excessive play in the steering wheel.

Figure 7–5 A U-joint used to change the angle of the steering shaft. Many times these U-joints become worn and can cause "jerky" or loose steering.

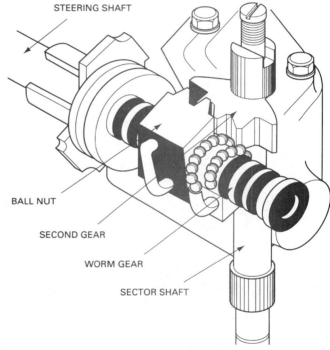

Figure 7–7 Recirculating (moving) steel balls reduce the friction between the worm gear teeth and the gear nut. (Courtesy of Ford Motor Company)

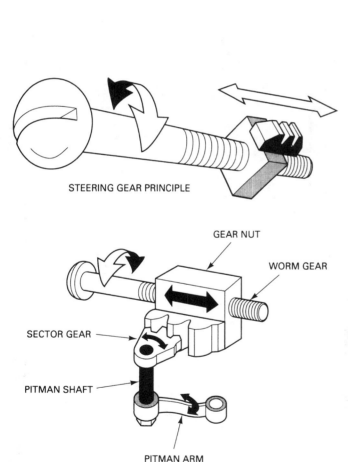

Figure 7–6 As the steering wheel is turned, the nut moves up or down on the threads shown using a bolt to represent the worm gear and the nut representing the gear nut that meshes with the teeth of the sector gear. (Courtesy of Ford Motor Company)

from the series of ball bearings placed between the input worm shaft and the ball nut (see Figures 7–7 and 7–8).

As the steering wheel turns, the worm shaft rolls inside a set of ball bearings. The movement of the bearings causes the ball nut to move. The ball nut has gear teeth that mesh with the gear teeth of the sector shaft. The sector gear and shaft rotate and move the pitman arm. The pitman arm is connected to steering linkage that moves and steers the front wheels. The rotating steel balls in the

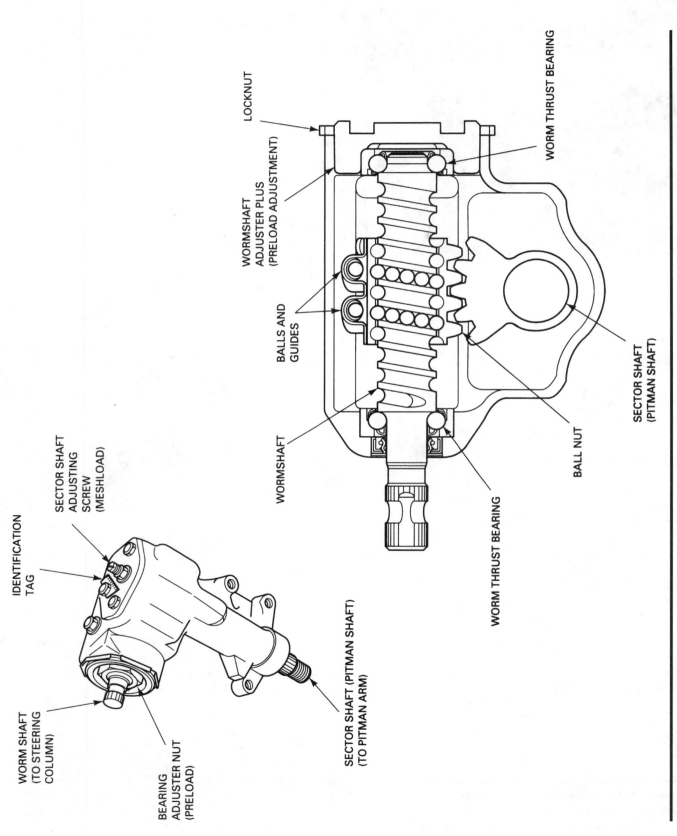

Figure 7-8 Recirculating ball steering gear showing worm shaft support bearings, worm shaft adjuster, and related parts. (Courtesy of Ford Motor Company)

LOCKNUT

WORM THRUST BEARING

WORMSHAFT ADJUSTER PLUS (PRELOAD ADJUSTMENT)

BALLS AND GUIDES

SECTOR SHAFT (PITMAN SHAFT)

WORMSHAFT

BALL NUT

WORM THRUST BEARING

IDENTIFICATION TAG

SECTOR SHAFT ADJUSTING SCREW (MESHLOAD)

WORM SHAFT (TO STEERING COLUMN)

BEARING ADJUSTER NUT (PRELOAD)

SECTOR SHAFT (PITMAN SHAFT) (TO PITMAN ARM)

ball nut reduce friction by rolling in the nut along the groove in the steering shaft and through a ball guide.

Standard (non-power-assisted) steering gears are lubricated by gear lubricant, usually SAE 80W-90. See Chapter 8 for adjustment locations and procedures.

Steering Gear Ratio When the steering wheel is turned, the front wheels turn on their steering axis. If the steering wheel is rotated 20 degrees and results in the front wheels rotating 1 degree, then the steering gear ratio is 20:1 (read as "20 to 1"). The front wheels usually are able to rotate through 60 degrees to 80 degrees of rotation. The steering wheel, therefore, has to rotate 20 times the number of degrees that the wheels move.

$20 \times 60° = 1200°$, or about three full revolutions ($360°$ = 1 full turn) of the steering wheel

$20 \times 80° = 1600°$, or over four revolutions of the steering wheel

A vehicle that turns three complete revolutions from full left to full right is said to have three turns "lock to lock."

A high ratio, such as 22 to 1 (22:1), means that the steering wheel must be rotated 22 degrees to move the front wheels 1 degree. This high ratio means that the steering wheel is easier to turn than with a lower ratio such as 14:1. The 14:1 ratio is considered to be "faster" than the 22:1 ratio. This fast ratio allows the front wheels to be turned with less movement of the steering wheel, yet more force may be required to turn the wheel. This is considered by some to be more "sporty."

Most steering gears and some rack and pinion steering gears feature a **variable ratio.** This feature causes the steering ratio to decrease as the steering wheel is turned from the on-center position. The high on-center ratio (such as 16:1) provides good steering feel at highway speeds, while the reduced off-center ratio (13:1) provides fewer steering wheel turns during turning and parking. The ratio is accomplished by changing the length of the gear teeth on the sector gear. See Figure 7–9 for an example of the teeth of a constant ratio sector gear.

See Figure 7–10 for an example of the teeth on a variable rate sector gear.

The sector gear meshes with the ball nut inside the steering gear, as shown in Figure 7–11.

Steering Linkage

Steering linkage relays steering forces from the steering gear to the front wheels. Most conventional steering linkages use the **parallelogram**-type design. A parallelogram is a geometric box shape where opposite sides are parallel and equal distance. A parallelogram-type linkage uses two **tie rods** (left and right), a **center link** (between the tie rods), and an **idler arm** on the passenger side and a **pitman arm** attached to the steering gear output shaft (pitman shaft) (see Figure 7–12).

As the steering wheel is rotated, the pitman arm is moved. The pitman arm attaches to a center link. At either end of the center link are inboard (inner) tie rods, adjusting sleeves, and outboard (outer) tie rods connecting to the steering arm, which moves the front wheels. The passenger side of all these parts is supported and held horizontal by an idler arm which is bolted to the frame. The center link may be known by several names, including:

Center link

Connecting link

Connecting rod

Relay rod

Intermediate rod

Drag link (usually a truck term only)

Other types of steering linkage often used on light trucks and vans include the **cross-steer linkage** and the **Haltenberger linkage.** See Figure 7–13 on page 145 for a comparison of parallelogram, cross-steer, and Haltenberger-type steering linkage arrangements.

> **NOTE:** Many light trucks, vans, and some luxury cars use a steering dampener attached to the linkage. A *steering dampener* is similar to a shock absorber, and it absorbs and dampens sudden motions in the steering linkage (see Figure 7–14 on page 146).

Connections between all steering component parts are constructed of small ball and socket joints. These joints allow movement side to side to provide steering of both front wheels as well as allow the joints to move up and down, which is required for normal suspension travel.

It is important that all these joints be lubricated with chassis grease through a grease fitting, also called a **zerk fitting,** at least every six months or per the vehicle manufacturer's specifications.

Frequently Asked Question ???

Why Is a Grease Fitting Sometimes Called a Zerk Fitting?

In 1922 the *zerk* fitting was developed by Oscar U. Zerk, an employee of the Alamite Corporation, a manufacturer of pressure lubrication equipment. A zerk or grease fitting is also known as an *Alamite fitting.*

Figure 7–10 Variable ratio steering gear sector shaft. Notice the larger center gear tooth.

Figure 7–9 Constant ratio steering gear sector shaft. Notice that all three gear teeth are the same size.

Some vehicles come equipped with sealed joints and do not require periodic servicing. Some vehicles come from the factory with plugs that need to be removed and replaced with grease fittings and then lubricated.

Tie Rod Ends

Tie rod ends connect the steering linkage to the steering knuckles and to other steering linkage components. Conventional tie rod ends use a hardened steel ball stud assembled into a hardened steel and thermoplastic bearing. An internal preload spring limits the ball stud end play and helps compensate for ball and socket wear. See Figure 7–15 on page 146 for two designs of tie rod ends.

Since the early 1980s, the Ford Motor Company has used tie rod ends that use rubber bonded to the steel ball stud. Since there is no sliding friction inside the tie rod end, no lubrication is needed or required. This type

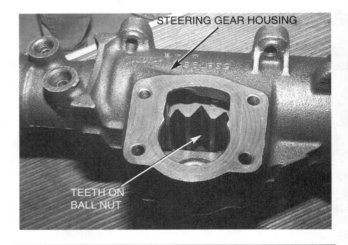

Figure 7–11 The sector gear meshes with the gear teeth on the ball nut.

Figure 7–12 Steering movement is transferred from the pitman arm that is splined to the sector shaft (pitman shaft), through the center link and tie rods to the steering knuckle at each front wheel. The idler arm supports the passenger side of the center link and keeps the steering linkage level with the road. (Courtesy of Dana Corporation)

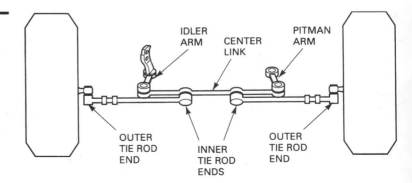

PARALLELOGRAM STEERING LINKAGE

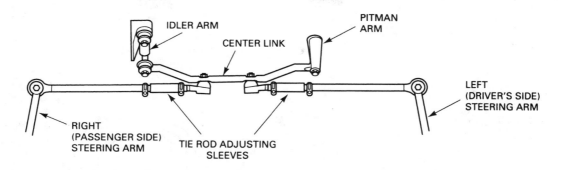

CROSS-STEER LINKAGE

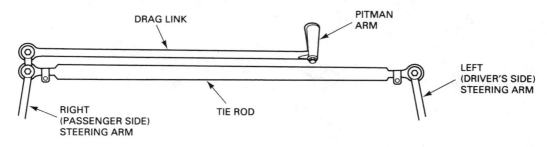

HALTENBERGER LINKAGE

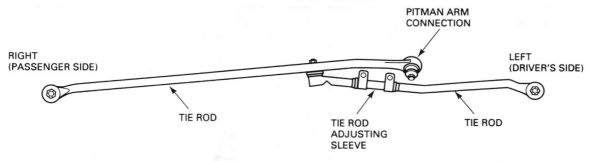

Figure 7–13 Types of steering linkage: Parallelogram steering linkage is commonly used on most rear-wheel-drive passenger cars and light trucks. The cross-steer and Haltenberger linkage designs are used on some trucks and vans.

of tie rod end is called **RBS (rubber bonded socket)** (see Figure 7–16 on page 147).

Most RBS tie rods have a hole in the location where a grease fitting normally is installed, but the hole is too large for a grease fitting. The purpose of the hole is to allow air to escape. **RBS tie rods should never be lubricated.** Applying grease to RBS tie rods could cause the rubber bonding between the housing and the ball stud to deteriorate.

Because the rubber bonding is elastic, the tie rods should be installed only when the front wheels are straight ahead. If the wheels are turned when an RBS tie rod is installed, the rubber inside of the joint would have a tendency to self-steer away from the straight-ahead position because rubber has a "memory." This

elastic force could cause the vehicle to lead or pull toward one side if not properly installed.

Manual Rack and Pinion Steering

A rack and pinion steering unit consists of a **pinion gear** that is in mesh with a flat gear called a **rack.** The ends of the rack are connected to the front wheels through tie rods. Turning the steering wheel rotates the pinion and causes the rack to move left and right in the housing. The rack housing attaches to the body or frame of the vehicle by rubber bushings to help isolate noise and vibration. The steering forces act in a straight line and provide direct steering action with no lost motion (see Figures 7–17 and 7–18 on page 147).

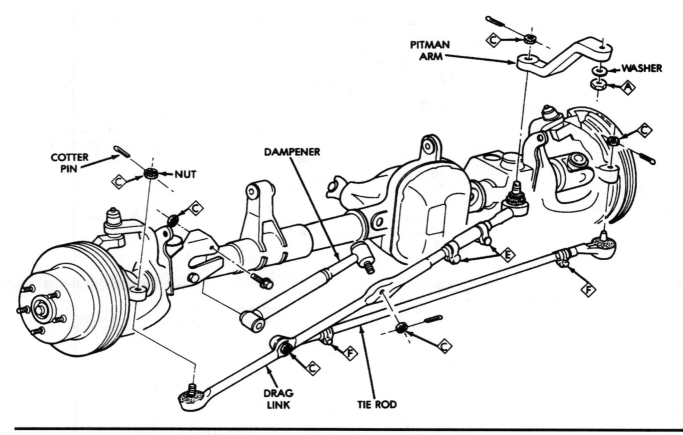

Figure 7–14 A typical steering dampener attaches to the frame or axle of the vehicle at one end and the steering linkage at the other end. (Courtesy of Chrysler Corporation)

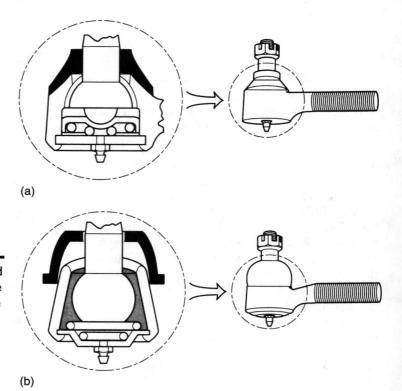

Figure 7–15 (a) A dual bearing design with a preload spring. The use of two bearing surfaces allows for one surface for rotation (for steering) and another surface for pivoting (to allow for suspension up-and-down movement). (b) The nylon wedge bearing type allows for extended lube intervals. Wear is automatically compensated for by the tapered design and spring loaded bearing. (Courtesy of Dana Corporation)

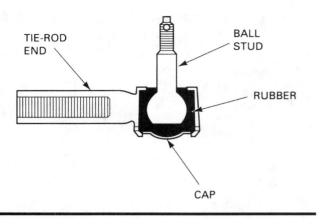

Figure 7–16 Rubber bonded socket (RBS)–type tie rod end. (Courtesy of Ford Motor Company)

Figure 7–16 Rubber bonded socket (RBS)–type tie rod end. (Courtesy of Ford Motor Company)

Figure 7–17 Basic principle and parts of a rack and pinion steering unit. (Courtesy of Dana Corporation)

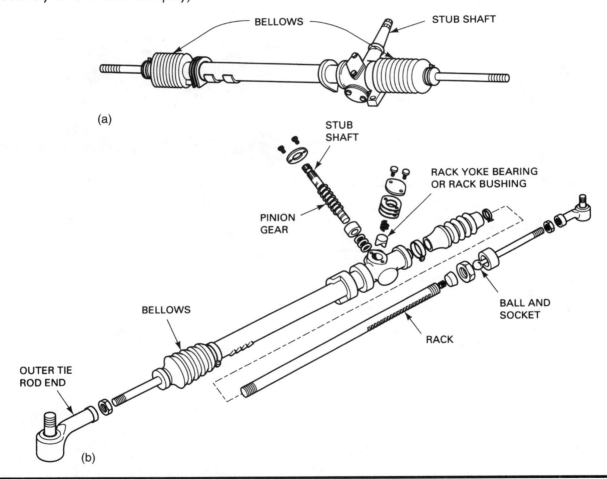

Figure 7–18 (a) Typical manual rack and pinion steering unit. (b) Exploded view of a manual rack and pinion assembly. The rubber bellows (boots) prevent dirt and moisture from getting around the rack and pinion gears. (Courtesy of Moog)

Rack and pinion steering is lightweight and small. The major disadvantage of a rack and pinion steering system is that there is more steering wheel feedback to the driver than with a conventional steering gear with parallelogram linkage. The inner tie rod ends attach to the rack with a ball and socket joint. This ball and socket is retained to the end of the rack by a soft pin, a roll pin, or a stacked flange (see Figure 7–19).

■ POWER STEERING PUMPS

Power-assisted steering, commonly called power steering, allows the use of faster steering ratios and reduced steering wheel turning force. A typical power steering system requires only 2 to 3.5 lb. (0.9 to 1.6 kg) of effort to turn the steering wheel.

Inner Tie Rod End (Ball Socket)

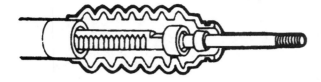

Soft Pin (Drill)

Roll Pin

Staked End

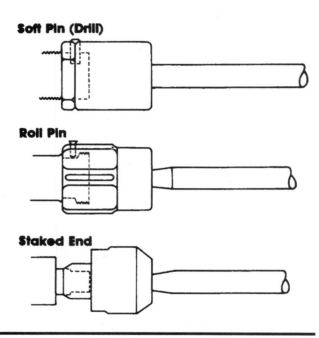

Figure 7–19 Inner tie rod ends use a ball and socket–type joint to allow the tie rod to move up and down as the wheels move up and down. Three methods used to retain the ball socket to the end of the rack include a soft metal pin (that can be easily drilled out), a roll pin (that must be pulled out), or a staked end. (Courtesy of Moog)

Most power steering systems use an engine-driven hydraulic pump. Power steering hydraulic pumps are usually belt driven from the front crankshaft pulley of the engine, as shown in Figures 7–20 and 7–21.

Some power steering (PS) pumps are driven from the camshaft, such as on General Motor's Quad Four engines. The power steering pump delivers a constant flow of hydraulic fluid to the power steering gear or rack. A typical power steering pump requires less than 1/2 horsepower, which is less than 1 percent of engine power while driving straight ahead. Even while parking at low speed, the power steering requires only about 3 horsepower while providing high hydraulic pressures. Typical pressures generated by a power steering system include:

Straight ahead	less than 200 psi (1400 kPa)
Cornering	about 450 psi (3100 kPa)
Parking	(maximum)750–1400 psi (5200–10,000 kPa)

Figure 7–20 This power steering pump uses a reservoir that is attached to the pump. Other designs of power steering pumps use a reservoir that is remotely mounted away from the pump housing itself.

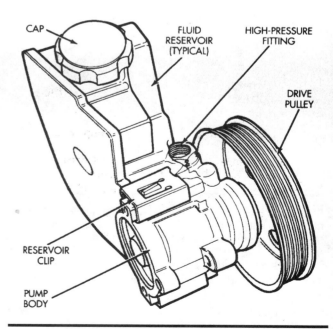

Figure 7–21 This power steering pump uses a reservoir that is clipped to the pump body. (Courtesy of Chrysler Corporation)

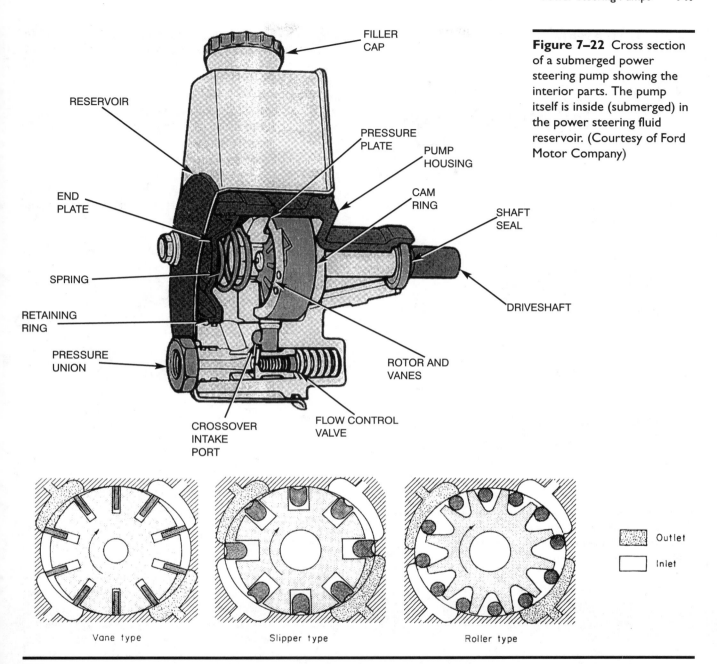

Figure 7–22 Cross section of a submerged power steering pump showing the interior parts. The pump itself is inside (submerged) in the power steering fluid reservoir. (Courtesy of Ford Motor Company)

FILLER CAP

RESERVOIR

PRESSURE PLATE

PUMP HOUSING

CAM RING

END PLATE

SHAFT SEAL

SPRING

DRIVESHAFT

RETAINING RING

PRESSURE UNION

ROTOR AND VANES

CROSSOVER INTAKE PORT

FLOW CONTROL VALVE

Vane type

Slipper type

Roller type

Outlet

Inlet

Figure 7–23 Three types of power steering pumps.

The power steering pump drive pulley is usually fitted to a chrome-plated shaft with a press fit. The shaft is applied to a **rotor** with **vanes** that rotate between a **thrust plate** and a **pressure plate** (see Figure 7–22).

Some power steering pumps are of the slipper or roller design instead of the vane type (see Figures 7–23 and 7–24). When the engine starts, the drive belt rotates the power steering pump pulley and the rotor assembly inside the power steering pump (see Figure 7–25).

With a vane-type pump, centrifugal force and hydraulic pressure push the vanes of the rotor outward into contact with the **pump ring.** The shape of the pump ring causes a change in the volume of fluid between the vanes. As the volume increases, the pressure is decreased in the space between the vanes and draws in

fluid from the pump reservoir. When the volume between the vanes decreases, the pressure is increased and flows out the pump discharge port (see Figure 7–26).

Flow Control Valve Operation

Between the pump discharge ports and the outlet fitting of the pump is the **flow control valve.** The operation of the flow control valve is determined by the needs of the power steering gear or rack. When engine speed is increased, the volume and pressure of the power steering pump increase. As the fluid flows through the outlet fitting orifice, a difference in pressure is created (see Figure 7–27 on page 151).

The pressure drop causes the flow control valve to move toward its lower pressure end. This movement

ROLLER TYPE

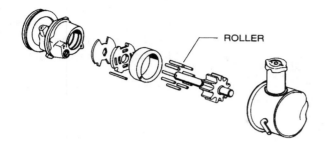

VANE TYPE

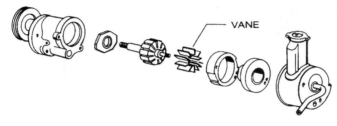

SLIPPER TYPE

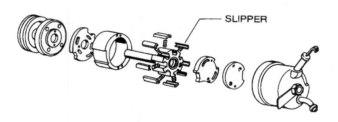

Figure 7–24 Exploded views of the three types of power steering pumps showing the design and shape of the pumps and reservoirs. (Courtesy of Moog)

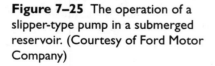

Figure 7–25 The operation of a slipper-type pump in a submerged reservoir. (Courtesy of Ford Motor Company)

opens a passage to allow the same fluid to return to the pump inlet. The opening increases as the pump speed increases to control the amount of fluid going to the power steering gear. Recirculating the power steering fluid back through the pump reduces the power required by the pump and keeps the fluid temperature from increasing, while still supplying adequate flow for proper power steering operation.

When the steering wheel is turned all the way to the left or right, the valve in the steering gear is closed off, shutting off flow from the pump. This causes the pressure to rise rapidly. A pressure relief valve inside the flow control valve acts as a safety valve to prevent the pressure from building higher. The pump is still trying to pump more power steering fluid into the system with no place to flow. A small steel ball is pushed off its seat at a prede-

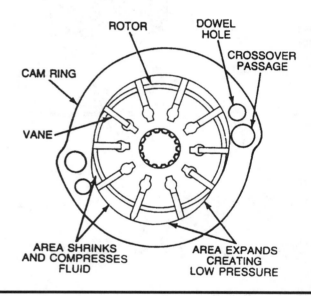

Figure 7–26 Fluid is drawn into the pump around the vanes as volume increases and the pressure is lowered. As the vanes rotate around toward a smaller volume, the pressure of the fluid trapped between the vanes increases approaching the discharge port. (Courtesy of Chrysler Corporation)

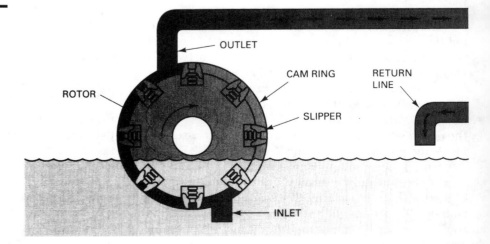

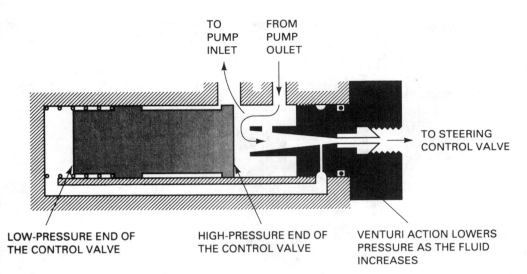

TO PUMP INLET
FROM PUMP OUTLET

TO STEERING CONTROL VALVE

LOW-PRESSURE END OF THE CONTROL VALVE

HIGH-PRESSURE END OF THE CONTROL VALVE

VENTURI ACTION LOWERS PRESSURE AS THE FLUID INCREASES

Figure 7–27 Operating principle of a flow control valve in position to return most of the fluid to the reservoir.

termined pressure. The fluid flows past the ball and through holes in the sides of the valve and into the low pressure intake section of the pump. The spring pressure working against the steel ball determines the maximum power steering pressure before pressure relief starts; this varies according to the size and type of vehicle, ranging from 750 to 1400 psi (5200 to 10,000 kPa). See Figure 7–28.

As the driver turns the steering wheel, the pressure demands of the gear are increased as determined by basic hydraulic principles:

1. A pump can pump a fluid into an open hose outlet and little pressure is developed.
2. To build pressure, the fluid flow needs to be restricted so that the pump volume exceeds the leakage in the system.

> **HINT:** Think of a garden hose without a nozzle. A full stream of water can flow out and the pressure is the same as the source pressure. Now place your finger over the end of the hose and the stream flow becomes restricted. The pressure is raised; at the same time, the volume of water actually leaving the end of the hose is reduced.

■ INTEGRAL POWER STEERING GEAR OPERATION

With power-assisted steering, the driver effort is always in proportion to the force necessary to turn the front wheels. The hydraulic pump pressurizes the power steering fluid and sends it through a pressure hose to the steering gear. Manual steering is available if the engine stops running or if there is a loss of hydraulic assist.

Power steering gears are usually called **integral gears** because the power piston or actuator for the power assist is inside (integrated into) the steering gear box. All integral power steering gears mount to the frame or sub-

frame of the vehicle. The driver's steering effort is transferred from the steering wheel through the steering column and intermediate shafts to the **stub shaft** or **input shaft** of the gear assembly (see Figure 7–29 on page 153). The stub shaft connects to the rotary (spool) valve inside the steering gear and directs and controls the flow of pressurized power steering fluid within the gear assembly. The stub shaft is sometimes called the **spool shaft.** The other end of the valve is connected to the worm gear through the **torsion bar.** When the front wheels are straight ahead, the rotary (spool) valve in the steering gear equalizes system pressure on both sides of the rack piston. The valve and housing cylinder are always full of power steering fluid to lubricate the internal parts and to dampen road shocks that would normally be transmitted to the driver through the steering wheel. The stub shaft and valve sleeve are held in a central position by the torsion bar. The fluid flows from the inlet port through the valve to the outlet port and back to the pump. No area of the steering gear is under high pressure in this position. The pressure in the system while driving straight is usually less than 150 psi (1000 kPa). See Figure 7–30 on page 153.

As the driver applies force to the steering wheel, the resistance of the tires on the road surface creates a resistive force and the torsion bar twists. This causes a change between the input shaft and the control sleeve, which restricts the flow of power steering fluid and directs the high-pressure fluid to one end of the piston in the gear housing. This high pressure forces the piston to move the sector gear and assists the turning effort. Figure 7–31 on page 154 shows the control valve passages and gear during a left turn; Figure 7–32 on page 155 shows a right turn.

The power steering fluid in the opposite end of the gear housing is forced out through the return outlet of the control valve and back to the pump reservoir.

When the driver stops applying steering effort, the valve sleeve and input shaft are returned to a centered position by the torsion bar. When this occurs, pressure is equalized on both sides of the piston. The normal forces on

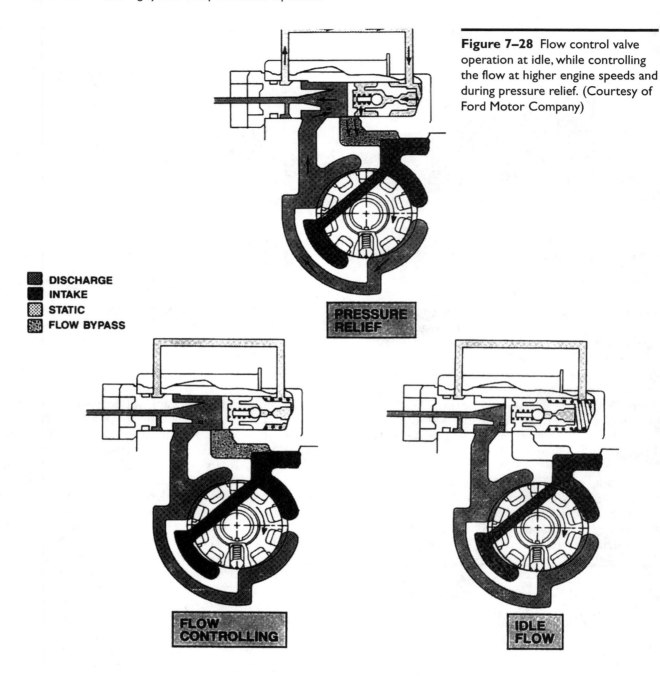

Figure 7–28 Flow control valve operation at idle, while controlling the flow at higher engine speeds and during pressure relief. (Courtesy of Ford Motor Company)

- ■ DISCHARGE
- ■ INTAKE
- ▨ STATIC
- ▨ FLOW BYPASS

PRESSURE RELIEF

FLOW CONTROLLING

IDLE FLOW

the tires due to the steering geometry tend to return the front wheels to the straight-ahead position. *Only when the torsion bar is being twisted does power assist occur.*

The torsion bar diameter determines the "feel" of the steering the driver senses. A thick bar requires more force to twist: The steering feels stiffer and is favored by many as being more "sporty." A smaller-diameter torsion bar twists easily and directs hydraulic fluid to the steering gear when very little force is applied to the steering wheel. This results in very easy steering and a "light" steering feel.

After the torsion bar twists enough to direct the hydraulic fluid to the steering gear, the combination of steering gear ratio and orifice size of the outlet fitting determines the ease of steering.

NOTE: The orifice size and gear ratios are carefully matched for each vehicle to achieve the proper steering effort for each particular vehicle. This is one reason why power steering components should not be interchanged. Just because they fit does not mean that the parts are matched correctly to the vehicle.

■ POWER RACK AND PINION STEERING

Since the 1980s, the power rack and pinion–type steering has been the most used in passenger cars as well as in light trucks and vans. Its light weight and small size

Frequently Asked Question ???

How Did Older Power Steering Systems Work?

Some earlier power steering systems used a conventional steering gear box and power piston connected to the steering linkage (see Figure 7–33 on page 156). This type of steering is called **linkage-type** power steering. The control valve for linkage-type power steering is connected to the pitman arm and controls the fluid to the power piston (see Figure 7–34 on page 156).

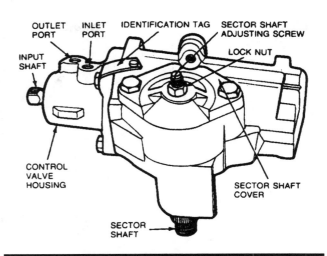

Figure 7–29 Typical integral power steering gear. (Courtesy of Ford Motor Company)

mean it can be mounted in a variety of locations. There are two basic designs of rack and pinion used today: the end-take-off and the center-take-off (see Figures 7–35 and 7–36 on pages 156–157).

The power steering pump supplies pressurized hydraulic fluid to the top, or "hat," section of the unit. The steering column attaches to the stub shaft and turns a rotary spool valve just as in a conventional integral power steering gear (see Figure 7–37 on page 157).

The spool valve assembly directs the pressurized fluid to one side of the rack piston, as shown in Figure 7–38 on page 157.

Fluid from the other side of the rack piston returns to the spool valve area as "return" fluid (see Figure 7–39 on page 158). Many power rack and pinion steering units use a power steering fluid cooler or simply extend the length of the lines to provide cooling as shown in Figure 7–40 on page 158.

Teflon seals are used to seal the spool valve passages in the rack housing bore, as shown in Figure 7–41 on page 159.

As the steering wheel is turned, the spool valve also turns inside the housing. Often the Teflon seals will wear grooves in the aluminum housing. When this occurs, the seals can no longer seal when cold (see the Frequently Asked Question, "What Can Be Done about Hard Steering Only When Cold?" in Chapter 8).

Some power rack and pinion steering units use a variable ratio rack (see Figure 7–42 on page 160). By changing the rack gear tooth pitch (spacing), easier steering effort can be achieved in city-type parking and driving maneuvers, while higher effort with reduced power assist can be provided during highway driving.

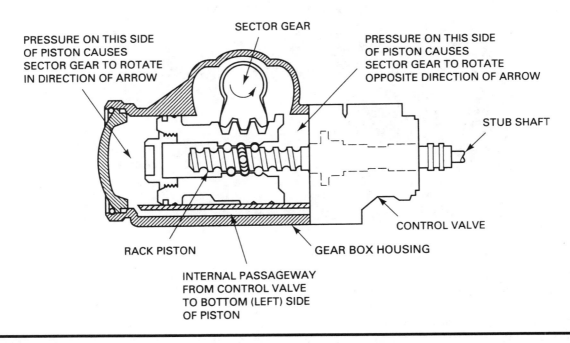

Figure 7–30 Forces acting on the rack piston of an integral power steering gear.

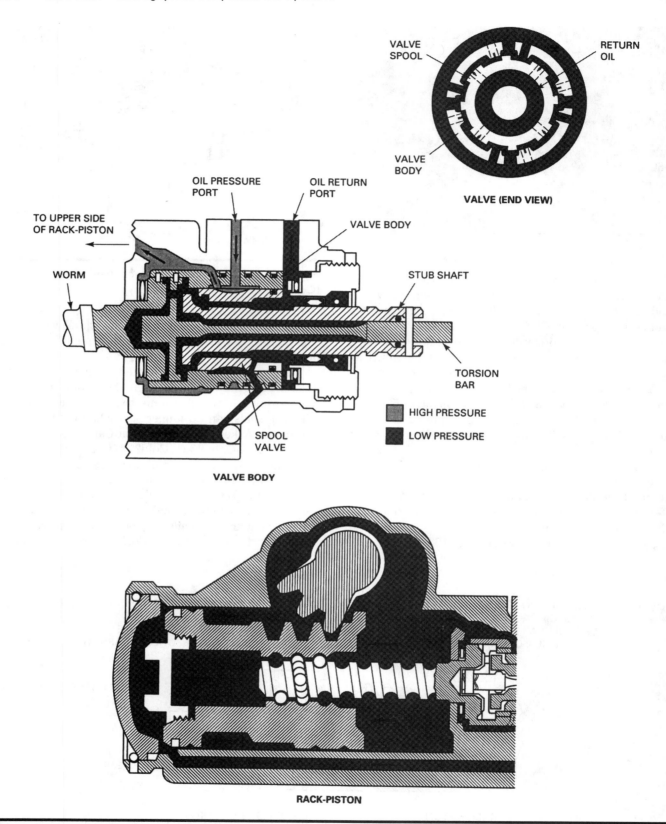

Figure 7–31 The flow of power steering fluid and the position of the control valve and rack piston during a left turn.

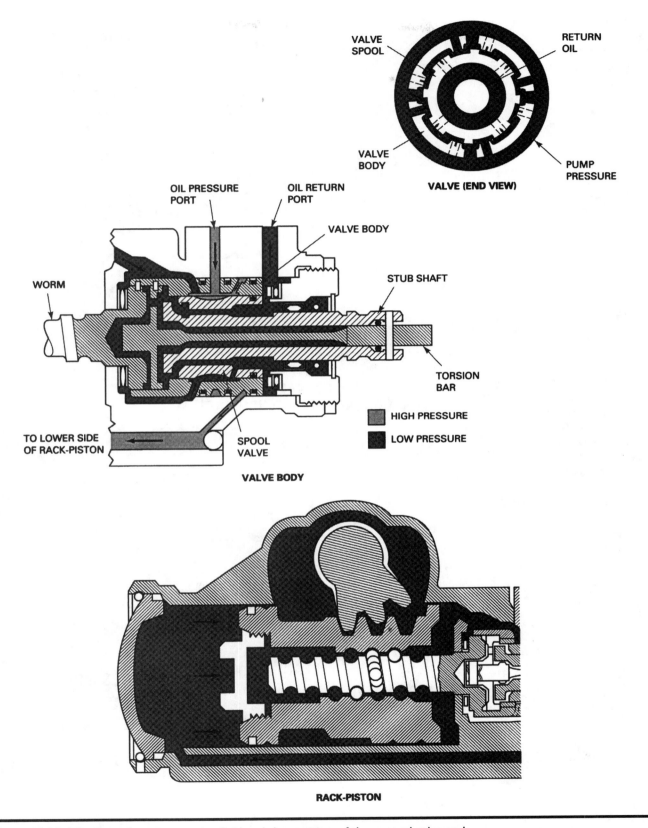

Figure 7–32 The flow of power steering fluid and the position of the control valve and rack piston during a right turn.

Figure 7–33 Linkage-type power steering. The power cylinder exerts force directly to the center link. (Courtesy of Ford Motor Company)

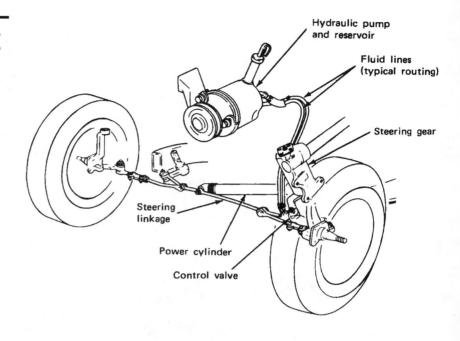

Figure 7–34 Operation of a typical linkage-type power steering system. A conventional steering gear is used. The pitman arm is connected to the control valve that directs the flow of pressurized power steering fluid to one side or the other of the power cylinder. (Courtesy of Ford Motor Company)

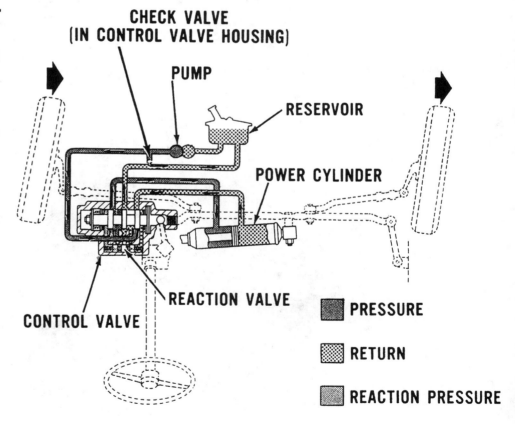

Figure 7–35 Typical end-take-off power rack and pinion steering gear. The tie rods connect to the ends of the rack with ball sockets.

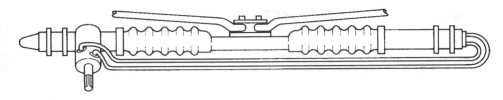

Figure 7–36 Typical center-take-off power rack and pinion steering gear. The long tie rods bolt to the center of the rack with step bolts that allow the outward ends of the tie rods to move freely up and down.

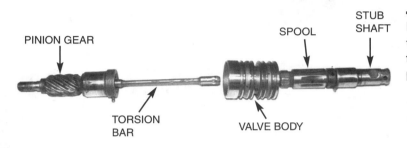

PINION GEAR

SPOOL

STUB SHAFT

TORSION BAR

VALVE BODY

Figure 7–37 The stub shaft is pinned to the torsion bar and twists to direct the power steering fluid in a power rack and pinion steering gear. This photo shows a disassembled control valve assembly.

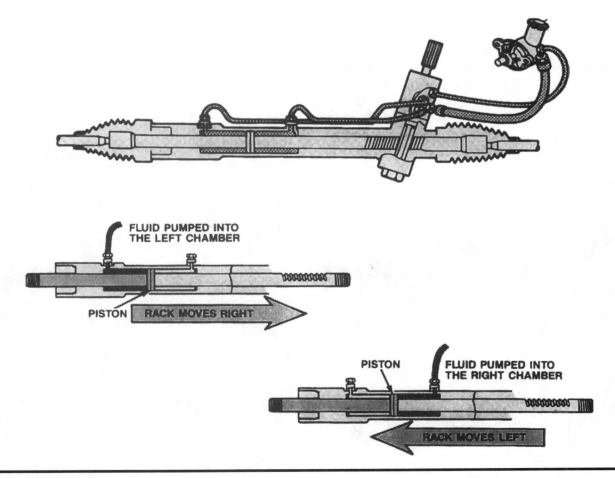

FLUID PUMPED INTO THE LEFT CHAMBER

PISTON   RACK MOVES RIGHT

PISTON

FLUID PUMPED INTO THE RIGHT CHAMBER

RACK MOVES LEFT

Figure 7–38 The control valve directs power steering fluid to one side or the other of the piston on the rack. Power steering fluid from the other side of the rack piston returns to the valve assembly where it is directed back to the pump reservoir. (Courtesy of Ford Motor Company)

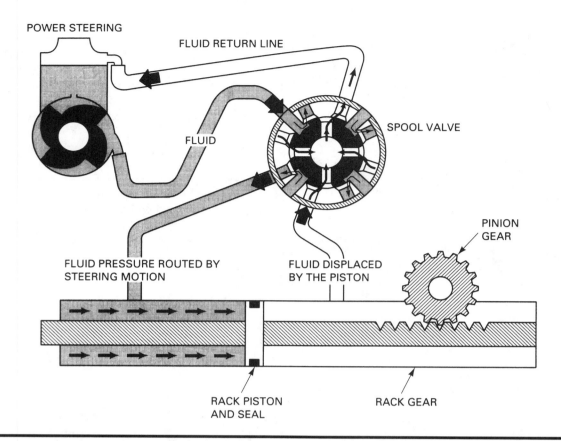

Figure 7–39 An end view of the control (spool) valve showing how pressurized power steering fluid is routed to one side of the rack piston. (Courtesy of Ford Motor Company)

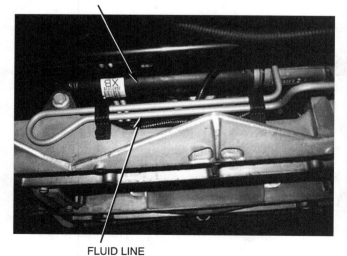

Figure 7–40 Many power rack and pinion steering units use extra length of fluid line to help cool the power steering fluid. Some units use a small radiator in the line.

Frequently Asked Question ???

What Do the Terms Understeer and Oversteer Mean?

The terms *understeer* and *oversteer* refer to vehicle handling while rounding a corner. Understeer is when the driver must turn the steering wheel more and more while turning a corner, such as turning onto an exit ramp from the expressway. Race car drivers and technicians refer to this as *pushing* or *plowing*. Most vehicles are designed to handle with some understeer because it gives the driver more control of the vehicle.

Oversteer means that when the vehicle enters a curve, the vehicle feels as if the steering wheel was turned too far. To counteract this feeling, the driver often must move the steering wheel in the opposite direction of the curve. Race car drivers and technicians usually refer to this type of handling as *loose*.

- *Understeer*—more steering is needed around a curve, "pushing."
- *Oversteer*—the steering wheel needs to be turned in the opposite direction as initially turned, "loose."

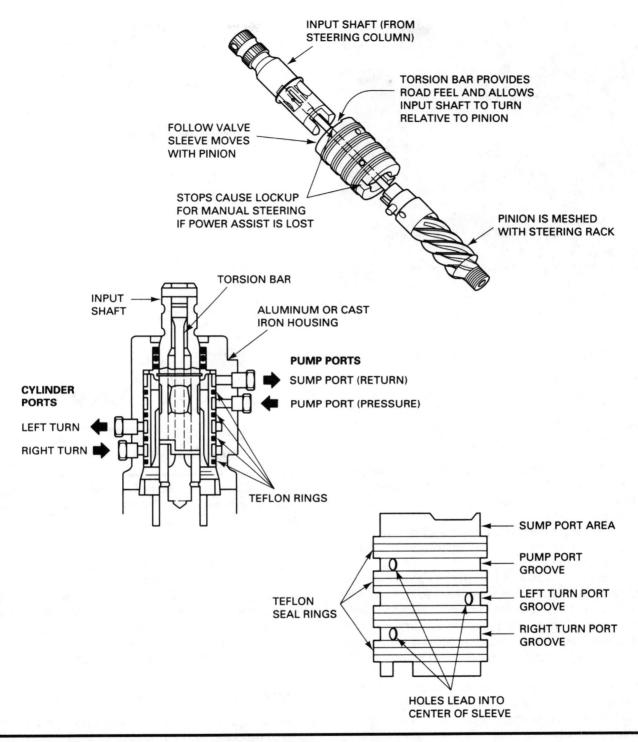

INPUT SHAFT (FROM
STEERING COLUMN)

TORSION BAR PROVIDES
ROAD FEEL AND ALLOWS
INPUT SHAFT TO TURN
RELATIVE TO PINION

FOLLOW VALVE
SLEEVE MOVES
WITH PINION

STOPS CAUSE LOCKUP
FOR MANUAL STEERING
IF POWER ASSIST IS LOST

PINION IS MESHED
WITH STEERING RACK

TORSION BAR

INPUT
SHAFT

ALUMINUM OR CAST
IRON HOUSING

PUMP PORTS

SUMP PORT (RETURN)

PUMP PORT (PRESSURE)

CYLINDER
PORTS

LEFT TURN

RIGHT TURN

TEFLON RINGS

SUMP PORT AREA

PUMP PORT
GROOVE

TEFLON
SEAL RINGS

LEFT TURN PORT
GROOVE

RIGHT TURN PORT
GROOVE

HOLES LEAD INTO
CENTER OF SLEEVE

Figure 7–41 A cutaway view of the "hat," or control valve area, of a typical power rack and pinion steering unit. The Teflon sealing rings provide a seal between the housing and the rotating control valve. A leak between any two of these Teflon seals can cause hard steering because the pressurized power steering fluid can cross channels. (Courtesy of Ford Motor Company)

Variable Effort Power Steering

Variable effort power steering uses an electrical control solenoid or motor to control the power steering pump output volume. The purpose of controlling the fluid volume out of the pump is to provide easy steering at low speeds for ease of maneuvering such as parking, yet provide increased road feel at highway speeds. There are two basic designs being used today, the **variable effort** and the **two-flow** or **switched** type. Figure 7–43 shows a variable system that uses an electric stepper motor–type actuator.

Both systems control the outlet flow rate between the power steering pump and the steering gear, or rack and pinion assembly.

A variable effort steering controller uses the vehicle speed either from the engine computer or from a separate sensor located in the transmission. Most variable effort power steering systems allow maximum power assist at low speeds and reduce power steering pump flow volume at speeds of about 20 mph (30 km/h) to increase road feel.

Some variable effort power steering systems *gradually* change the amount of power steering assist as the vehicle speed changes, thereby providing a less noticeable change in steering effort. Besides vehicle speed, many *variable effort* systems use a **steering wheel rotation sensor** to measure rapid steering wheel movement (see Figure 7–44).

The system can provide maximum power assist when the drive needs it, such as in a defensive driving maneuver, to avoid an accident.

The default mode of most variable effort power steering is to allow maximum power assist at all speeds. If, for example, the electrical connector were to become disconnected from the output actuator of the power steering

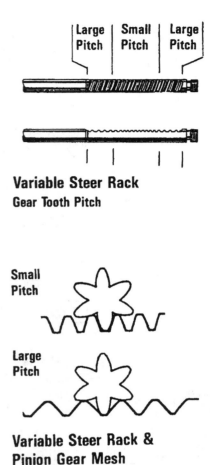

Figure 7–42 Variable ratio in a rack and pinion steering unit is accomplished by varying the gear tooth pitch on the rack. (Courtesy of TRW Inc.)

pump, the orifice size is set to its largest and maximum steering assist is available. Since most variable power steering systems use the vehicle speed sensor input for changing the amount of assist, a fault in the vehicle speed sensor would cause full power assist all the time. A fault in the vehicle speed sensor or circuit would also affect the speedometer and speed (cruise) control.

Front Steer versus Rear Steer

Front steer, also called **forward steer,** is the term used to describe a vehicle that has the steering gear in front of the front wheel centerline. Steering gear located in this position improves handling and directional stability, especially when the vehicle is heavily loaded.

Front-steer vehicles usually produce an understeer effect that makes the vehicle feel very stable while cornering. If the steering gear linkage is located behind the wheels, it is called **rear steer** and the cornering forces are imposed on the steering in the direction of the turn. This is an oversteer effect. It tends to make the steering easier and makes the vehicle feel less stable.

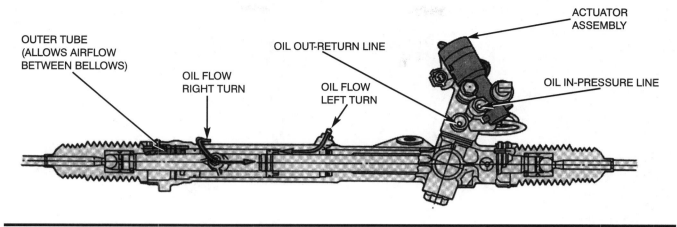

Figure 7–43 Typical variable effort power rack and pinion steering unit with electrical actuator assembly. (Courtesy of Ford Motor Company)

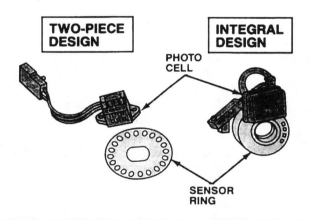

Figure 7–44 Two designs of steering angle sensors. These sensors can be used by the controller (computer) to vary the amount of power steering assist and for electronic ride control if the vehicle is equipped. (Courtesy of Ford Motor Company)

Most front-wheel-drive vehicles are rear steering, with the rack and pinion steering unit attached to the bulkhead or subframe behind the engine (see Figure 7–45).

Power Steering Computer Sensor

Many power steering systems, especially those using smaller four- and six-cylinder engines, use a sensor switch for the computer. The typical power steering sensor switch is a pressure switch that completes an electrical circuit to the computer (controller) whenever the power steering pressure exceeds a certain point, usually about 300 psi (2070 kPa). As the pressure of the power steering increases, the load on the engine increases. The computer senses this pressure increase and increases idle speed and fuel delivery to compensate for the increased load. The computer also senses other loads, including air-conditioning and the gear selector, to help maintain a stable idle regardless of the load on the engine.

Four-Wheel Steering

Some vehicles are equipped with a system that steers all four wheels. There are four terms that are commonly used when discussing four-wheel steering:

1. *Passive rear steering.* Passive means that no steering wheel input is needed to cause a rear-wheel steer effect. As the vehicle corners, forces on the suspension system allow a change in the rear toe angle. It is this slight change in toe of the rear wheels that contributes to a slight steering effect. This variation in the suspension mountings is due to the rubber (elastomeric materials) deformation. This change in suspension geometry is called **elastokinematics** (see Chapter 11 for details on toe and related alignment angles).
2. *Active rear steering.* Active means that either a mechanical, electrical, or hydraulic mechanism moves the wheels to change rear toe.
3. *Same-phase steering.* Same-phase steering means that the front and rear wheels are steered in the same direction. Same-phase steering improves steering response, especially during rapid-lane-change-type maneuvers (see Figure 7–46).
4. *Opposite-phase steering.* Also called **negative-phase mode,** opposite-phase steering is when the front wheels and rear wheels are steered in the opposite direction (see Figure 7–46). Opposite-phase steering will quickly change the vehicle's direction, but may cause a feeling of oversteering.

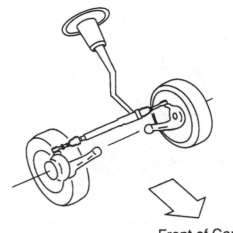

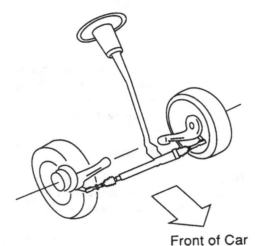

Front of Car

REAR STEER

Front of Car

FRONT STEER

Figure 7–45 In a rear-steer vehicle, the steering linkage is behind the centerline of the front wheels, whereas the linkage is in front on a front-steer vehicle. (Courtesy of Dana Corporation)

Opposite-phase steering is best at low speeds; same-phase steering is best for higher speed handling and lane-change maneuvers. Figure 7–47 shows several types of four-wheel steering systems.

Electric Steering

Electric power-assisted rack and pinion steering saves weight and horsepower, improves fuel economy, and offers faster steering system response than hydraulic-assisted steering. The steering effort can be tailored to provide a different feel for different types of vehicles. For example, the electronics can be changed to provide less power assist for sporty vehicles and more power assist for luxury vehicles (see Figure 7–48).

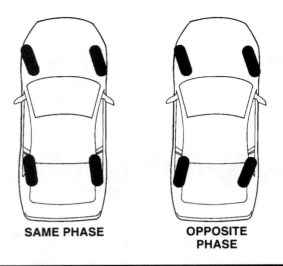

SAME PHASE   **OPPOSITE PHASE**

Figure 7–46 Opposite-phase four-wheel steer is usually used only at low vehicle speed to help in parking maneuvers. Same-phase steering helps at higher speed and may not be noticeable by the average driver.

■ SUMMARY

1. Steering wheel movement from the driver is transmitted through the steering column to the intermediate shaft and flexible coupling to the steering gear or rack and pinion unit.

2. Most conventional steering gears use a recirculating ball nut–type design.

3. The most common steering linkage used with a conventional steering gear is the parallelogram type. Other types commonly used on light trucks and vans are the cross-steer and Haltenberger linkage.

4. Most steering system components use greaseable ball and socket joints to allow for suspension travel and steering. Some manufacturers use rubber bonded sockets (RBS) that are not to be greased.

5. In a rack and pinion steering unit, the stub shaft connects to the small pinion gear. The pinion gear meshes with gear teeth cut into a long shaft called a *rack*.

6. Power steering pumps supply hydraulic fluid to the steering gear, power piston assembly, or power rack and pinion unit. The spool (rotary) valve controls and directs the high-pressure fluid to the power piston to provide power-assisted steering.

7. Variable effort power steering systems use electronically controlled valves to reduce hydraulic fluid flow to the gear at higher speeds to improve road feel.

8. Rack and pinion steering units work in a straight line without the need for multiple steering linkage components.

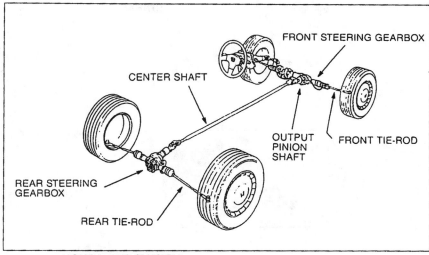

HONDA 4WS SYSTEM

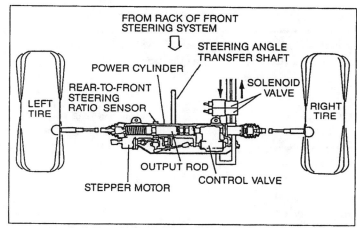

MAZDA REAR STEERING SYSTEM

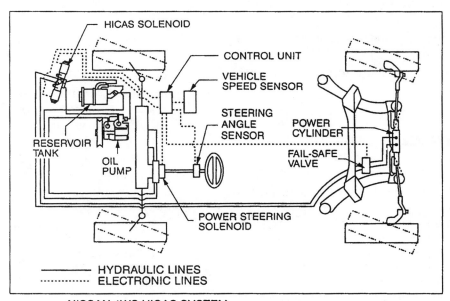

NISSAN 4WS HICAS SYSTEM

Figure 7–47 Several four-wheel steer vehicles have been sold including the Honda 4-WS system, the Mazda rear-steer system, and the Nissan HICAS system. (Courtesy of Dana Corporation)

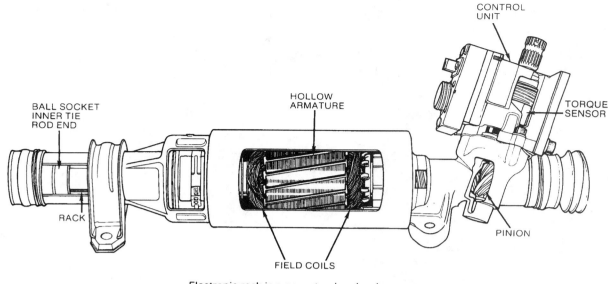

Electronic rack is a new steering development.

Figure 7–48 A typical electronic rack and pinion steering saves the weight and expense of a hydraulic pump and high-pressure lines. (Courtesy of Dana Corporation)

■ REVIEW QUESTIONS

1. List all the parts that move when the steering wheel is turned with a conventional steering unit and a rack and pinion steering unit.

2. Explain how recirculating steel balls reduce friction in a recirculating ball–type steering gear.

3. Describe the hydraulic fluid flow from the pump through the flow control valve and to the steering gear.

4. Explain how power steering fluid is directed and controlled by the spool (rotary) valve.

5. List the two types of variable effort power steering.

■ ASE CERTIFICATION-TYPE QUESTIONS

1. Which type of steering linkage component must not be lubricated?
   a. Zerk fittings
   b. Alamite fittings
   c. RBS tie rods
   d. Ball guides

2. What steering component is between the intermediate shaft from the steering column and the stub shaft of the steering gear on a rack and pinion unit?
   a. Pitman shaft
   b. Flexible coupling
   c. Sector shaft
   d. Tie rod

3. Normal straight-ahead power steering pump pressure is
   a. Less than 200 psi (1400 kPa)
   b. About 450 psi (3100 kPa)
   c. 800–1000 psi (5500 kPa)
   d. Up to 1400 psi (10,000 kPa)

4. Which part of the power steering pump limits the maximum pressure?
   a. Flow control valve
   b. Pressure relief valve
   c. Spool valve
   d. Rotary valve

5. What part in the power steering gear assembly (either conventional or rack and pinion) determines the steering "feel"?
   a. Torsion bar
   b. Flow control valve
   c. Spool valve
   d. Rotary valve

6. Which is *not* an input device to the electronic controller that controls the variable effort actuator?
   a. Vehicle speed sensor
   b. Brake switch
   c. Steering wheel rotation sensor
   d. Ignition switch

7. Power-assisted steering uses hydraulic fluid under pressure to move the
   a. Rack piston
   b. Pitman arm
   c. Spool valve
   d. Torsion bar

8. The upper limit pressure of a power steering system can reach up to 1400 psi during
   a. Gentle turns at fast speeds
   b. Straight-ahead driving
   c. Turning against the steering stops
   d. Just as the steering wheel is rotated by the driver

9. A drag link is a term used to describe which steering component?
   a. An idler arm
   b. A center link
   c. A pitman arm
   d. A stabilizer bar

10. On some vehicles, the air-conditioning is shut off if the power steering system pressure exceeds a certain level to reduce the load on the engine during turning.
   a. True
   b. False

Steering System Diagnosis and Service

When the steering wheel is turned as far as possible, the steering should *not* stop inside the steering gear! Forces exerted by the power steering system can do serious damage to the steering gear if absorbed by the steering gear rather than the steering stop.

NOTE: Many rack and pinion steering units are designed with a rack travel limit internal stop and do not use an external stop on the steering knuckle or control arm.

The proper operation of the steering system is critical to the safe operation of any vehicle. Proper lubrication and inspection on a regular basis is important.

■ UNDER-VEHICLE LUBRICATION

Keeping all joints equipped with a grease fitting that is properly greased is necessary for long life and ease of steering (see Figure 8–1).

During a chassis lubrication, do not forget to put grease on the *steering stop*, if so equipped. **Steering stops** are the projections or built-up areas on the control arms of the front suspension designed to limit the steering movement at full lock (see Figure 8–2).

Figure 8–1 Greasing a tie rod end. Some joints do not have a hole for excessive grease to escape, and excessive grease can destroy the seal.

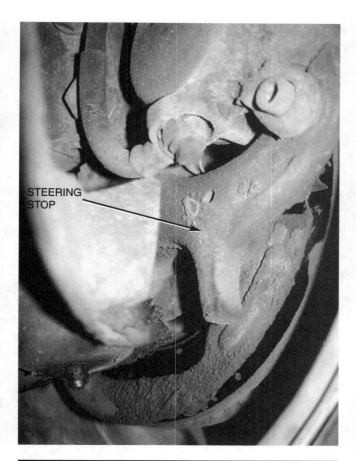

STEERING
STOP

Figure 8–2 Part of steering linkage lubrication is applying grease to the steering stops. If these stops are not lubricated, a grinding sound may be heard when the vehicle hits a bump when the wheels are turned all the way one way or the other. This often occurs when driving into or out of a driveway that has a curb.

Most steering stops are designed so that the lower control arm hits a small section of the body or frame when the steering wheel is turned to the full "lock" position. Steering stops should be lubricated to prevent a loud grinding noise when turning while the vehicle is going over a bump. This noise is usually noticeable when turning into or out of a driveway.

■ DRY PARK TEST

Since many steering (and suspension) components do *not* have *exact* specifications for replacement purposes, it is extremely important that the beginning service technician work closely with an experienced veteran technician. While most technicians can determine when a steering component such as a tie rod end is definitely in need of replacement, marginally worn parts are often hard to spot and can lead to handling problems. One of the most effective, yet easy to perform, steering component inspection methods is called the **dry park test.**

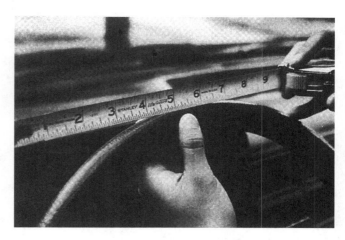

Figure 8–3 Checking for free play in the steering.

Excessive play in the steering wheel can be caused by worn or damaged steering components. Looseness in the steering components usually causes free play in the steering. Free play refers to the amount of movement of the steering wheel required to cause movement of the front wheels. The exact cause of free play in the steering should be determined if the free play exceeds 2" (5 cm) for parallelogram-type steering linkage, or 3/8" (1 cm) for rack and pinion steering (see Figure 8–3).

This simple test is performed with the vehicle on the ground or on a drive-on ramp-type hoist, moving the steering wheel back and forth *slightly* while an assistant feels for movement at each section of the steering system. The technician can start checking for any looseness in the steering linkage starting either at the outer tie rod ends and working toward the steering column, or from the steering column toward the outer tie rod ends. It is important to check each and every joint and component of the steering system including:

Figure 8–4 The flexible coupling on the left came from a twenty-year-old Chevrolet pickup truck. Note how the splines are worn or corroded away compared with the replacement coupling on the right. This repair greatly improved the steering by eliminating over 2″ of free play at the steering wheel.

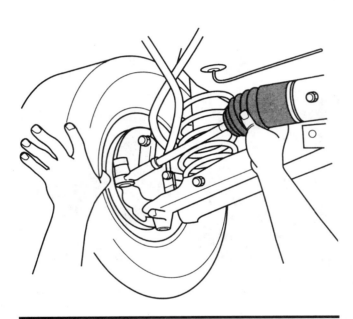

Figure 8–5 All joints should be felt during a dry park test. Even inner tie rod ends (ball socket assemblies) can be felt through the rubber bellows on many rack and pinion steering units.

1. The intermediate shaft and flexible coupling (see Figure 8–4)
2. All steering linkage joints, including the inner tie rod end ball socket (see Figure 8–5)
3. Steering gear mounting and rack and pinion mounting bushings (see Figure 8–6)

■ COMMON WEAR ITEMS

On a vehicle equipped with a conventional steering gear and parallelogram linkage, as shown in Figure 8–7, typical items that wear first, second, etc., include:

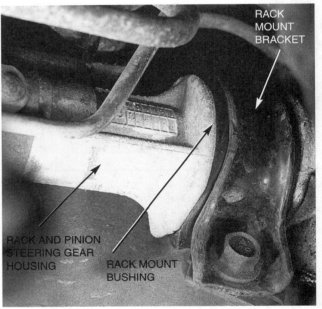

Figure 8–6 A rack mounting bushing. Often these rubber bushings are worn or softened by fluid leaks from the engine or transaxle.

Steering Component	Estimated Mileage to Wear Out*
1. Idler arm	40,000–60,000 miles (60,000–100,000 km)
2. Outer tie rod ends (replaced in pairs only)	60,000–100,000 miles (100,000–160,000 km)
3. Inner tie rod ends	80,000–120,000 miles (130,000–190,000 km)
4. Center link	90,000–130,000 miles (140,000–180,000 km)
5. Pitman arm	100,000–150,000 miles (160,000–240,000 km)

*Mileage varies greatly due to different road conditions and levels of vehicle maintenance; this chart should be used as a guide only.

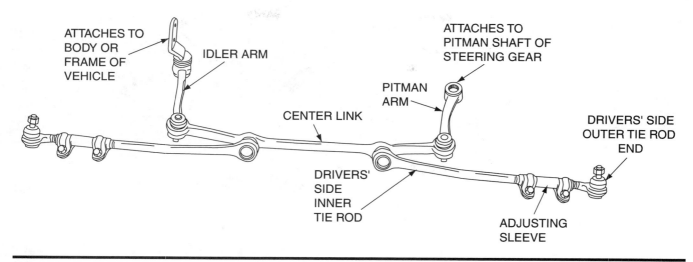

Figure 8–7 Typical parallelogram steering linkage. The center link can also be named the relay rod, drag link, or connecting link.

Note that there are overlapping mileage intervals for several components. Also note that the mileage interval for an idler arm is such that by the time other components are worn, the idler arm may need to be replaced a second time.

For vehicles that use rack and pinion–type steering systems, the list is shorter because there are fewer steering components and the forces exerted on a rack and pinion system are in a straight line. The first to wear are usually outer tie rod ends (one or both) followed by the inner tie rod ball and socket joints, usually after 60,000 miles (100,000 km) or more. Intermediate shaft U-joints usually become worn and can cause steering looseness after 80,000 miles (130,000 km) or more.

> **HINT:** Experienced front-end technicians can often guess the mileage of a vehicle simply by careful inspection of the steering linkage. For example, if the idler arm is a replacement part and again needs to be replaced, and the outer tie rods also need replacement, then the vehicle probably has at least 60,000 miles and usually more! When inspecting a used vehicle for possible purchase, perform a careful steering system inspection. This is one area of the vehicle where it is difficult to hide long or hard service.

■ UNDER-VEHICLE INSPECTION

After checking the steering system components as part of a dry park test, hoist the vehicle and perform a thorough part-by-part inspection:

1. Inspect each part for damage due to an accident or bent parts due to the vehicle's hitting an object in the roadway.

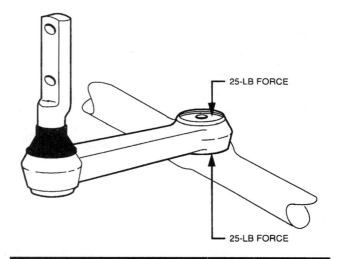

Figure 8–8 To check an idler arm, most vehicle manufacturers specify that 25 lb. of force be applied by hand up and down to the idler arm. The idler arm should be replaced if the total movement (up and down) exceed 1/4″ (6 mm). (Courtesy of Dana Corporation)

> **CAUTION:** Never straighten bent steering linkage; always replace with new parts.

2. Idler arm inspection is performed by using *hand* force of 25 lb. (110 N-m) up and down on the arm. If the *total* movement exceeds 1/4″ (6 mm), the idler arm should be replaced (see Figure 8–8).

3. All other steering linkage should be tested *by hand* for any vertical or side-to-side looseness. Tie rod ends use ball and socket joints to allow for freedom of movement for suspension travel and transmit steering forces to the front wheels. It is, therefore, normal for tie rods to rotate in their sockets when the toe rod sleeve is rocked. **End play in any tie**

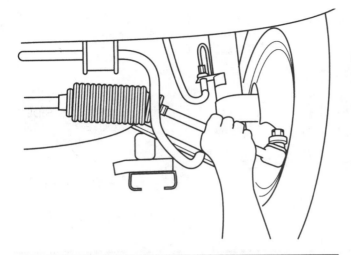

Figure 8–9 Steering system component(s) should be replaced if any noticeable looseness is detected when moved by hand.

rod should be zero. Many tie rods are spring loaded to help keep the ball and socket joint free of play as the joint wears. Eventually, the preloaded spring cannot compensate for the wear, and end play occurs in the joint (see Figures 8–9 and 8–10).

4. All steering components should be tested with the wheels in the straight-ahead position. If the wheels are turned, some apparent looseness may be noticed due to the angle of the steering linkage.

CAUTION: Do not turn the front wheels of the vehicle while suspended on a lift to check for looseness in the steering linkage. The extra leverage of the wheel and tire assembly can cause a much greater force being applied to the steering components than can be exerted by hand alone. This extra force may cause some apparent movement in good components that may not need replacement.

Figure 8–10 All joints should be checked by hand for any lateral or vertical play. (Courtesy of Moog)

TECH TIP ✔

The Killer B's

The three B's that can cause steering and suspension problems include bent, broken, or binding components. Always inspect each part under the vehicle for each of the killer B's.

■ STEERING LINKAGE REPLACEMENT
Parallelogram Type

When replacing any steering system component, it is best to replace all defective and marginally good components at the same time. Use the following guidelines.

Parts that can be replaced individually:

Idler arm

Center link

Pitman arm

Intermediate shaft

Intermediate shaft U-joint

Parts that should be replaced in pairs only:

Outer tie rod ends

Inner tie rod ends

Idler arm (if there are two on the same vehicle, such as GM's Astro van)

Replacing steering system components involves these steps:

Step 1 Hoist the vehicle safely with the wheels in the straight-ahead position. Remove the front wheels, if necessary, to gain access to the components.

Step 2 Loosen the retainer nut on tapered components, such as tie rod ends. Use a tie rod removal puller

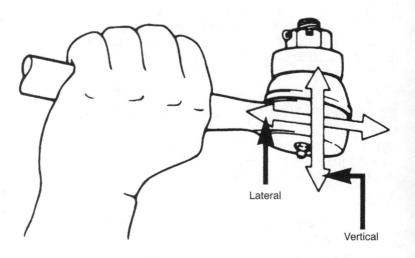

Lateral

Vertical

Diagnostic Story

Bump Steer

Bump steer, or *orbital steer*, is used to describe what happens when the steering linkage is not level: The front tires turn inward or outward as the wheels and suspension move up and down. (Automotive chassis engineers call it *roll steer*.) The vehicle's direction is changed *without moving the steering wheel* whenever the tires move up and down over bumps, dips in the pavement, or even over gentle rises!

This author experienced bump steer once and will never forget the horrible feeling of not having control of the vehicle. After replacing an idler arm and aligning the front wheels, everything was okay until about 40 mph (65 km/h); then the vehicle started darting from one lane of the freeway to another. Because there were no "bumps" as such, bump steer was not considered as a cause. Even when holding the steering wheel perfectly still and straight ahead, the vehicle would go left, then right. Did a tie rod break? It certainly felt exactly like that's what happened. I slowed down to below 30 mph and returned to the shop.

After several hours of checking everything, including the alignment, I discovered that the idler arm was not level with the pitman arm. This caused a pull on the steering linkage whenever the suspension moved up and down. As the suspension compressed, the steering linkage pulled inward on the tie rod on that side of the vehicle. As the wheel moved inward (toed in), it created a pull just as if the wheel had been turned by the driver.

This is why all steering linkage must be parallel with the lower control. The reason for the bump steer was that the idler arm bolted to the frame, which was slotted vertically. I didn't pay any attention to the location of the original idler arm and simply bolted the replacement to the frame. After raising the idler arm back up where it belonged (about 1/2″ [13 mm]), the steering problem was corrected.

Other common causes of bump steer are worn or deteriorated rack mounting bushings, noncentered steering linkage or a bent steering linkage. If the steering components are not level, any bump or dip in the road will cause the vehicle to steer one direction or the other (see Figure 8–11).

Always check the steering system carefully whenever a customer complains about any "weird" handling problem.

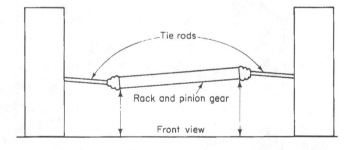

Figure 8–11 If a rack and pinion or any other steering linkage system is not level, the front tires will be moved inward and/or outward whenever the wheels of the vehicle move up or down.

(also called a *taper breaker)*, as shown in Figure 8–12, or use hammers to slightly deform the taper, as shown in Figure 8–13.

> **CAUTION:** Vehicle manufacturers often warn not to use a tapered *pickle fork* tool to separate tapered parts. The wedge tool can tear the grease seal and damage both the part being removed and the adjoining part.

Pitman arms require a larger puller to remove the pitman arm from the splines of the pitman shaft (see Figures 8–14 and 8–15 on pages 172–173).

Step 3 Replace the part using the hardware and fasteners supplied with the replacement part. *Do not reuse* the pre-crimped torque prevailing nuts used at the factory as original equipment on many tie rod ends.

> **CAUTION:** Whenever tightening the nuts of tapered parts such as tie rods, *DO NOT* loosen after reaching the proper assembly torque to align the cotter key hole. If the cotter key does not fit, *TIGHTEN* the nut further until the hole lines up for the cotter key. See Figure 8–16 on page 173.

When replacing tie rod ends, use the adjusting sleeve to adjust the total length of the tie rod to the same position and length as the original. Measure the original length of the tie rods and assemble the replacement tie rod(s) to the same overall length (see Figure 8–17 on page 173).

When positioning the tie rod end(s), check that the stud is centered in the socket, as shown in Figure 8–18 on page 174. This permits maximum steering linkage movement without getting into a bind if the steering linkage is pivoted beyond the angle the tie rod end can move in the socket.

> **NOTE:** To ensure a proper wheel alignment, install the adjusting sleeve with an equal number of threads showing at each end of the sleeve. Some manufacturers also specify a *minimum* of three threads showing at each end. If the sleeve itself is corroded or bent, it should be replaced along with either or both of the tie rod ends (inner and outer). See Figure 8–19 on page 174.

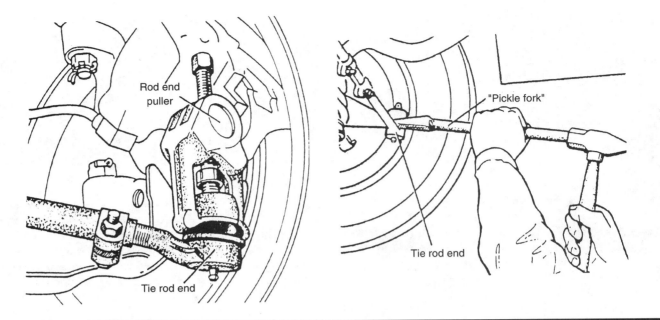

Figure 8-12 The preferred method for separating the tie rod end from the steering knuckle is to use a puller such as the one shown. A "pickle fork"–type tool should only be used if the tie rod end is going to be replaced. A pickle fork–type tool can damage or tear the rubber grease boot.

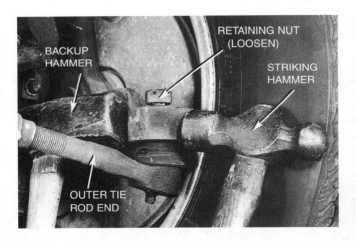

Figure 8-13 Two hammers being used to disconnect a tie rod end from the steering knuckle. One hammer is used as a backing for the second hammer. Notice that the attaching nut has been loosened, but not removed. This prevents the tie rod end from falling when the tapered connection is knocked loose.

Rack and Pinion

Inner tie rod end assemblies used on rack and pinion steering units require special consideration and often special tools. The inner tie rod end, also called a **ball socket assembly,** should be replaced whenever there is any noticeable free play in the ball and socket joint. Another test of this joint is performed by disconnecting

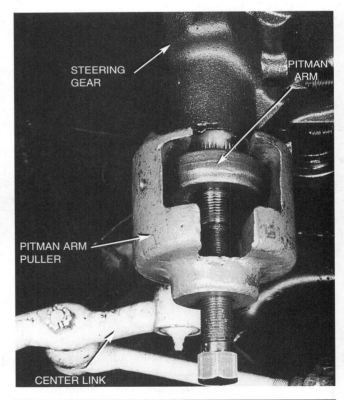

Figure 8-14 Using a puller to remove the pitman arm from the pitman shaft.

Figure 8–15 Pitman arm and pitman shaft indexing splines.

Figure 8–16 Align the hole in the tie rod end with the slot in the retaining nut. If the holes do not line up, always tighten the nut further (never loosen) until the hole lines up.

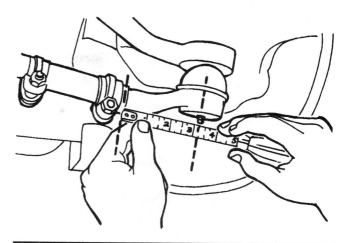

Figure 8–17 Replacement tie rods should be installed the same length as the original. Measure from the edge of the tie rod sleeve to the center of the grease fitting. When the new tie rod is threaded to this dimension, the toe setting will be close to the original. (Courtesy of Dana Corporation)

the outer tie rod end and measuring the effort required to move the tie rod in the socket, as shown in Figures 8–20 and 8–21. This is called the **articulation test.**

> **NOTE:** The articulation test is to be used on metal-to-metal ball socket assemblies. Low friction (polished ball and plastic liner–type joints) may require less effort to move and still be serviceable.

The inner tie rod assemblies are attached to the end of the steering rack by one of several methods.

Staked This method is common on Saginaw-style rack and pinion steering units found on General Motors and many Chrysler vehicles. Use two wrenches to remove, as shown in Figure 8–22.

The flange around the outer tie rod must be restaked to the flat on the end of the rack, as shown in Figures 8–23 and 8–24.

Figure 8–18 All tie rod ends should be installed so that the stud is in the center of its operating range as shown. (Courtesy of Dana Corporation)

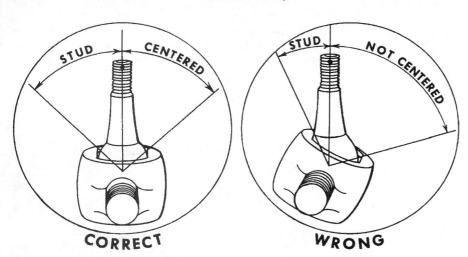

CORRECT   WRONG

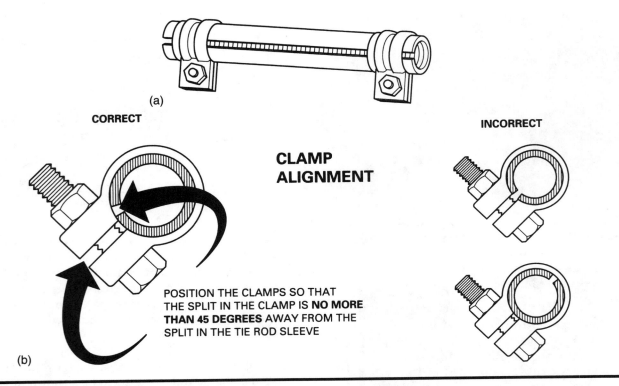

CLAMP
ALIGNMENT

CORRECT

INCORRECT

POSITION THE CLAMPS SO THAT
THE SPLIT IN THE CLAMP IS **NO MORE
THAN 45 DEGREES** AWAY FROM THE
SPLIT IN THE TIE ROD SLEEVE

Figure 8–19 (a) Tie rod adjusting sleeve. (Courtesy of Dana Corporation) (b) Be sure
to position the clamp correctly on the sleeve. (Courtesy of Ford Motor Company)

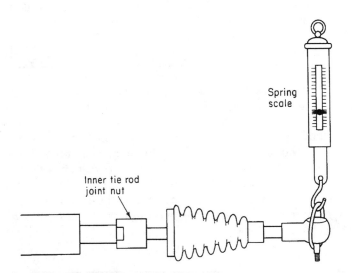

Figure 8–20 An articulation test uses a spring scale to
measure the amount of force needed to move the tie
rod in the ball socket assembly. Most manufacturers
specify a minimum of 1 lb. (4.4 N) of force and a
maximum of 6 lb. (26 N).

Figure 8–21 This worn ball socket assembly had no
resistance to movement. This joint caused a noise and
looseness in the steering.

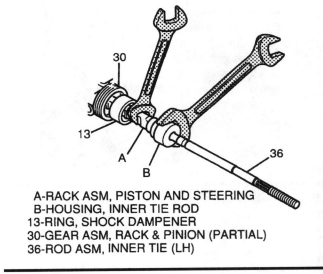

A-RACK ASM, PISTON AND STEERING
B-HOUSING, INNER TIE ROD
13-RING, SHOCK DAMPENER
30-GEAR ASM, RACK & PINION (PARTIAL)
36-ROD ASM, INNER TIE (LH)

Figure 8–22 Removing a staked inner tie rod assembly
requires two wrenches: one to hold the rack and the other
to unscrew the joint from the end of the steering rack.
(Courtesy of Oldsmobile)

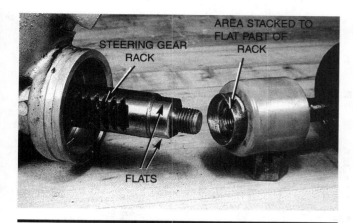

Figure 8–23 As the inner tie rod end is unthreaded from the rack, the staked socket housing bends.

Riveted or Pinned This method is commonly found on Ford vehicles (see Figure 8–25). Some roll pins require a special puller, or the pin can be drilled out. Many styles use an aluminum rivet. A special, very deep socket or a large open-end wrench can usually be used to shear the aluminum rivet by unscrewing the socket assembly from the end of the rack while the rack and pinion unit is still in the vehicle.

> **CAUTION:** While shearing of the aluminum rivet is done in many repair shops, care should be used to prevent damaging the threads of the rack while unscrewing the socket. For best results without the possibility of damage, always remove the rivet or pin using factory-authorized tools and procedures (see Figure 8–26 on page 177).

It is necessary to remove both bellows (boots) when servicing the passenger side ball socket assembly to allow access to the rack teeth on the driver side. The rack teeth must be held with a wrench whenever removing a ball socket assembly to avoid damage to the pinion gear or steering gear housing. The bellows and outer tie rod have to be transferred from the old to the replacement ball socket assembly (see Figure 8–27 on page 178).

> **HINT:** When replacing a rack and pinion assembly, specify a *long rack* rather than a *short rack*. A short rack does not include the bellows (boots) or inner tie rod ends (ball socket assemblies). The labor and cost required to exchange or replace these parts usually make it easier and less expensive to replace the entire steering unit.

Center Take-Off Racks Use bolts to secure the inner tie rods to the rack, as shown in Figure 8–28 on page 179.

Always follow the instructions that come with the replacement part(s). When securing the ball socket to

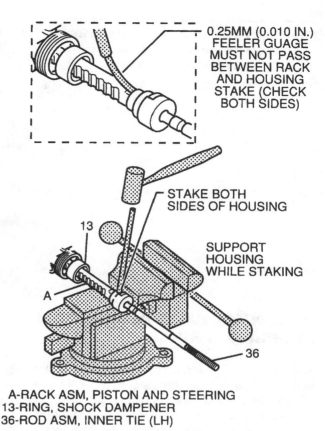

A-RACK ASM, PISTON AND STEERING
13-RING, SHOCK DAMPENER
36-ROD ASM, INNER TIE (LH)

Figure 8–24 When the inner tie rod end is reassembled, both sides of the housing must be staked down onto the flat shoulder of the rack. (Courtesy of Oldsmobile)

the rack, drill a new hole and install the supplied drill pin. If drilling is not required, install a hollow roll pin or stake, per the manufacturer's instructions. Lock the set screw, if so equipped (see Figure 8–29 on page 180).

TECH TIP

Jounce/Rebound Test

All steering linkage should be level and "work" at the same angle as the suspension arms (as shown in Figure 8–28). A simple test to check these items is performed as follows:

1. Park on a hard, level surface with the wheels straight ahead and the steering wheel in the *unlocked* position.
2. Bounce (jounce) the vehicle up and down at the front bumper while watching the steering wheel. The steering wheel should *not* move.

The steering wheel should *not* move during this test. If the steering wheel moves while the vehicle is being bounced, look for a possible bent steering linkage, suspension arm, or steering rack (see Figure 8–30 on page 181).

Figure 8–25 Many inner tie rod ends (ball socket assemblies) are pinned or riveted to the end of the rack. (Courtesy of Dana Corporation)

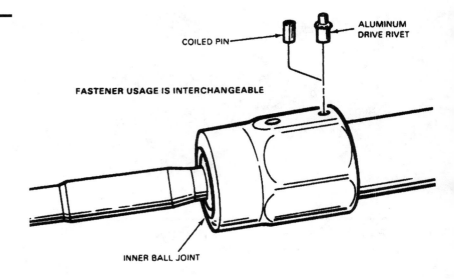

■ MANUAL RACK AND PINION STEERING SERVICE

Manual rack and pinion steering units are lightweight and provide quick, accurate steering (see Figure 8–31).

But because the movement of the front wheels is connected in a straight line directly to the rack, some feedback to the driver is often noticeable when the front tires hit a bump or dip in the road. This feedback can often jerk the steering wheel out of the driver's hands if the driver is not aware that this can happen, especially when driving up into a driveway or a curb when one wheel hits the bump first.

This type of steering wheel kickback should not be considered a fault in the steering mechanism. There are four basic areas of concern the technician should inspect when working with a manual rack and pinion steering unit:

1. Inspect the outer tie rod ends for wear.
2. Inspect the inner tie rod ends for wear.
3. Check for cracked or split rack boots and the air **breather tube** that connects the air chambers of the bellows. (As the rack moves, it compresses one bellows and expands the opposite bellows. The connecting tube allows air to move freely between the two bellows.)

> **NOTE:** The purpose of the bellows is to prevent water and dirt from getting into the rack housing. Dirt acts as an abrasive and can cause excessive wear to the rack and pinion gear teeth as well as possible wear to the rack support bushings. It is very important that cut or damaged rack bellows (boots) be replaced as soon as possible to prevent wear that can cause a steering problem.

4. Carefully inspect and replace the rack support bushings as needed.

Rack Friction Adjustment

Another less common service procedure on a manual rack and pinion steering unit is adjustment of the **rack bearing.** This bearing is also called a **yoke bushing.** This bearing pushes on the rack on the side opposite the pinion gear. The purpose of this bearing is to keep the rack teeth in proper mesh with the pinion gear teeth.

As a turning force is applied to the pinion gear by the driver turning the steering wheel, a bending force is applied to the rack. This force tends to force the rack out of engagement with the pinion. The rack bearing is spring loaded and is designed to keep constant pressure exerted on the rack itself. *If the pressure on the rack is too loose, the vehicle may wander. If the pressure is too great, hard steering may result.* See Chapter 12 for additional information on the correction of vehicle handling problems. The adjustment of this spring pressure is controlled in two ways:

1. Adjustment of the adjuster plug that screws into the rack housing. This plug is adjusted on most steering racks by tightening the nut until snug, then backing off the plug about 60 degrees (or one "flat" of the hex-shaped plug). This method is commonly used on General Motors and other types of vehicles that use the Saginaw rack and pinion unit. The Ford procedure is similar but uses a special socket that is marked to show how much to back off on the adjuster plug. When the plug bottoms, position the socket to align the "0" mark with a scribe mark on the housing. Then back off (turn the socket counterclockwise) until the second mark on the socket aligns with the scribe mark.

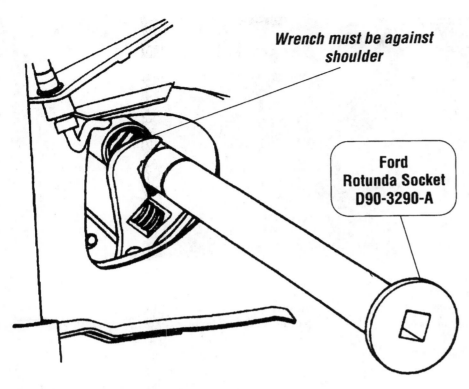

Wrench must be against shoulder

Ford
Rotunda Socket
D90-3290-A

Figure 8–26 Special tools are often necessary to remove the inner tie rod end (ball socket assemblies) while still on the vehicle. Notice that a wrench is being used to hold the rack teeth to prevent damage to the rack or the pinion gear as the inner tie rod socket is being unthreaded from the end of the rack. Special pullers make the job of removing the roll pin easy. (Courtesy of Ford Motor Company)

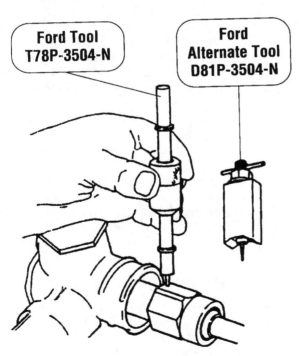

Ford Tool
T78P-3504-N

Ford
Alternate Tool
D81P-3504-N

2. The use of shims between a cover plate and the rack housing is another common rack bearing adjustment method used on many Ford vehicles and others. These thin metal shims are used to change the tension the rack bearing spring exerts on the rack (see Figure 8–32 on page 181).
   a. Adding shims *decreases* the rack spring tension (decreases the rack friction).

   b. Subtracting (removing) shims *increases* the rack spring tension (increases the rack friction).

   To determine the exact thickness of shim needed to provide the proper tension, remove the cover, all shims, and the gasket. Reinstall just the cover (without the gasket) and measure the gap with a feeler (thickness) gauge between the cover and the housing.

Figure 8–27 The worn ball socket assembly (bottom) is being replaced on a Ford Taurus. The bellows (boot) and outer tie rod end have already been transferred to the new inner ball socket assembly.

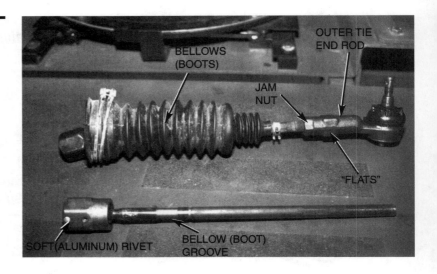

Add 0.005″ to 0.006″ (5–6 thousandths of an inch) to the feeler gauge reading. This total is the thickness of shims necessary to provide the proper tension.

Consult the service manual for the correct procedure for the vehicle on which you are working.

■ POWER STEERING DIAGNOSIS AND TROUBLESHOOTING

Power steering systems are generally very reliable, yet many problems are caused by not correcting simple service items such as:

1. **A loose, worn, or defective power steering pump drive belt,** including serpentine belts. This can cause jerky steering and belt noise, especially when turning.

> **NOTE:** Do not guess at the proper belt tension. Always use a belt tension gauge or observe the marks on the tensioner (see Figures 8–33 and 8–34 on page 182). Always apply force on the pump at the proper location to prevent damage to the pump (see Figures 8–35 and 8–36 on page 182).

2. **A bent or misaligned drive pulley,** usually caused by an accident or improper reassembly of the power steering pump after an engine repair procedure. This can cause a severe grinding noise whenever the engine is running and may sound like an engine problem.
3. **Low or contaminated power steering fluid,** usually caused by a slight leak at the high-pressure hose or defective inner rack seals on a power rack and pinion power steering system (see Figure 8–37 on page 183). This can cause a loud whine and a lack of normal power steering assist.

4. **Broken or loose power steering pump mounting brackets.** In extreme cases, the pump mounting bolts can be broken. These problems can cause jerky steering. It is important to inspect the pump mounting brackets and hardware carefully when diagnosing a steering-related problem. The brackets tend to crack at the adjustment points and pivot areas. Tighten all the hardware to ensure the belt will remain tight and not slip, causing noise or a power-assist problem.
5. **Underinflated tires.**
6. **Engine idle speed below specifications.**
7. **A defective power steering pressure switch.** If this switch fails, the computer will not increase engine idle speed while turning (see Figure 8–38 on page 183).
8. **Internal steering gear mechanical binding.**

As part of a complete steering system inspection and diagnosis, a steering wheel turning effort test should be performed (see Figure 8–39 on page 183). The power steering force, as measured by a spring scale during turning, should be less than 5 lb. (2.3 kg).

Power Steering Fluid

The correct power steering fluid is *critical* to the operation and service life of the power steering system! The *exact* power steering fluid to use varies as to vehicle manufacturer. There are even differences within the same company because of various steering component suppliers.

> **NOTE:** Remember, multiple-purpose power steering fluid does not mean *all*-purpose power steering fluid. Always consult the power steering reservoir cap, service manual, or owner's manual for the exact fluid to be used in the vehicle being serviced.

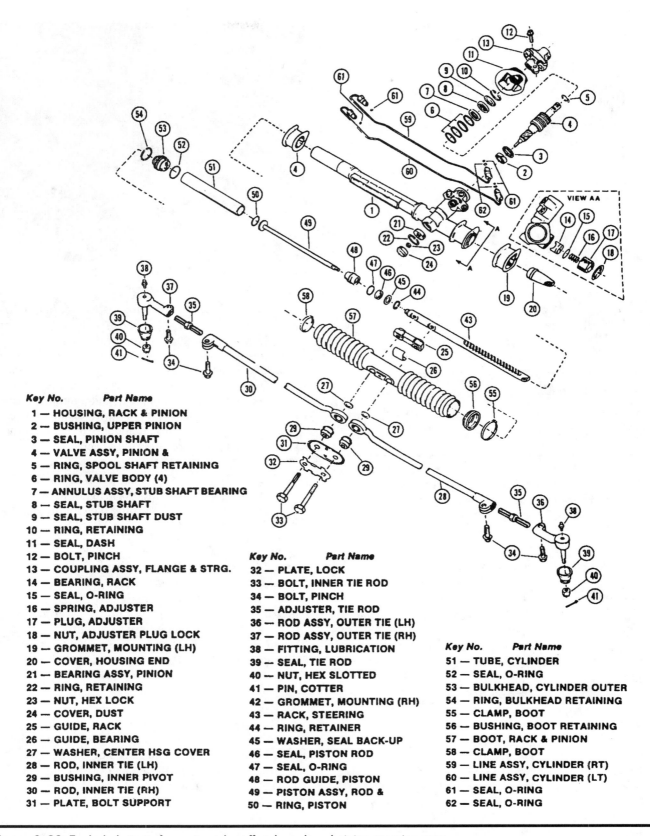

Key No.	Part Name
1 —	HOUSING, RACK & PINION
2 —	BUSHING, UPPER PINION
3 —	SEAL, PINION SHAFT
4 —	VALVE ASSY, PINION &
5 —	RING, SPOOL SHAFT RETAINING
6 —	RING, VALVE BODY (4)
7 —	ANNULUS ASSY, STUB SHAFT BEARING
8 —	SEAL, STUB SHAFT
9 —	SEAL, STUB SHAFT DUST
10 —	RING, RETAINING
11 —	SEAL, DASH
12 —	BOLT, PINCH
13 —	COUPLING ASSY, FLANGE & STRG.
14 —	BEARING, RACK
15 —	SEAL, O-RING
16 —	SPRING, ADJUSTER
17 —	PLUG, ADJUSTER
18 —	NUT, ADJUSTER PLUG LOCK
19 —	GROMMET, MOUNTING (LH)
20 —	COVER, HOUSING END
21 —	BEARING ASSY, PINION
22 —	RING, RETAINING
23 —	NUT, HEX LOCK
24 —	COVER, DUST
25 —	GUIDE, RACK
26 —	GUIDE, BEARING
27 —	WASHER, CENTER HSG COVER
28 —	ROD, INNER TIE (LH)
29 —	BUSHING, INNER PIVOT
30 —	ROD, INNER TIE (RH)
31 —	PLATE, BOLT SUPPORT

Key No.	Part Name
32 —	PLATE, LOCK
33 —	BOLT, INNER TIE ROD
34 —	BOLT, PINCH
35 —	ADJUSTER, TIE ROD
36 —	ROD ASSY, OUTER TIE (LH)
37 —	ROD ASSY, OUTER TIE (RH)
38 —	FITTING, LUBRICATION
39 —	SEAL, TIE ROD
40 —	NUT, HEX SLOTTED
41 —	PIN, COTTER
42 —	GROMMET, MOUNTING (RH)
43 —	RACK, STEERING
44 —	RING, RETAINER
45 —	WASHER, SEAL BACK-UP
46 —	SEAL, PISTON ROD
47 —	SEAL, O-RING
48 —	ROD GUIDE, PISTON
49 —	PISTON ASSY, ROD &
50 —	RING, PISTON

Key No.	Part Name
51 —	TUBE, CYLINDER
52 —	SEAL, O-RING
53 —	BULKHEAD, CYLINDER OUTER
54 —	RING, BULKHEAD RETAINING
55 —	CLAMP, BOOT
56 —	BUSHING, BOOT RETAINING
57 —	BOOT, RACK & PINION
58 —	CLAMP, BOOT
59 —	LINE ASSY, CYLINDER (RT)
60 —	LINE ASSY, CYLINDER (LT)
61 —	SEAL, O-RING
62 —	SEAL, O-RING

Figure 8–28 Exploded view of a center take-off-style rack and pinion steering gear assembly. (Courtesy of Oldsmobile)

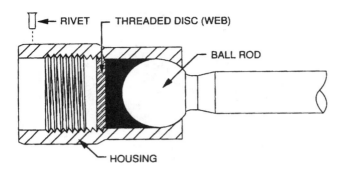

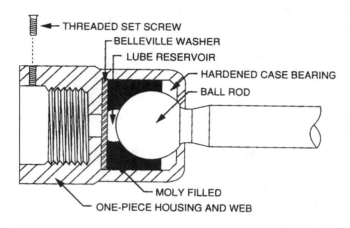

Figure 8–29 After replacing an inner tie rod end, the socket assembly should be secured with a rivet or set screw depending on the style of the replacement part. (Courtesy of Dana Corporation)

The fluids *generally* recommended include:

Manufacturer	Fluid Recommendation
Acura	Honda-specific power steering fluid
AMC/Eagle	Power steering fluid
AMC/Renault	Dexron III
Audi	Dexron III
BMW	Dexron III
Chrysler, Dodge, Plymouth	Power steering fluid
Colt/Champ	Dexron III
Fiat	Dexron III
Ford	ATF, type F
General Motors	Power steering fluid
Honda	Honda-specific power steering fluid
Hyundai	Dexron III
Isuzu	Dexron III
Jaguar	Dexron III
Jeep	Power steering fluid
Mazda	ATF: some use type F; some use Dexron III
Mercedes-Benz	Power steering fluid
Mitsubishi	Dexron III
Nissan/Datsun	Dexron-type ATF
Peugeot	Power steering fluid
Porsche	Dexron III
Saab	Power steering fluid
Subaru	Some use power steering fluid; all others Dexron III
Toyota	Dexron III
Volkswagen	Power steering fluid
Volvo	ATF: some use Dexron III; some use type F

TECH TIP

The "Pinky" Test

Whenever diagnosing any power steering complaint, check the level *and* condition of the power steering fluid. Often this is best accomplished by putting your finger ("pinky") down into the power steering fluid reservoir and pulling it out to observe the texture and color of the fluid (see Figure 8–40 on page 184).

A common problem with some power rack and pinion units is the wearing of grooves in the housing by the Teflon sealing rings of the spool (control) valve. When this wear occurs, aluminum particles become suspended in the power steering fluid, giving it a grayish color and thickening the fluid.

Normally clear power steering fluid that is found to be grayish in color and steering that is difficult when cold are clear indications as to what has occurred and why the steering is not functioning correctly.

> **NOTE:** Some vehicles use power steering reservoir caps with *left-hand threads*. Always clean the top of the cap and observe all directions and cautions. Many power steering pump reservoirs and caps have been destroyed by technicians attempting to remove a cap in the wrong direction using large pliers.

The main reason for using the specified power steering fluid is the compatibility of the fluid with the materials used in seals and hoses of the system. Using the wrong fluid (substituting clear power steering fluid for ATF, or vice versa) can lead to seal or hose deterioration and/or failure and fluid leaks. Honda-specific fluid contains friction-reducing additives.

> **NOTE:** Always use the power steering fluid recommended by the manufacturer. The correct fluid to use is usually imprinted on or near the power steering reservoir fill cap. See Figure 8–41 on page 185.

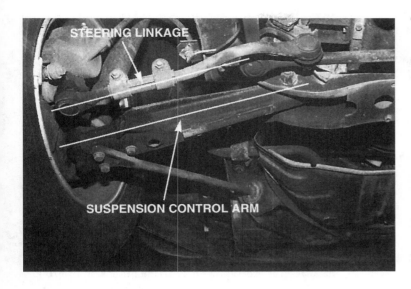

Figure 8–30 The steering and suspension arms must remain parallel to prevent the up and down motion of the suspension from causing the front wheels to turn inward or outward.

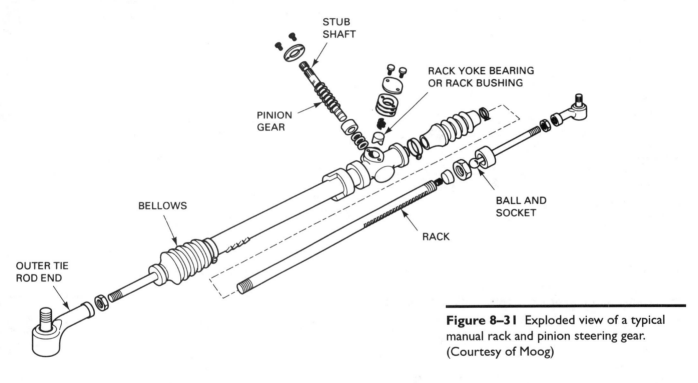

Figure 8–31 Exploded view of a typical manual rack and pinion steering gear. (Courtesy of Moog)

Figure 8–32 Shims used to adjust the tension of the rack bearing.

1. Check the power steering fluid level. Bleed the air from the system by turning the steering wheel lock-to-lock with the engine running (see details later in this chapter).
2. Check the condition and tension of the drive belt. If in doubt, replace the belt.
3. Inspect the condition of all hoses, checking for any soft hose or places where the hose could touch another component.
4. Check the tightness of all mounting bolts of the pump and gear.

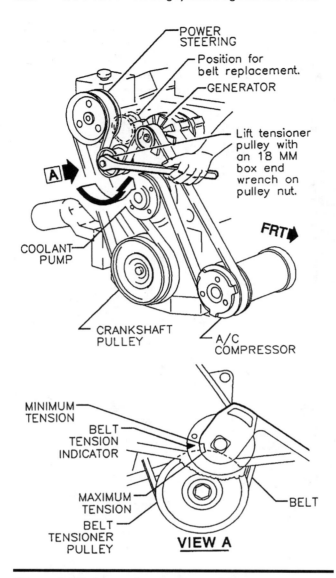

Figure 8–33 A typical service manual illustration showing the method to use to properly tension the accessory drive belt.

Figure 8–36 Service manual illustrations such as this can be used to locate the adjustment point for the power steering pump belt. (Courtesy of Chrysler Corporation)

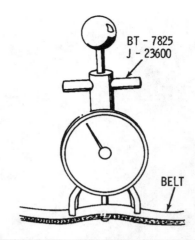

Figure 8–34 Belt tension gauge. Most vehicle manufacturers recommend a maximum tension of 170 lb. (760 N) for a new belt and a minimum tension of 90 lb. (400 N) for a used belt. (Courtesy of Oldsmobile)

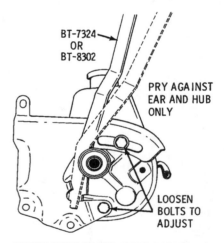

PULLEY REMOVED FOR PHOTO PURPOSE

Figure 8–35 Most power steering pumps have a designated area to be used to pry against to tension the drive belt. Never pry against the sheet metal or plastic reservoir. (Courtesy of Oldsmobile)

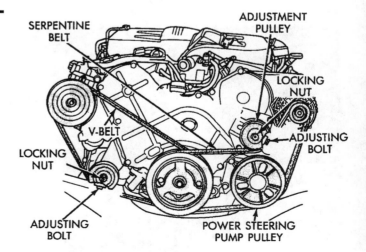

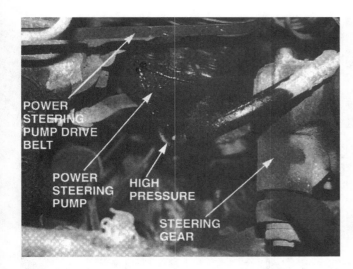

Figure 8–37 This power steering system leak could be the result of a leaking high-pressure hose or a defective seal in the pump. The entire area may have to be cleaned before the exact location of the leak can be determined.

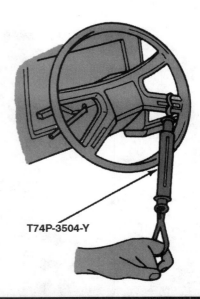

Figure 8–39 When diagnosing a hard steering complaint, use a spring scale to actually measure the amount of force needed to turn the steering wheel. The results should then be compared with specifications or compared with the effort of a similar vehicle. (Courtesy of Ford Motor Company)

Power Steering Fluid Flushing Procedure

Whenever there is any power steering service performed, such as replacement of a defective pump or steering gear or rack and pinion unit, the entire system should be flushed. If all of the old fluid is not flushed from the system, small pieces of a failed bearing or rotor could be circulated through the system. These metal particles can block paths in the control valve and cause failure of the new power steering pump or gear assembly.

> **NOTE:** Besides flushing the old power steering fluid from the system and replacing with new fluid, many technical experts recommend installing a filter in the low-pressure return line as an added precaution against serious damage from debris in the system. Power steering filters are commonly available through vehicle dealer parts departments, as well as aftermarket sources from local auto supply stores.

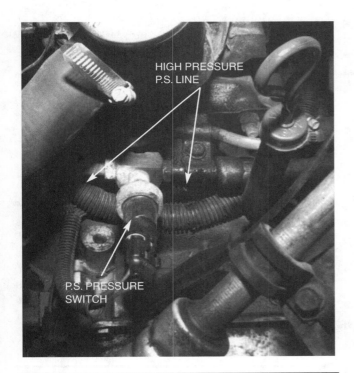

Figure 8–38 Typical power steering pressure switch installed in the high-pressure line. This switch closes and signals the engine computer whenever the pressure exceeds 300 psi (2000 kPa) indicating a moderate turn or a parking situation. The engine computer then raises the engine speed enough to compensate for the power steering load.

Two people are needed to flush the system, using the following steps:

Step 1 Remove the low-pressure return hose from the pump and plug the line fitting on the pump.

Figure 8–40 A check of the power steering fluid should include inspecting not only the level but the condition and color of the fluid, which could indicate a possible problem with other components in the steering system.

Step 2 Place the low-pressure return hose into an empty container.

Step 3 Fill the pump reservoir with fresh fluid and start the engine.

Step 4 As the dirty old power steering fluid is being pumped into the container, keep the reservoir full of clean fluid while the assistant turns the steering wheel full lock one way to full lock the other way.

> **CAUTION:** Never allow the pump reservoir to run dry of power steering fluid. Severe internal pump damage can result.

Step 5 When the fluid runs clean, stop the engine and reattach the low-pressure return hose to the pump reservoir.

Step 6 Restart the engine and fill the reservoir to the full mark. Turn the steering wheel back and forth avoiding the stops one or two times to bleed any trapped air in the system.

Bleeding Air out of the System

If the power steering fluid is pink (if ATF is the power steering fluid) or tan (if clear power steering fluid is used), there may be air bubbles trapped in the fluid. Stop the engine and allow the air to burp out to the surface for several minutes. Lift the vehicle off the ground, then rotate the steering wheel. This method prevents the breakup of large air bubbles into thousands of smaller bubbles that are more difficult to bleed out of the system.

> **NOTE:** To help rid the power steering system of unwanted trapped air, some vehicle manufacturers recommend cranking or starting the engine and turning the wheels from stop to stop *with the tires off the ground.* Do not crank the engine for longer than fifteen seconds to prevent starter damage due to overheating. Allow the starter to cool at least thirty seconds before cranking again.

Figure 8–41 Some power steering fluid is unique to the climate, such as this cold climate fluid recommended for use in General Motors vehicles when temperatures are low.

Sometimes trapped air just cannot be bled out of the system using ordinary methods. This trapped air makes the pump extremely noisy; this noise sometimes convinces the technician that the pump itself is defective. If the power steering system has been opened for repairs and the system drained, trapped air may be the cause.

See Figure 8–42 for adapters that can be built by the technician to use a vacuum pump to remove air from the power steering system. A vacuum pump from a diagnostic tester or an air-conditioning machine can be used.

Step 1 Fill the power steering fluid to the correct level.

Step 2 Connect the vacuum pump and start evacuating using 15″ Hg of vacuum.

Step 3 Start the engine and allow it to idle for fifteen minutes. Turn the wheels full left and full right every five minutes.

Step 4 Release the vacuum and refill the fluid level if necessary. If the noise is still present, repeat the procedure.

Hose Inspection

Both high-pressure and low-pressure return hoses should be inspected as part of any thorough vehicle inspection. While the low-pressure return hose generally feels softer than the high-pressure, neither should feel *spongy*. A soft, spongy hose should always be replaced (see Figure 8–43).

When replacing any power steering hose, make certain that it is routed the same as the original and does not interfere with any accessory drive belt, pulley, or other movable component such as the intermediate steering shaft.

Pressure Testing

Using a power steering pressure tester:

1. Disconnect the pressure hose at the pump.
2. Connect the leads of the tester to the pump and the disconnected pressure line (see Figure 8–44).

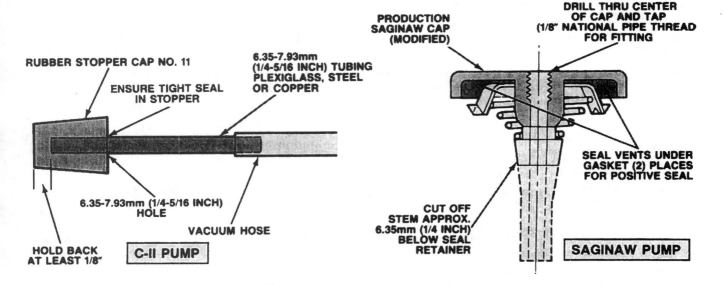

Figure 8–42 A vacuum adapter for a power steering pump can be made from a modified filler cap or a rubber stopper. (Courtesy of Ford Motor Company)

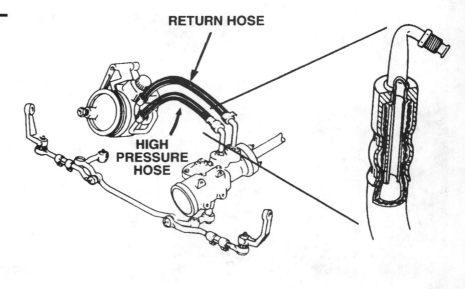

Figure 8–43 Inspect both high-pressure and return power steering hoses. Make sure the hoses are routed correctly and not touching sections of the body to prevent power steering noise from being transferred to the passenger compartment. (Courtesy of Moog)

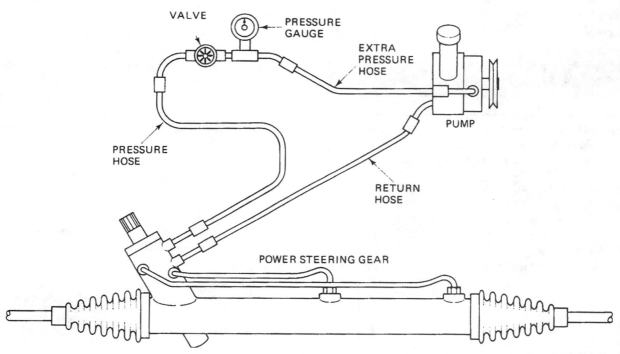

Figure 8–44 A drawing showing how to connect a power steering analyzer to the system.

3. Open the valve on the tester.
4. Start the engine. Allow the power steering system to reach operating temperatures.
5. The pressure gauge should register 80–125 psi (550–860 kPa). If the pressure is greater than 200 psi (1400 kPa), check for restrictions in the system including the operation of the poppet valve located in the inlet of the steering gear.
6. Fully close the valve three times. (Do not leave the valve closed for more than 5 seconds!) All three readings should be within 50 psi (345 kPa) of each

other and the peak pressure higher than 1000 psi (6900 kPa).
7. If the pressure readings are high enough *and* within 50 psi (345 kPa) of each other, the pump is okay.
8. If the pressure readings are high enough, yet not within 50 psi (345 kPa) of each other, the flow control valve is sticking.
9. If the pressure readings are less than 1000 psi (6900 kPa), replace the flow control valve and recheck. If the pressures are still low, replace the

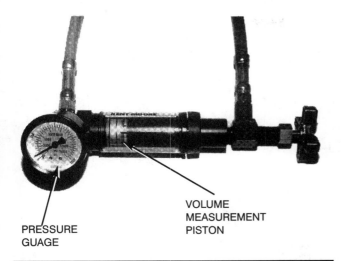

PRESSURE
GUAGE

VOLUME
MEASUREMENT
PISTON

Figure 8–45 A power steering analyzer that measures both pressure and volume. The shut-off valve at the right is used to test the maximum pressure of the pump.

rotor and vanes in the power steering pump. (See the overhaul procedure later in this chapter.)

10. If the pump is okay, turn the steering wheel to both stops. If the pressure at both stops is not the same as the maximum pressure, the steering gear (or rack and pinion) is leaking internally and should be repaired or replaced.

Many vehicle manufacturers recommend using a pressure gauge that also measures volume as shown in Figure 8–45.

Knowing the volume flow in the system provides information to the technician in addition to that of the pressure gauge. Many manufacturers' diagnostic procedures specify volume measurements and test results that can help pinpoint flow control or steering gear problems. Always follow the vehicle manufacturer's recommended testing procedures. Pressure and volume measurements specified by the manufacturer usually fall within the following ranges below.

Pump Service

Some power steering pump service can usually be performed without removing the pump, including:

1. Replacing the high-pressure and return hoses
2. Removing and cleaning the flow control valve assembly (see Figure 8–46)

Most power steering pump service requires the removal of the pump from the engine mounting and/or removal of the drive pulley.

> **NOTE:** Most replacement pumps are not equipped with a pulley. The old pulley must be removed and installed on the new pump. The old pulley should be carefully inspected for dents, cracks, or warpage. If the pulley is damaged, it must be replaced.

The pulley must be removed and installed with a pulley removal and installation tool (see Figure 8–47 on page 189).

> **CAUTION:** Do not hammer the pump shaft or pulley in an attempt to install the pulley. The shock blows will damage the internal components of the pump.

After removing the pump from the vehicle and removing the drive pulley, disassemble the pump according to the manufacturer's recommended procedure. See Figures 8–48 through 8–52 on pages 190–191 for the disassembly and reassembly of a typical power steering pump.

Clean all parts in power steering fluid. Replace any worn or damaged parts and all seals.

■ STEERING GEAR DIAGNOSIS

Steering gears are usually very reliable and long-lasting components. Often accidents cause damage to the gear itself or to the steering linkage that binds or flexes. Often the problem is not in the steering gear itself but rather is caused by other factors including:

1. Loose mounting bolts that hold the steering gear onto the frame
2. A loose or binding steering wheel, steering column, intermediate shaft between the bulkhead and the steering gear, or a defective flexible coupling (rag joint)

Typical Power Steering Pressures and Volume

Steering Action	Pressure psi (kPa)	Volume* (gal/min) (l/min)
Straight ahead no steering	Less than 200 psi (1000 kPa)	2.0 to 3.3 gpm (10 to 15 lpm)
Slow cornering	300–450 psi (2000–3000 kPa)	Within 1 gpm (4 lpm) of straight ahead
Full turn at stops	750–1450 psi** (5200–10,000 kPa)	Less than 1 gpm (4 lpm)

*Volume is determined by orifice size in the outlet of the pump and is matched to the steering gear.
**Upper limit pressure is determined by the calibration of the pressure relief valve.

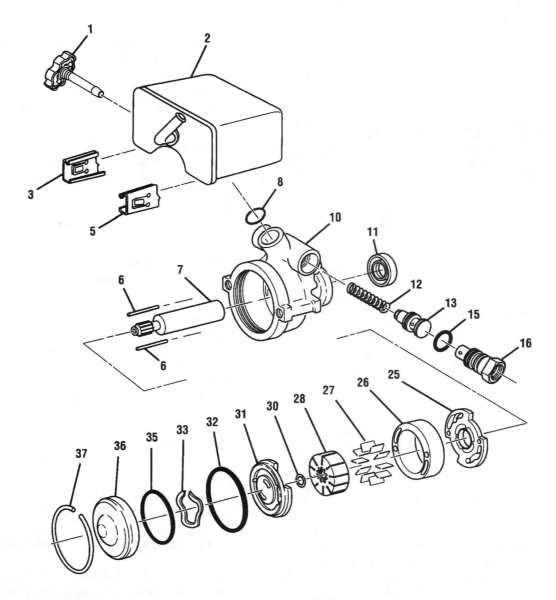

Key No.	Part Name
1 -	CAPSTICK ASM, RESERVOIR
2 -	RESERVOIR ASM, HYD PUMP (TYPICAL)
3 -	CLIP, RESERVOIR RETAINING (LH)
5 -	CLIP, RESERVOIR RETAINING (RH)
6 -	PIN, PUMP RING DOWEL
7 -	SHAFT, DRIVE
8 -	SEAL, O-RING
10 -	HOUSING ASM, HYD PUMP
11 -	SEAL, DRIVE SHAFT
12 -	SPRING, FLOW CONTROL
13 -	VALVE ASM, CONTROL
15 -	SEAL, O-RING

Key No.	Part Name
16 -	FITTING, O-RING UNION
25 -	PLATE, THRUST
26 -	RING, PUMP
27 -	VANE
28 -	ROTOR, PUMP
30 -	RING, SHAFT RETAINING
31 -	PLATE, PRESSURE
32 -	SEAL, O-RING
33 -	SPRING, PRESSURE PLATE
35 -	SEAL, O-RING
36 -	COVER, END
37 -	RING RETAINING

Figure 8–46 Typical power steering pump showing the order of assembly. The high-pressure (outlet) hose attaches to the fitting (#16). The flow control valve can be removed from the pump by removing the fitting. (Courtesy of Oldsmobile)

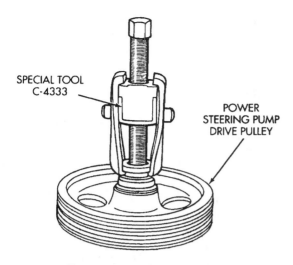

Remove Drive Pulley (Typical)

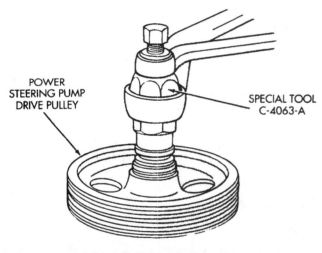

Install Drive Pulley (Typical)

Figure 8–47 Typical tools required to remove and install a drive pulley on a power steering pump. Often these tools can be purchased at a relatively low cost from automotive parts stores and will work on many different makes of vehicles. (Courtesy of Chrysler Corporation)

To make sure that the steering problem is in the steering gear, the first step is to remove the pitman arm from the pitman shaft. With the pitman arm removed from the pitman shaft of the steering gear, move the steering linkage through its normal left and right travel. Listen as well as feel for any binding or scraping that may indicate a steering linkage or suspension problem such as defective ball joints or MacPherson strut bearings and/or bushings. Once the problem has been isolated to the steering gear, the unit can then be removed from the vehicle and repaired or replaced. There are four choices that the technician and the customer or owner of the vehicle need to select.

1. Replace the steering gear with a new unit available from the dealer. This is the most expensive option. The cost is usually very high, but the core usually need not be turned into the dealer.
2. Purchase a remanufactured steering gear. This costs less than a new unit, but requires that the core (the old steering gear) be returned.
3. Purchase a replacement unit from a salvage or wrecking yard from the same or similar model vehicle that has the same Hollander interchange number (see Figure 8–53 on page 191).
4. Rebuild the existing steering gear. Although this may be the least expensive, it does require a higher level of skill. After disassembly of the unit, the damage to the unit may require more money in replacement parts than the cost of a rebuilt unit.

■ STEERING GEAR OVERHAUL AND SERVICE

Off the Vehicle

Always follow manufacturer's recommended procedures whenever removing or servicing any steering gear. Figures 8–54 through 8–75 on pages 192–200 show typical integral power steering gear disassembly and reassembly.

On the Vehicle

Manual steering gears usually require SAE 80W-90 gear lube as the lubricant specified by most manufacturers. The fluid level should be one of the first items to check if hard steering is the problem. Power steering gears use the power steering fluid that is circulated throughout the system as its lubricant.

Pitman Shaft Seal Replacement A pitman shaft seal that is leaking is a major source of power steering leaks. Full system pressure is applied to these seals. (The Saginaw 800 series gear uses two seals—one single lip seal and a double lip seal toward the pitman arm end of the pitman shaft.) To replace the pitman seals on a typical steering gear, follow these steps.

Step 1 Hoist the vehicle safely and remove the pitman arm nut and pitman arm from the pitman shaft using an appropriate puller.

Step 2 Remove the snap ring and backup washer.

Step 3 Lower the vehicle and place the oil drain pan under the steering gear.

Step 4 Start the engine and rotate the steering wheel to the stops. (On vehicles equipped with the Saginaw 800–type steering gear, turn to the left stop. On vehicles equipped with the Saginaw 605–type steering gear, turn to the right stop. (If the type of gear is unknown, turn the wheel first to the left and then to the right.) The high pressure will blow the seals out of the steering gear and into the drain pan. Immediately

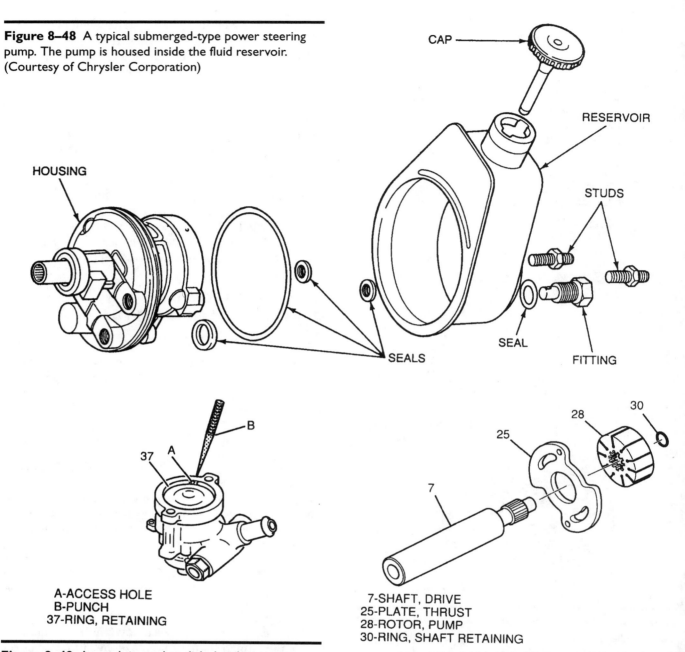

Figure 8–48 A typical submerged-type power steering pump. The pump is housed inside the fluid reservoir. (Courtesy of Chrysler Corporation)

HOUSING

CAP

RESERVOIR

STUDS

SEAL

FITTING

SEALS

A-ACCESS HOLE
B-PUNCH
37-RING, RETAINING

7-SHAFT, DRIVE
25-PLATE, THRUST
28-ROTOR, PUMP
30-RING, SHAFT RETAINING

Figure 8–49 A punch is used to dislodge the retaining ring. (Courtesy of Oldsmobile)

Figure 8–50 The drive shaft attaches to the drive pulley at one end and splined to the pump rotor at the other end. The vanes are placed in the slots of the rotor. (Courtesy of Oldsmobile)

stop the engine to prevent a complete loss of power steering fluid.

Step 5 Clean the seal pocket and pitman shaft. Use a crocus cloth, if necessary, to remove any surface rust or corrosion.

NOTE: If the pitman shaft has more severe rust than can be smoothed with crocus cloth, the pitman shaft should be replaced. Rust pitting can easily damage the new replacement seals: The lips of the seals will not be able to seal the high-pressure power steering fluid.

Step 6 Install both seals, backup washers, and snap ring using the appropriate seal installer (see Figure 8–76 on page 200).

Step 7 Before installing the pitman arm, refill the power steering reservoir and start the engine. Turn the steering wheel to bleed air from the system and check for leaks.

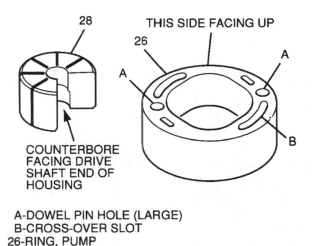

A-DOWEL PIN HOLE (LARGE)
B-CROSS-OVER SLOT
26-RING, PUMP
28-ROTOR, PUMP

Figure 8–51 The pump ring *must* be installed correctly. If it is installed upside down, the internal passages will not line up and the pump will have no output. (Courtesy of Oldsmobile)

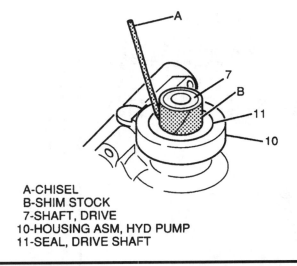

A-CHISEL
B-SHIM STOCK
7-SHAFT, DRIVE
10-HOUSING ASM, HYD PUMP
11-SEAL, DRIVE SHAFT

Figure 8–52 The shaft seal must be chiseled out. A thin metal shim stock should be used to protect the shaft from damage. (Courtesy of Oldsmobile)

Step 8 Install the pitman arm and double-check for leaks. Turn the steering wheel until bleeding is completed.

Over-Center Adjustment As mileage accumulates, some wear occurs in the steering gear. When this wear occurs, more and more clearance (lash) develops between the pitman shaft teeth (sector gear teeth) and the teeth on the rack piston. The result of this normal wear is excessive steering wheel free play. A common customer complaint is that the steering feels loose even though all steering linkage is okay.

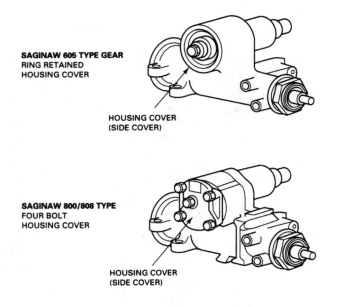

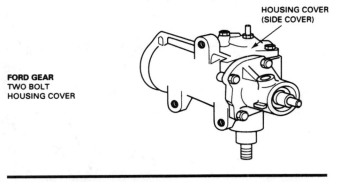

Figure 8–53 The Saginaw 605–type steering gear can be easily identified by its round side cover (which is usually on the top!). The Saginaw 800/808 uses a four-bolt side cover with the adjustment screw in the center. The two-bolt Ford gear uses two side cover bolts plus the adjustment screw in the center of the side cover.

> **NOTE:** This play in the steering can also cause a steering wheel shimmy that an alignment or normal steering linkage parts replacement cannot solve.

An *over-center* adjustment, also known as *sector lash* adjustment, may be necessary to correct this common problem.

> **CAUTION:** While many automotive manufacturers and experts recommend the following procedure, some vehicle manufacturers do not recommend that this procedure be performed with the steering gear installed in the vehicle.

With the front wheels in the straight-ahead position, visually check for lack of movement of the pitman shaft when the steering wheel is being moved

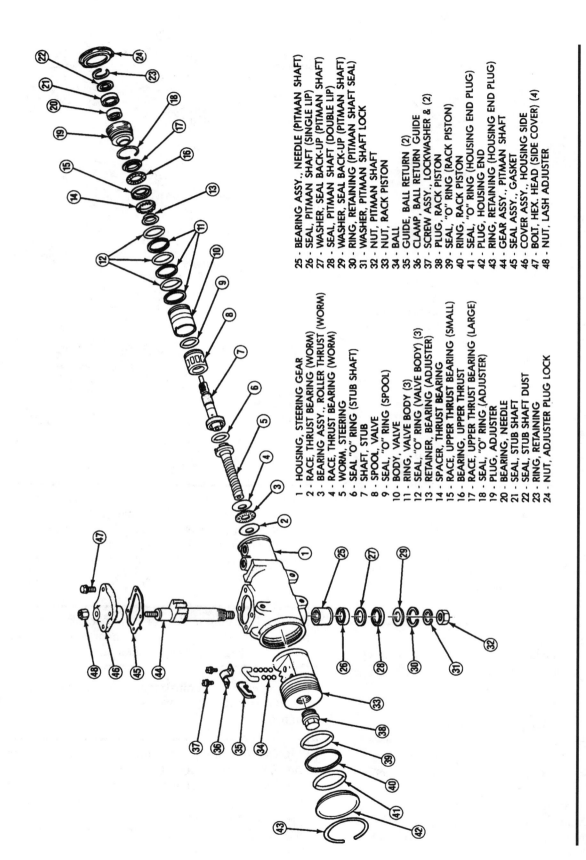

1 - HOUSING, STEERING GEAR
2 - RACE, THRUST BEARING (WORM)
3 - BEARING ASSY., ROLLER THRUST (WORM)
4 - RACE, THRUST BEARING (WORM)
5 - WORM, STEERING
6 - SEAL "O" RING (STUB SHAFT)
7 - SHAFT, STUB
8 - SPOOL, VALVE
9 - SEAL, "O" RING (SPOOL)
10 - BODY, VALVE
11 - RING, VALVE BODY (3)
12 - SEAL, "O" RING (VALVE BODY) (3)
13 - RETAINER, BEARING (ADJUSTER)
14 - SPACER, THRUST BEARING
15 - RACE, UPPER THRUST BEARING (SMALL)
16 - BEARING, UPPER THRUST
17 - RACE, UPPER THRUST BEARING (LARGE)
18 - SEAL, "O" RING (ADJUSTER)
19 - PLUG, ADJUSTER
20 - BEARING, NEEDLE
21 - SEAL, STUB SHAFT
22 - SEAL, STUB SHAFT DUST
23 - RING, RETAINING
24 - NUT, ADJUSTER PLUG LOCK

25 - BEARING ASSY., NEEDLE (PITMAN SHAFT)
26 - SEAL, PITMAN SHAFT (SINGLE LIP)
27 - WASHER, SEAL BACK-UP (PITMAN SHAFT)
28 - SEAL, PITMAN SHAFT (DOUBLE LIP)
29 - WASHER, SEAL BACK-UP (PITMAN SHAFT)
30 - RING, RETAINING (PITMAN SHAFT SEAL)
31 - WASHER, PITMAN SHAFT LOCK
32 - NUT, PITMAN SHAFT
33 - NUT, RACK PISTON
34 - BALL
35 - GUIDE, BALL RETURN (2)
36 - CLAMP, BALL RETURN GUIDE
37 - SCREW ASSY., LOCKWASHER & (2)
38 - PLUG, RACK PISTON
39 - SEAL, "O" RING (RACK PISTON)
40 - RING, RACK PISTON
41 - SEAL, "O" RING (HOUSING END PLUG)
42 - PLUG, HOUSING END
43 - RING, RETAINING (HOUSING END PLUG)
44 - GEAR ASSY., PITMAN SHAFT
45 - SEAL ASSY., GASKET
46 - COVER ASSY., HOUSING SIDE
47 - BOLT, HEX. HEAD (SIDE COVER) (4)
48 - NUT, LASH ADJUSTER

Figure 8-54 An exploded view of a Saginaw 800/808 steering gear showing all of its internal parts. Saginaw supplies steering gears to General Motors, Chrysler, and other vehicle manufacturers. (Courtesy of Chrysler Corporation)

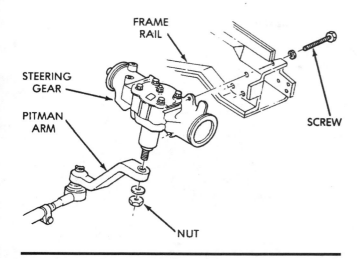

Figure 8–55 Most steering gears are bolted to the frame of the vehicle. (Courtesy of Chrysler Corporation)

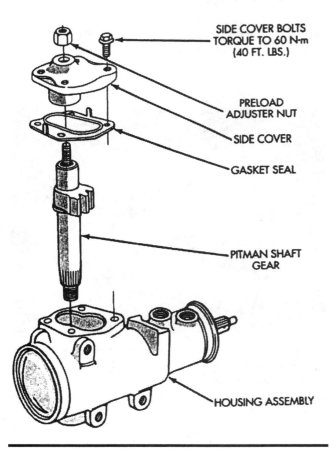

Figure 8–56 After removing the pitman arm and centering the steering gear, the side cover can be removed along with the pitman shaft as an assembly by removing the side cover bolts. (Courtesy of Chrysler Corporation)

slightly back and forth. (Move the steering wheel about 2 1/2″ [1 cm] total.) If the pitman arm does not move when the stub shaft (intermediate shaft) is being moved, then there is clearance (lash) in the steering gear itself.

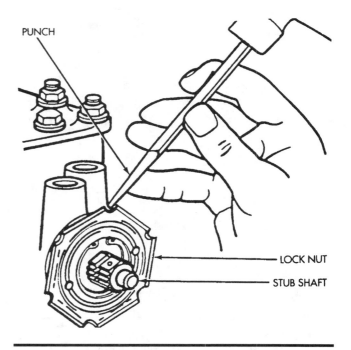

Figure 8–57 The adjuster plug is removed by first loosening the lock nut. The stub shaft is turned to force the end plug from the housing. (Courtesy of Chrysler Corporation)

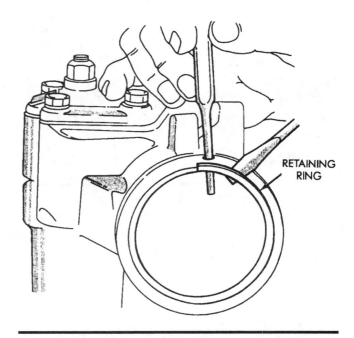

Figure 8–58 Insert a punch through a hole in the housing to release the snap ring that holds the end plug to the steering gear housing. (Courtesy of Chrysler Corporation)

If there is excessive clearance in the steering gear, then carefully perform the following steps:

Step 1 Drive the vehicle into the work area, keeping the front wheels straight ahead. Stop the engine.

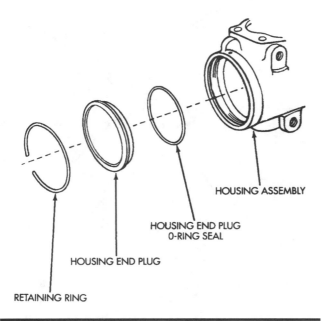

HOUSING ASSEMBLY

HOUSING END PLUG
O-RING SEAL

HOUSING END PLUG

RETAINING RING

Figure 8–59 After removing the snap ring, the stub shaft is turned to force the end plug from the housing. The housing end cap and O-ring seal can then be removed. (Courtesy of Chrysler Corporation)

Figure 8–60 Two different housing end caps. The larger one on the right is from a Saginaw model 808, while the smaller cap is from a Saginaw model 800. Except for this difference in diameter of housing, the units look identical. When a seal kit is purchased for a Saginaw 800/808 steering gear, both sizes of seals and O-rings are included.

Step 2 Loosen the over-center adjustment locking nut. This usually requires a 5/8″ socket or box end wrench.

NOTE: Most steering gears use conventional right-hand threads. The smaller Saginaw 605 steering gear uses left-hand threads. The 605 can be identified by the round side cover that is retained by a snap ring, instead of four bolts as in the 800 series.

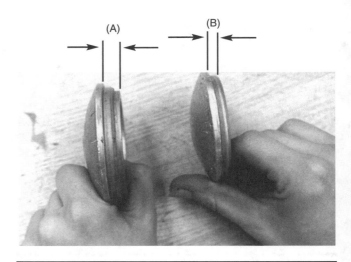

(A) (B)

Figure 8–61 The thickness of the end caps varies according to exact application. If the travel of the steering linkage is not properly matched to the steering stops on the steering linkage, the rack piston can hit the end cap and create a serious leak.

END PLUG RACK PISTON

Figure 8–62 After removal of the housing end cap, the end plug must be removed from the end of the rack piston.

Step 3 After loosening the lock nut, turn the adjusting screw clockwise until resistance is felt; then turn the screw back (counterclockwise) 1/4 turn (90 degrees). (With the Saginaw 605 gear, turn the adjusting screw counterclockwise to remove excessive clearance.)

NOTE: It is very important that the adjusting screw be loosened 1/4 turn to ensure proper clearance between the sector gear and the rack piston teeth.

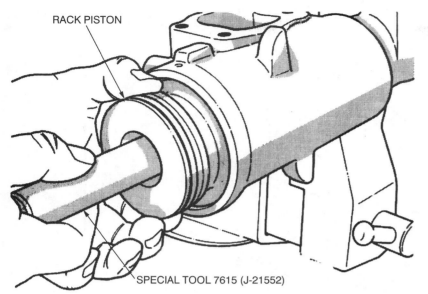

RACK PISTON

SPECIAL TOOL 7615 (J-21552)

Figure 8–63 A dowel rod or special tool is used to hold the ball bearings from falling out of the rack piston as it is being removed by turning the stub shaft. (Courtesy of Chrysler Corporation)

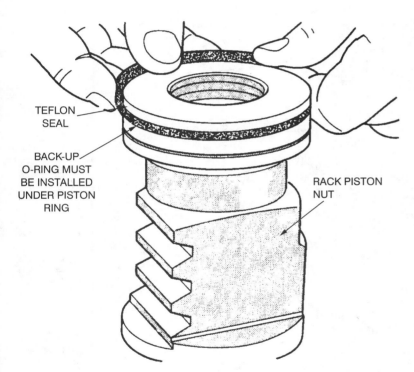

TEFLON SEAL

BACK-UP O-RING MUST BE INSTALLED UNDER PISTON RING

RACK PISTON NUT

Figure 8–64 The rack piston Teflon seal should be replaced whenever servicing the inside of the steering gear assembly. (Courtesy of Chrysler Corporation)

Step 4 While holding the adjusting screw to keep it from moving, tighten the locking nut.

Step 5 The adjustment procedure is complete, but the vehicle should be driven before returning it to the customer. This step is very important! For example, if the adjusting screw did move when the lock nut was tightened, the steering might feel too tight and might not return properly after turning a corner. *Carefully test-drive the vehicle, being especially careful when turning the first time.*

If the steering is too tight, repeat the adjustment procedure or remove the steering gear for repair.

■ POWER RACK AND PINION SERVICE

Service work on most power rack and pinion steering units is limited to replacement of outer and inner tie rod ends, boots (bellows), and rack mounting bushings. Whenever there is an internal leak or excessive wear is determined, the most economical repair is usually the replacement of the entire unit as an assembly. There are certain advantages and disadvantages to any repair procedure; they are summarized here.

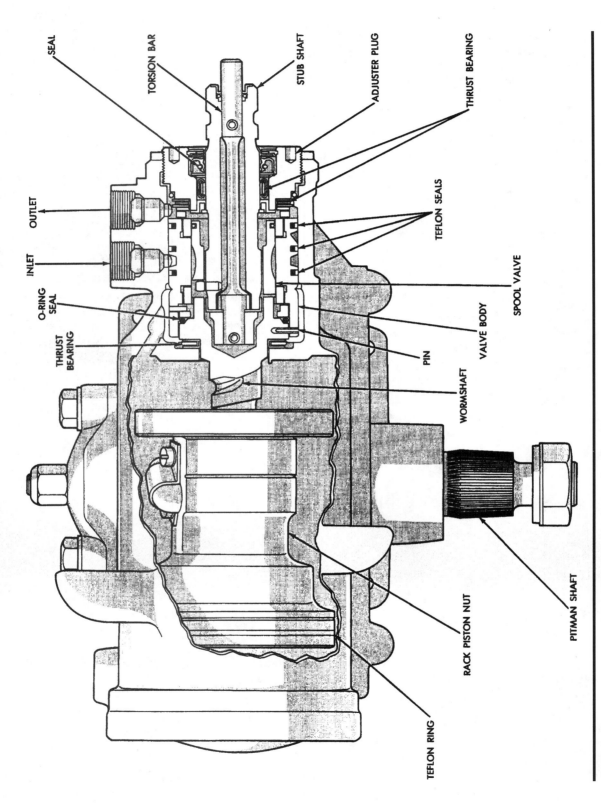

SEAL

TORSION BAR

STUB SHAFT

ADJUSTER PLUG

THRUST BEARING

OUTLET

INLET

TEFLON SEALS

O-RING SEAL

THRUST BEARING

SPOOL VALVE

VALVE BODY

PIN

WORMSHAFT

RACK PISTON NUT

TEFLON RING

PITMAN SHAFT

Figure 8–65 Cross section of a typical integral power steering gear assembly showing the relationship of the control valve, rack piston, worm shaft, and pitman shaft. (Courtesy of Chrysler Corporation)

What Can Be Done about Hard Steering Only When Cold?

Many technicians are asked to repair hard steering that occurs only when the vehicle is cold, usually first thing in the morning. After a couple of minutes, normal steering effort returns. As the vehicle gets older, the problem tends to get worse at higher and higher temperatures until the steering remains hard to turn even when warm. This condition occurs when the Teflon sealing rings wear grooves in the aluminum spool valve area of the rack and pinion steering unit (see Figure 8–77 on page 201).

During cold weather, these Teflon seals are stiff and power steering fluid leaks past them. This leakage causes hydraulic pressure to be applied to both sides of the power piston on the rack. With hydraulic pressure on both sides, instead of one side only, the steering wheel is *extremely* difficult to turn. The seals on the spool valve that are leaking determine in which direction it is hard to steer. Some vehicles are hard to steer in only one direction, while other vehicles may be difficult to steer in both directions.

As the power steering fluids heat up, the Teflon seals become more pliable and seal correctly, thereby restoring proper steering effort.

> **NOTE:** Power steering fluid leaking past the Teflon sealing rings of the spool valve will not create an external power steering fluid leak. This leakage around the spool valve area is simply an internal leak that causes the power steering fluid, under pressure, to be applied to the wrong end of the rack piston. External power steering fluid leaks are commonly caused by leaking seals on the rack itself.

There are several methods that many technicians use to "cure" morning sickness.

Method 1 Replace the entire rack and pinion steering unit. While this is the most expensive method, it is also the most commonly used repair (see Figures 8–78, 8–79, and 8–80 on pages 201–202).

Method 2 Replace the Teflon sealing rings with lap-joint Teflon seals. This procedure is usually performed when the vehicle is under warranty and involves the removal, disassembly, and replacement of the seal followed by reinstallation in the vehicle.

Method 3 Flush the old power steering fluid and refill with new fluid. This method is the least expensive and is recommended for mild cases of cold-weather reduction of power assist.

Method 4 Use an additive and power steering repair kits, available from aftermarket manufacturers. One method changes the calibration of the flow control at the pump to help compensate for the internal leakage that occurs at the spool valve area of the rack. Flushing the system and using special additives is another commonly available method.

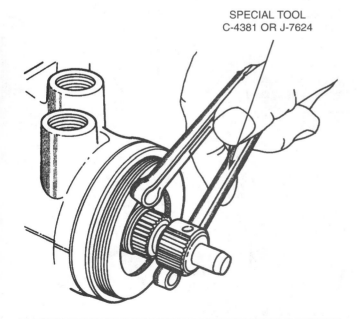

Figure 8–66 The adjuster plug is removed from the housing by using a tool that fits into two holes. (Courtesy of Chrysler Corporation)

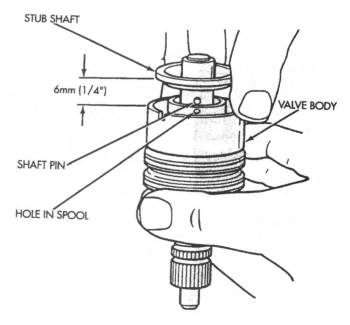

Figure 8–67 Removing the stub shaft from the valve assembly. (Courtesy of Chrysler Corporation)

Problem	Replacement Rack	Rack Repair
Hard steering when cold (spool valve seals normally wear grooves into the housing).	Includes the cost of the remanufactured unit. Usually a manufacturer's warrantee. Usually available units. R and R rack and realign the wheels.	Remove the rack and reseal the spool valve with lap-joint-type Teflon seals (dealer item). No warranty.
Leaking power steering fluid from the boots (bellows).	A remanufactured rack usually is tested for leaks before shipment. Manufacturer's warranty. R and R the rack and realign the wheels.	Remove and disassemble the rack. Replace both the rack seals including the hard to reach inner rack seal. If the rack is discovered to have rust pits, the entire unit needs to be replaced because the rust pits can damage the new seals.
Hard steering, binding.	A remanufactured rack usually includes all new bearings and seals. Most have manufacturer's warrantee.	Repair usually is not possible due to cracked or excessively worn housing. Replacement parts may be hard to find or secure.

See Figures 8–81 through 8–90 on pages 202–205 for the basic steps necessary for disassembly and reassembly of a typical power rack and pinion steering gear unit.

Steering System Troubleshooting Guide

Problem	Possible Cause/Correction
Excessive play or looseness in the steering.	Worn idler arm or other steering linkage, loose or defective front-wheel bearings. Perform a dry park test.
Hard steering.	Low tire pressure, loose or defective power steering belt, defective power rack (spool valve wear), binding steering gear, incorrect wheel alignment, lack of lubrication of the steering linkage, stocking control valve in the pump, steering column binding.
Steering wheel fails to return after a turn.	Steering gear adjusted too tightly, steering linkage is bent or binding, lack of lubrication of the steering linkage, incorrect wheel alignment, kinked power steering return hose.
Hissing noise in the rack and pinion.	May be normal valve noise. Compare with similar vehicles.
Growl noise in the power steering pump.	Restriction in the power steering hoses or steering gear. Defective or worn pump.
Groan noise in the power steering pump.	Air in the system. Bleed or use a vacuum pump to remove.
Temporary reduction of power steering assist when cold	Worn rack housing.

Figure 8–68 The spool valve can be removed from the valve body. The tolerance between these two parts is so close that if you hold the spool valve in your hand for just a minute, the heat of your body is enough to expand the valve so that it will not fit back into the valve body until it has cooled to room temperature. Dirt and contamination in this area are the causes of many power steering problems. (Courtesy of Chrysler Corporation)

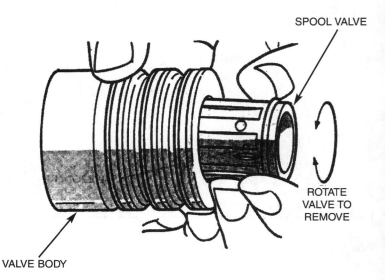

SPOOL VALVE

ROTATE VALVE TO REMOVE

VALVE BODY

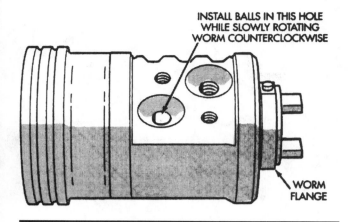

Figure 8–69 The steel ball bearings can be fed back into the gear nut. (Courtesy of Chrysler Corporation)

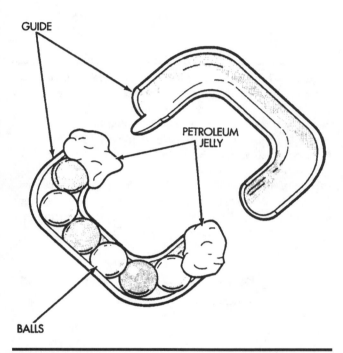

Figure 8–70 Petroleum jelly is used to hold the steel balls in the ball guide as it is assembled back onto the rack piston. (Courtesy of Chrysler Corporation)

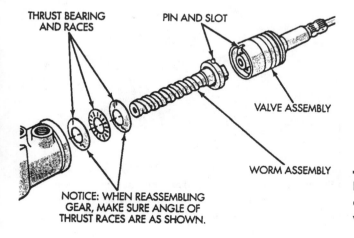

NOTICE: WHEN REASSEMBLING GEAR, MAKE SURE ANGLE OF THRUST RACES ARE AS SHOWN.

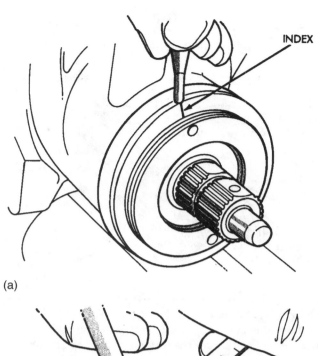

(a)

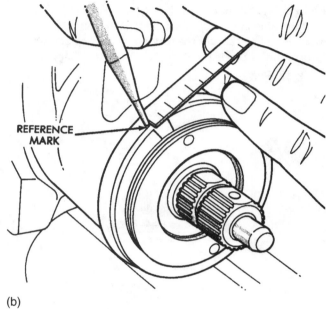

(b)

Figure 8–72 (a) After reassembling the rack piston and control valve, tighten the adjuster plug until it firmly bottoms in the housing. Place an index mark on the housing inline with one of the holes in the adjuster plug. (Courtesy of Chrysler Corporation) (b) Measure back 1/2" (13 mm) and make a reference mark. (Courtesy of Chrysler Corporation)

Figure 8–71 When reassembling the steering gear, be careful to note the proper position of the thrust bearing washers. (Courtesy of Chrysler Corporation)

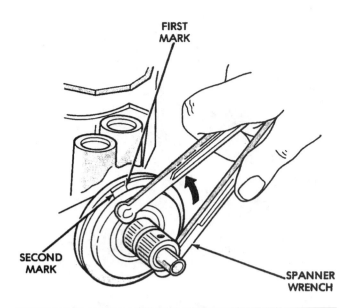

Figure 8–73 Rotate the adjuster plug counterclockwise until the hole in the adjuster plug aligns with the second mark. Install and secure the lock nut. This step adjusts the "worm thrust bearing preload." (Courtesy of Chrysler Corporation)

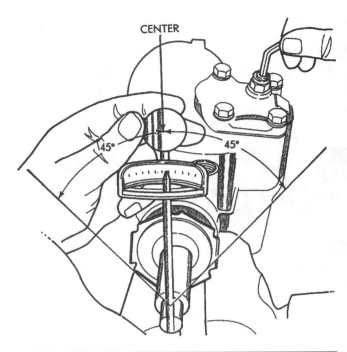

Figure 8–75 With the adjuster screw loosened, measure the amount of turning torque required to turn the stub shaft through a 90 degree arc 45 degrees either side from center. If the worm thrust bearing preload is okay, it should be within 4 to 10 in.-lb (0.45 to 1.13 N-m). Now turn the pitman shaft with an allen wrench (or screwdriver on some models) until there is an additional torque of 4 to 8 in.-lb for a total preload not to exceed 18 in. lb. (2 N-m). This is the last steering gear adjustment that needs to be made before reinstalling the steering gear in the vehicle. (Courtesy of Chrysler Corporation)

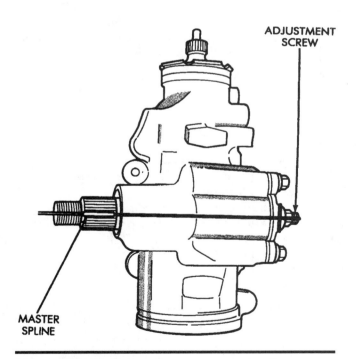

Figure 8–74 To make an "over-center preload" adjustment, start by rotating the stub shaft from stop to stop. Count the number of turns. Start at either stop and turn the stub shaft back one-half the total number of turns. This is the center of the gear travel, and the master spline on the pitman shaft should be aligned with the adjustment screw. (Courtesy of Chrysler Corporation)

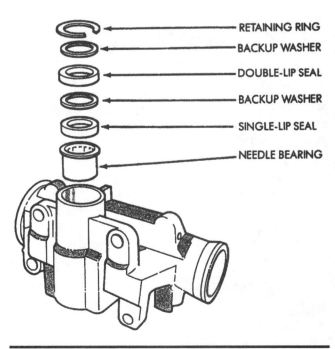

Figure 8–76 Pitman shaft seal and bearing locations in a typical Saginaw integral power steering gear. (Courtesy of Chrysler Corporation)

GROOVES CUT INTO
HOUSING BY SEALING RINGS

Figure 8–77 A section was cut from the housing of a power rack and pinion steering gear to show the wear grooves that have been cut into the aluminum.

Figure 8–78 Remanufacturers purchase these precut steel sleeves to be pressed into bored-out control valve housings to correct for the grooves created by the Teflon sealing rings.

Figure 8–79 A completed repair using a steel sleeve.

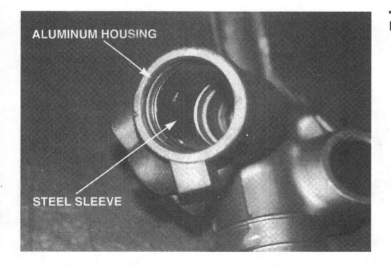

ALUMINUM HOUSING

STEEL SLEEVE

T E C H T I P

Pocket the Ignition Key to Be Safe

When replacing any steering gear such as a rack and pinion steering unit, be sure that no one accidentally turns the steering wheel! If the steering wheel is turned without being connected to the steering gear, the air bag wire coil (clock spring) can become off center. This can cause the wiring to break when the steering wheel is rotated after the steering gear has been replaced. To help prevent this from occurring, simply remove the ignition key from the ignition and put it in your pocket while servicing the steering gear.

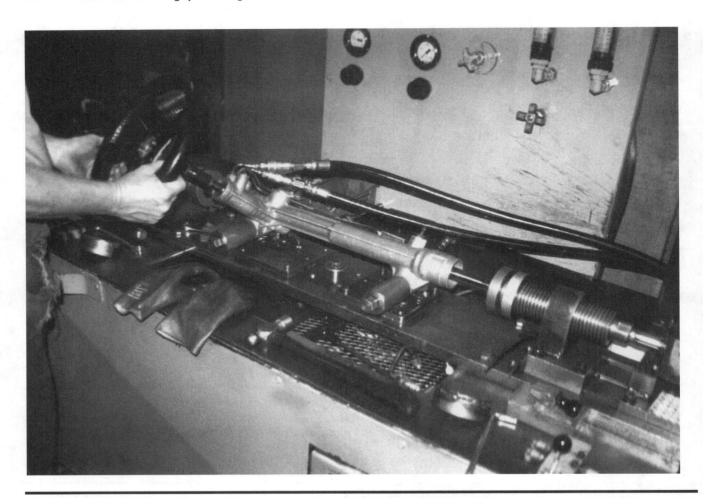

Figure 8–80 After a power rack and pinion is remanufactured, it is tested on a machine that tests the operation and measures the pressures and volume of power steering fluid through a test cycle.

Figure 8–81 The rack and pinion in this four-wheel-drive minivan looks easier to remove from the vehicle than most.

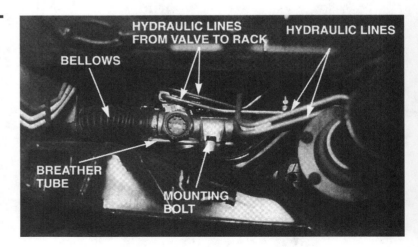

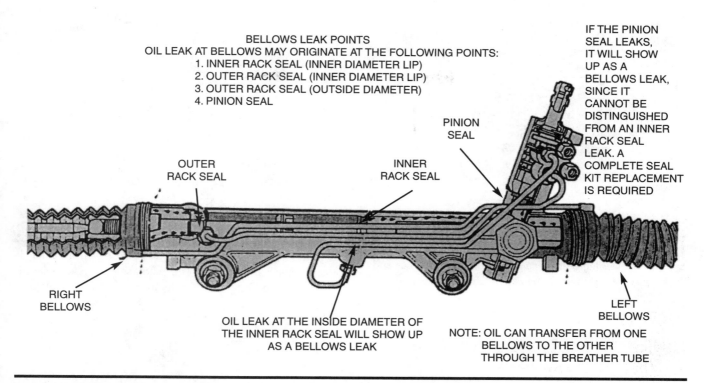

BELLOWS LEAK POINTS
OIL LEAK AT BELLOWS MAY ORIGINATE AT THE FOLLOWING POINTS:
1. INNER RACK SEAL (INNER DIAMETER LIP)
2. OUTER RACK SEAL (INNER DIAMETER LIP)
3. OUTER RACK SEAL (OUTSIDE DIAMETER)
4. PINION SEAL

IF THE PINION SEAL LEAKS, IT WILL SHOW UP AS A BELLOWS LEAK, SINCE IT CANNOT BE DISTINGUISHED FROM AN INNER RACK SEAL LEAK. A COMPLETE SEAL KIT REPLACEMENT IS REQUIRED

PINION SEAL

OUTER RACK SEAL

INNER RACK SEAL

RIGHT BELLOWS

OIL LEAK AT THE INSIDE DIAMETER OF THE INNER RACK SEAL WILL SHOW UP AS A BELLOWS LEAK

LEFT BELLOWS

NOTE: OIL CAN TRANSFER FROM ONE BELLOWS TO THE OTHER THROUGH THE BREATHER TUBE

Figure 8–82 Seeing the source of the leak helps to identify which seal or seals are leaking. (Courtesy of Ford Motor Company)

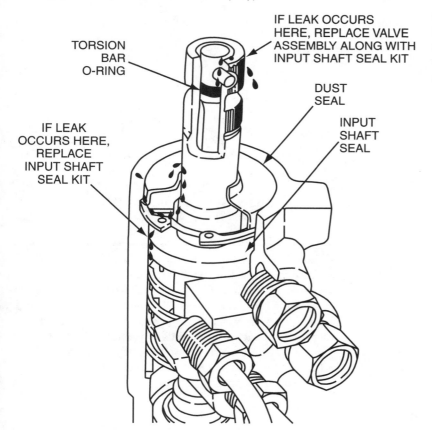

TORSION BAR O-RING

IF LEAK OCCURS HERE, REPLACE VALVE ASSEMBLY ALONG WITH INPUT SHAFT SEAL KIT

DUST SEAL

INPUT SHAFT SEAL

IF LEAK OCCURS HERE, REPLACE INPUT SHAFT SEAL KIT

Figure 8–83 A close-up view of the stub shaft area of the rack and pinion steering gear where hydraulic fluid leaks could occur. (Courtesy of Ford Motor Company)

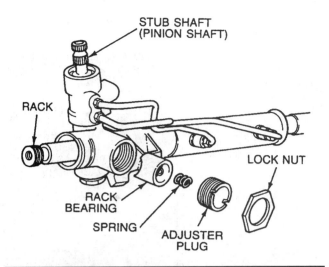

Figure 8–84 Start the disassembly of a typical power rack and pinion steering gear by removing the adjuster plug and rack bearing (bushing). (Courtesy of Chrysler Corporation)

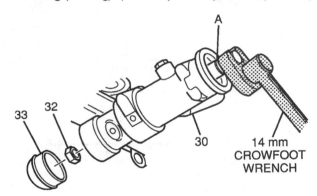

A-SHAFT, STUB
30-GEAR ASM, RACK & PINION (PARTIAL)
32-NUT, HEX LOCK
33-COVER, DUST

Figure 8–85 While holding the stub shaft, remove the pinion nut cover and nut. (Courtesy of Oldsmobile)

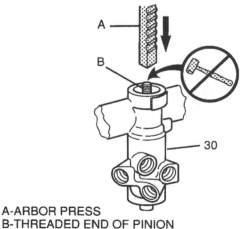

A-ARBOR PRESS
B-THREADED END OF PINION
30-GEAR ASM, RACK & PINION (PARTIAL)

Figure 8–86 After marking the center position of the gear, use a press or a vise to push the valve and pinion gear assembly out of the gear housing. (Courtesy of Oldsmobile)

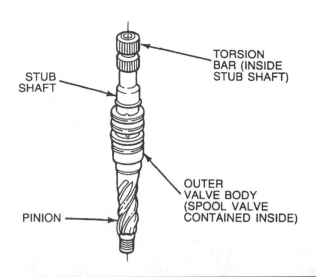

Figure 8–87 The rotary valve and pinion gear assembly is usually serviced as a unit. The Teflon seals can be replaced separately and reinstalled back into the housing. (Courtesy of Chrysler Corporation)

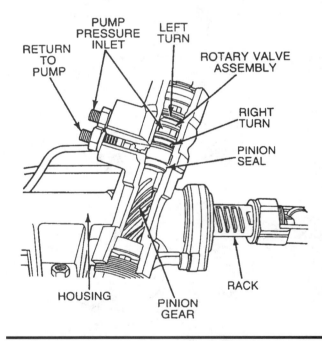

Figure 8–88 Cross section of assembled valve assembly. Be sure to center the rack before reinstalling the valve/pinion gear assembly. If the rack has been removed to replace the rack seals, be sure to cover the rack teeth with card stock or a plastic film to prevent the rack teeth from damaging the rack seals. Torque the pinion gear nut to specifications. Reinstall the rack bearing and adjuster plug; adjust to the manufacturer's specification. (Courtesy of Chrysler Corporation)

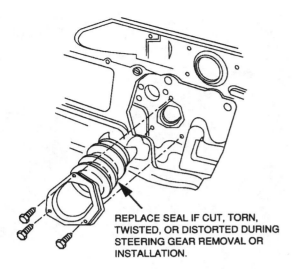

REPLACE SEAL IF CUT, TORN, TWISTED, OR DISTORTED DURING STEERING GEAR REMOVAL OR INSTALLATION.

Figure 8–89 Carefully inspect the coupling seal whenever replacing a rack and pinion steering gear. The seal connects to the bulkhead at one end and to the pinion shaft housing of the steering gear at the other end. If this seal touches the steering shaft, a noise that sounds like a defective rack and pinion steering gear can be heard. (Courtesy of Dana Corporation)

Figure 8–90 Refilling a power steering reservoir after replacing the rack and pinion steering gear assembly. Notice that the technician is pouring the clear power steering fluid from a container with the outlet toward the top. This is the correct way to pour from a container of this design because it allows air to get into the container, which allows the liquid in the container to flow smoothly.

■ SUMMARY

1. The **dry park test** is a very important test to detect worn or damaged steering parts. With the vehicle on the ground, have an assistant move the steering wheel back and forth while the technician feels for any looseness in every steering system part.

2. The steering system must be level side to side to prevent unwanted **bump steer.** Bump steer is when the vehicle's direction is changed when traveling over bumps or dips in the road.

3. The idler arm usually is the first steering system component to wear out in a conventional parallelogram-type steering system. Following the idler arm in wear are the tie rods, center link, and then the pitman arm.

4. Always use a belt tension gauge when checking, replacing, or tightening a power steering drive belt. The proper power steering fluid should always be used to prevent possible seal or power steering hose failure.

5. Power steering troubles can usually be diagnosed using a power steering pressure gauge. Lower than normal pump pressure could be due to a weak (defective) power steering pump or internal leakage inside the steering gear itself. If the pressure reaches normal when the shut-off valve on the gauge is closed, then the problem is isolated to being a defective gear.

6. Care should be taken when repairing or replacing any steering gear assembly to follow the manufacturer's recommended procedures exactly; do not substitute parts from one steering gear to another.

PHOTO SEQUENCE Dry Park Test

PS8–1 To perform a dry park test to determine the condition of the steering linkage and components, have an assistant sit behind the steering wheel with the vehicle on a drive-on-type hoist.

PS8–2 Raise the vehicle high enough to be able to test all of the steering components.

PS8–3 Have the assistant wiggle the steering wheel back and forth about 2″ (5 cm) during the test.

PS8–4 Start at one end of the steering system and work your way along the various components checking for any looseness or worn parts. Feeling the intermediate shaft helps find any free play that may be found in the shaft U-joints or coupler.

PS8–5 Feel for any looseness in the pitman arm or in the joint between the pitman arm and the center link.

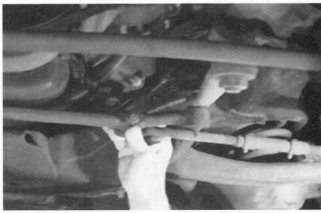

PS8–6 Feel for any looseness at the joint between the center link and the driver's side tie rod.

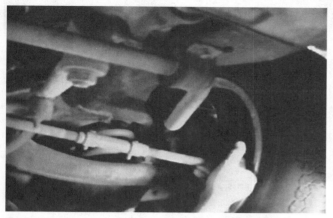

PS8–7 By grabbing the joint between the steering knuckle and the outer tie rod end, you will be able to feel any movement that you may not be able to see while the assistant continues to wiggle the steering wheel.

PS8–8 Continue with the steering linkage diagnosis by checking the joint between the idler arm and the center link.

PS8–9 The idler arm can be checked as a separate component by simply applying a 25-lb. force up and a 25-lb. force down on the idler arm. A good idler arm should have less than 1/4" of total play (1/8" up and 1/8" down). The assistant does not need to wiggle the steering wheel during this test procedure.

PS8–10 Any looseness in the idler arm where it connects to the frame can be easily determined by feeling the joint while the assistant turns the steering wheel back and forth.

PS8–11 Finally the passenger side outer tie rod end should be checked for any wear or looseness.

PS8–12 After the dry park inspection. Lower the vehicle and write down the results of the test on the repair order.

PHOTO SEQUENCE Rack and Pinion Steering Unit Replacement

PS8–13 The first step is to safely hoist the vehicle to a comfortable working position and remove both front wheels.

PS8–14 Remove the cotter key (if equipped) from the outer tie rod end attaching nut and remove the nut.

PS8–15 A tie rod end puller can be used to separate the tapered tie rod end from the tapered hole in the steering knuckle.

PS8–16 Many technicians simply use a large ball-peen hammer to separate the tie rod end from the steering knuckle by striking the steering knuckle.

PS8–17 After the tapered tie rod end has been separated, the tie rod can be disconnected.

Rack and Pinion Steering Unit Replacement—continued

PS8–18 Use a line (flare nut) wrench to loosen the hydraulic lines from the control valve area of the rack and pinion steering gear assembly.

PS8–19 Be sure to have a drain pan positioned under the vehicle when the lines are loosened from the rack, because power steering fluid will leak out.

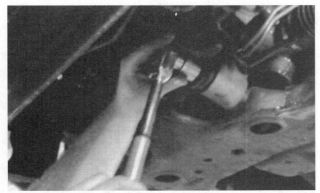

PS8–20 After the hydraulic lines have been removed from the control valve area of the rack and pinion, the rack attaching bolts can be removed.

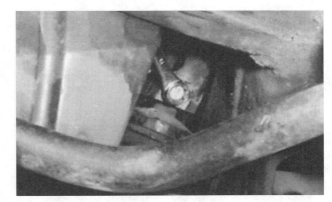

PS8–21 A pinch bolt holds the intermediate steering shaft to the stub shaft (input shaft) of the rack and pinion steering gear assembly. The bolt has to be completely removed before the two could be separated.

PS8–22 The rack and pinion steering gear assembly can now be removed from the vehicle. The replacement rack and pinion assembly is then reinstalled reversing the procedure.

Rack and Pinion Steering Unit Replacement—continued

PS8–23 After the rack and pinion assembly has been installed and the hydraulic lines have been attached and torqued to specifications, power steering fluid is added to the reservoir.

PS8–24 To bleed any trapped air from the power steering system, be sure the wheels are off the ground and simply turn the steering wheel from one side to the other with the engine off.

PS8–25 After turning the steering wheel several times to force the power steering fluid throughout the system, check the level of the fluid and add more if necessary.

PS8–26 Start the engine and turn the steering wheel from side to side checking for any leaks or abnormal noise. After this step, the vehicle can be lowered and the steering checked with the weight of the vehicle on the ground. After adjusting the alignment (toe) and a thorough test drive to verify proper operation, the vehicle can be returned to the owner.

■ REVIEW QUESTIONS

1. Describe how to perform a dry park test.
2. List the steering parts that should be replaced in pairs.
3. List five possible causes for hard steering.
4. Explain the procedure for flushing a power steering system.
5. Briefly describe adjustment and service procedures for a conventional power steering gear.
6. Briefly describe adjustment and service procedures for a power rack and pinion steering unit.

■ ASE CERTIFICATION-TYPE QUESTIONS

1. A dry park test to determine the condition of the steering components and joints should be performed with the vehicle
   a. On level ground
   b. On turn plates that allow the front wheels to move
   c. On a frame contact lift with the wheels off the ground
   d. Lifted off the ground about 2″ (5 cm)

2. Two technicians are discussing bump steer. Technician A says that an unlevel steering linkage can be its cause. Technician B says that if the steering wheel moves when the vehicle is bounded up and down, the steering linkage may be bent. Which technician is correct?
   a. Technician A only
   b. Technician B only
   c. Both Technician A and B
   d. Neither Technician A nor B

3. Two technicians are discussing the proper procedure for bleeding air from a power steering system. Technician A says that the front wheels of the vehicle should be lifted off the ground before bleeding. Technician B says that the steering wheel should be turned left and right with the engine off during the procedure. Which technician is correct?
   a. Technician A only
   b. Technician B only
   c. Both Technician A and B
   d. Neither Technician A nor B

4. A power steering pressure test is being performed, and the pressure is higher than specifications with the engine running and the steering wheel stationary in the straight-ahead position. Technician A says that a restricted high-pressure line could be the cause. Technician B says that internal leakage inside the steering gear or rack and pinion unit could be the cause. Which technician is correct?
   a. Technician A only
   b. Technician B only
   c. Both Technician A and B
   d. Neither Technician A nor B

5. A vehicle with power rack and pinion steering is hard to steer when cold (temporary loss of power assist when cold). The most likely cause is
   a. Leaking rack seals
   b. A defective or worn power steering pump
   c. Worn grooves in the housing by the spool valve seals
   d. Using the incorrect power steering fluid

6. Integral power steering gears use _____ for lubrication of the unit.
   a. SAE 80W-90 gear lube
   b. Chassis grease (NLGI #2)
   c. Power steering fluid in the system
   d. Molybdenum disulfide

7. Two technicians are discussing replacement of the pitman shaft seal on an integral power steering gear. Technician A says that the pitman arm must be removed before the old seal can be removed. Technician B says that the steering gear unit should be removed from the vehicle before the seal can be removed. Which technician is correct?
   a. Technician A only
   b. Technician B only
   c. Both Technician A and B
   d. Neither Technician A nor B

8. Two technicians are discussing the adjustment of the rack bearing (yoke bushing) on a rack and pinion steering unit. Technician A says that if the adjustment is too loose, the vehicle may wander. Technician B says that the rack friction is usually adjusted with an adjusting screw on the rack housing. Which technician is correct?
   a. Technician A only
   b. Technician B only
   c. Both Technician A and B
   d. Neither Technician A nor B

9. Which steering component should be replaced along with the same part on the other side of the vehicle? In other words, which part must be replaced as a pair?
   a. The idler arm
   b. The center link
   c. The pitman arm
   d. The ball socket assembly

10. Torque-prevailing nuts should be replaced with a new part if they are removed for service.
    a. True
    b. False

Suspension System Components and Operation

Objectives: After studying Chapter 9, the reader should be able to:

1. List various types of suspensions and their component parts.
2. Explain how coil, leaf, and torsion bar springs work.
3. Describe how front suspension components function to allow wheel movement up and down and provide for turning.
4. Discuss rear suspension function for both front-wheel-drive and rear-wheel-drive, as well as four-wheel-drive, vehicles.
5. Explain the function of electronic suspension systems.

Street-driven cars and trucks use a suspension system to keep the tires on the road and to provide acceptable riding comfort. A vehicle with a solid suspension, or no suspension, would bounce off the ground when the tires hit a bump. If the tires are off the ground, even for a fraction of a second, loss of control is possible. The purpose of the suspension is to provide the vehicle with:

1. A smooth ride
2. Accurate steering
3. Responsive handling

■ UNSPRUNG WEIGHT

A suspension system has to be designed to allow the wheels to move up and down quickly over bumps and dips without affecting the entire weight of the car or truck. In fact, the lighter the total weight of the components which move up and down, the better the handling and ride. This weight is called **unsprung** weight. The idea of very light weight resulted in magnesium wheels for racing cars which are very light, yet strong. Look-alike wheels, therefore, are often referred to as *mag* wheels. For best handling and ride, the unsprung weight should be kept as low as possible.

Sprung weight is the term used to identify the weight of the car or truck which does *not* move up and down and is supported or *sprung* by the suspension.

■ TYPES OF SUSPENSIONS

Early suspension systems on old horse wagons, buggies, and older vehicles used a solid axle for front and rear wheels (see Figure 9–1). If one wheel hit a bump, the other wheel was affected as shown in Figure 9–2.

Most vehicles today use a separate control arm–type of suspension for each front wheel which allows for movement of one front wheel without affecting the other front wheel. This type of suspension is called **independent** front suspension (see Figure 9–3).

Many rear suspensions also use independent-type suspension systems. Regardless of the design type of suspension, all suspensions use springs in one form or another.

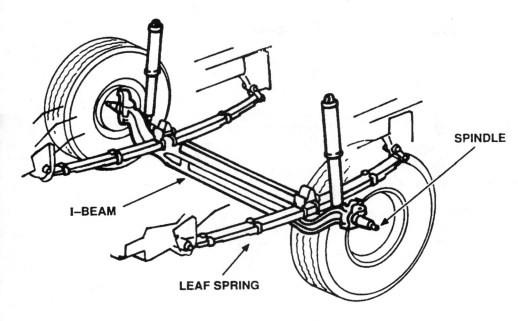

Figure 9–1 Solid I-beam axle with leaf springs. (Courtesy of Hunter Engineering Company)

SPINDLE

I–BEAM

LEAF SPRING

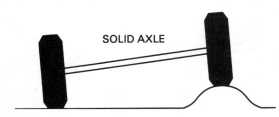

SOLID AXLE

Figure 9–2 When one wheel hits a bump or drops into a hole, both left and right wheels are moved. Because both wheels are affected, the ride is often harsh and feels stiff.

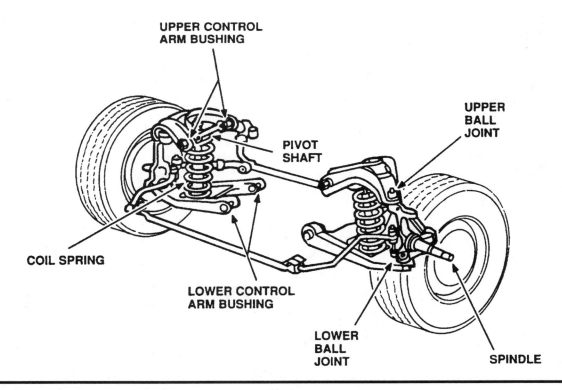

UPPER CONTROL ARM BUSHING

PIVOT SHAFT

UPPER BALL JOINT

COIL SPRING

LOWER CONTROL ARM BUSHING

LOWER BALL JOINT

SPINDLE

Figure 9–3 A typical independent front suspension used on a rear-wheel-drive vehicle. Each wheel can hit a bump or hole in the road *independently* without affecting the opposite wheel. (Courtesy of Hunter Engineering Company)

■ HOOKE'S LAW

Regardless of type, all suspensions use springs which share a common characteristic described by Hooke's Law. Robert Hooke (1635–1703), an English physicist, discovered the force characteristics of springs: *The deflection (movement or deformation) of a spring is directly proportional to the applied force.*

What this means is that when a coil spring (for example) is depressed 1″, it pushes back with a certain force (in pounds), such as 400 pounds. If the spring is depressed another inch, the force exerted by the spring is increased by another 400 pounds. The spring **rate** or force constant for this spring is therefore "400 lb per inch," usually symbolized by the letter **K.** Since the force constant is the force per unit of displacement (movement), it is a measure of the stiffness of the spring. The higher the spring rate (K), the stiffer the spring (see the example in Figure 9–4).

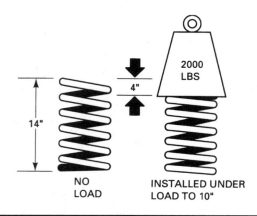

Figure 9–4 This spring was depressed 4 inches due to a weight of 2000 lb. This means that this spring has a spring rate (K) of 500 lb per inch (2000 ÷ 4″ = 500 lb/in.). (Courtesy of Moog)

■ COIL SPRINGS

Coil springs are made of special round spring steel wrapped in a helix shape. The strength and handling characteristics of a coil spring depend on:

1. Coil diameter
2. Number of coils
3. Height of spring
4. Diameter of the steel coil that forms the spring (See Figure 9–5.)

The spring rate (K) for coil springs is expressed by the formula:

$$K = \frac{Gd^4}{8ND^3}$$

where G = 11,250,000 (constant for steel)
d = diameter of wire
N = number of coils
D = diameter of the coil
Coil springs are used in front and/or rear suspensions.
The larger the diameter of the steel, the "stiffer" the spring.
The shorter the height of the spring, the stiffer the spring.
The fewer the coils, the stiffer the spring.
Springs are designed to provide desired ride and handling (see Figure 9–6).

The use of spacers between the coils of a coil spring is *not* recommended because the force exerted by the spacers on the springs can cause spring breakage. When a spacer is installed between the coils, the number of coils is reduced and the spring becomes stiffer. The force exerted on the coil spring at the contact points of the spacer can cause the spring to break.

Figure 9–5 The spring rate of a coil spring is determined by the diameter of the spring and the diameter of the steel used in its construction plus the number of coils and the free length (height). (Courtesy of Moog)

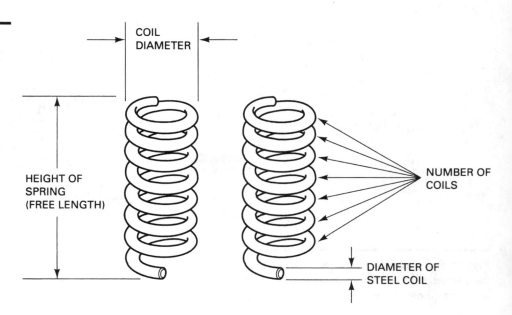

Variable-Rate Coil Springs

Many coil springs are designed to provide a variable spring rate. This means that as the spring is being compressed, the spring becomes stiffer. This allows for a smooth ride when bumps and dips in the road are small and provides load-carrying capacity and resistance to bottoming out when traveling over rough roads. A commonly used method of manufacturing a variable-rate coil spring is to vary the distance between the coils (see Figure 9–7).

As the coil spring compresses, the coils that are close together bottom out and the spring rate increases because the number of free coils is decreased.

Another way to achieve a variable-rate coil spring is to vary the diameter of the steel wire used to make the coil spring. The smaller diameter wire section compresses more easily than the thicker diameter section of wire (see the examples in Figure 9–8).

Coil Spring Mounting

Coil springs are usually installed in a **spring pocket** or **spring seat.** Hard rubber or plastic **cushions** or **insulators** are usually mounted between the coil spring and the spring seat (see Figure 9–9). The purpose of these insulators is to isolate and dampen road noise and vibration from the vehicle body. The type of end on the coil spring also varies and determines the style of spring mount (see Figures 9–10 and 9–11 on pages 216–217).

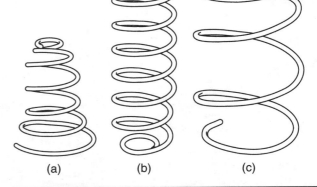

Figure 9–6 Most springs are identified by part number tags as shown. The spring used for the left front is usually different from the right front. Replacement springs may also be different side to side, and technicians should be careful not to switch springs during suspension service or repair.

(a) (b) (c)

Figure 9–7 Typical variable-rate coil spring. Note that the coils are closer together at the top and farther apart toward the bottom of the spring. (Courtesy of Dana Corporation)

Figure 9–8 (a) A semiconical spring used in rear suspension of passenger cars. It provides a variable rate because it is designed so that the coils bottom out on the spring seat. Because the coils can compress into itself (nest), it conserves space. (b) Typical application on the rear coil suspensions of light trucks. This cylindrical spring is manufactured from tapered wire to achieve a variable rate. (c) A shaped linear-rate spring conserves space by compressing into itself. This type of design permits body stylists to achieve an aerodynamic wedge-shaped vehicle.

Don't Cut Those Coil Springs!

Chassis service technicians are often asked to lower a vehicle. One method is to remove the coil springs and cut off 1/2 or more coils from the spring. While this *will* lower the vehicle, this method is generally not recommended:

1. A coil spring could be damaged during the cutting-off procedure, especially if a torch is used to do the cutting.
2. The spring will get stiffer when shortened, often resulting in a very harsh ride.
3. The amount the vehicle is lowered is *less* than the amount cut off from the spring, because as the spring is shortened, it becomes stiffer. The stiffer spring will compress less than the original.

Instead of cutting springs to lower a vehicle, there are two methods available that are preferred if the vehicle *must* be lowered:

1. There are replacement springs designed specifically to lower that model vehicle. A change in shock absorbers may be necessary because the shorter springs change the operating height of the stock (original) shock absorbers. Consult spring manufacturers for exact installation instructions and recommendations. See Figure 9–12.
2. There are replacement spindles designed to *raise* the location of the wheel spindle, thereby lowering the body in relation to the ground. Except for ground clearance problems, this is the method recommended by many chassis service technicians. Replacement spindles keep the same springs, shock absorbers, and ride, while lowering the vehicle without serious problems.

Figure 9–9 The spring cushion helps isolate noise and vibration from being transferred to the passenger compartment. (Courtesy of Moog)

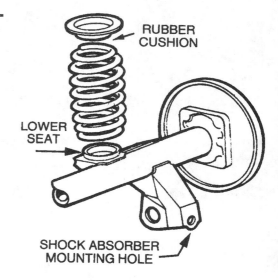

  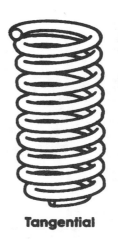 

| Square (Tapered) | Square (Untapered) | Tangential | Pig Tail |

Figure 9–10 Coil spring end types. (Courtesy of Moog)

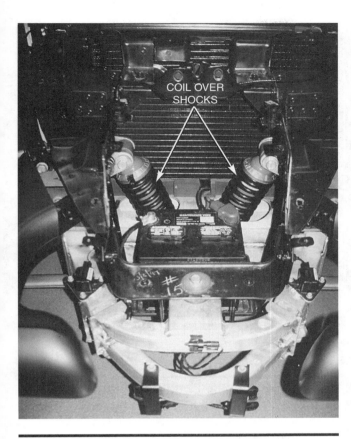

Figure 9–11 Coil springs can be used for many different types of front and rear suspensions. This is the front suspension of a Plymouth Prowler showing how the shock absorbers are inside the coil springs. This arrangement is called "coil-over-shocks."

Spring Coatings

All springs are painted or coated with epoxy to help prevent breakage. A scratch, nick, or pit caused by corrosion can cause a **stress riser** that can lead to spring failure. The service technician should be careful not to remove any of the protective coating. Whenever a service operation requires the spring to be compressed, always use a tool that will not scratch or nick the surface of the spring.

■ LEAF SPRINGS

Leaf springs are constructed of one or more strips of long, narrow spring steel. These metal strips, called leaves, are assembled with plastic or synthetic rubber insulators between the leaves, allowing freedom of movement during spring operation (see Figure 9–13).

The ends of the spring are rolled or looped to form eyes. Rubber bushings are installed in the eyes of the spring and act as noise and vibration insulators (see Figure 9–14).

The leaves are held together by a **center bolt**, also called a **centering pin** (see Figure 9–15).

Figure 9–12 The replacement coil spring (left) designed to lower a Ford Mustang is next to the original taller spring (right).

Figure 9–13 Typical leaf spring used on the rear of a rear-wheel-drive vehicle.

One end of a leaf spring is mounted to a hanger with a bolt and rubber bushings directly to the frame. The other end of the leaf spring is attached to the frame with movable mounting hangers called **shackles** (see Figure 9–16).

The shackles are necessary because as the spring hits a bump, the slightly curved spring (semi-elliptical) becomes longer and straighter, and the shackles allow this rearward movement. **Rebound** or **spring alignment** clips help prevent the leaves from separating whenever the leaf spring is rebounding from hitting a bump or rise in the roadway (see Figure 9–17).

Single leaf steel springs, called **monoleaf,** are used on some vehicles. A single or mono leaf spring is usually tapered to produce a variable spring rate. Leaf springs are used for rear suspensions on cars and many light trucks. A variable rate can be accomplished with a leaf spring suspension by providing contacts on the mount that effectively shorten the spring once it is compressed to a certain point. This provides a smoother ride when the load is light and still provides a stiffer spring when the load is heavy (see Figure 9–18).

To provide additional load-carrying capacity, especially on trucks and vans, auxiliary or *helper* leaves are commonly used. This extra leaf becomes effective only when the vehicle is heavily loaded, as shown in Figure 9–19.

Leaf springs are used on the front suspension of many four-wheel-drive trucks, especially medium and heavy trucks.

Composite Leaf Springs

Since the early 1980s, fiberglass-reinforced epoxy plastic leaf springs have been used on production vehicles. They save weight: An 8-lb. spring can replace a conventional 40-lb. steel leaf spring. The secret to making a

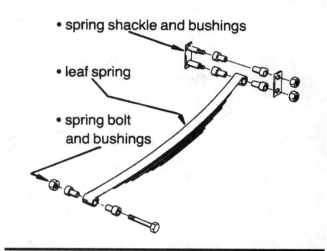

Figure 9–14 Rubber or urethane bushings are used between both ends of a leaf spring and the mounts to help isolate noise and vibration from getting into the passenger compartment. (Courtesy of Dana Corporation)

Figure 9–15 All multileaf springs use a center bolt to not only hold the leaves together but also to help retain the leaf spring in the center of the spring perch. (Courtesy of Moog)

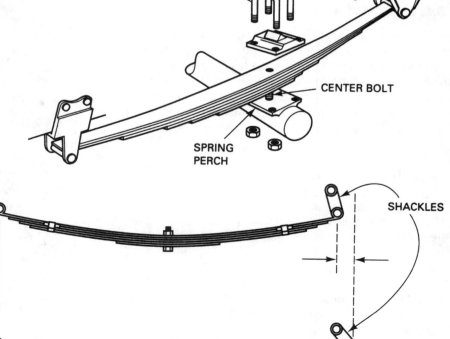

Figure 9–16 When a leaf spring is compressed, the spring flattens and becomes longer. The shackles allow for this lengthening.

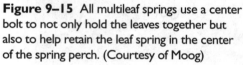

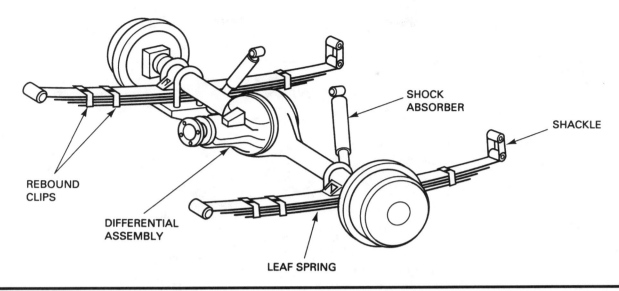

Figure 9–17 Typical rear leaf spring suspension of a rear-wheel-drive vehicle.

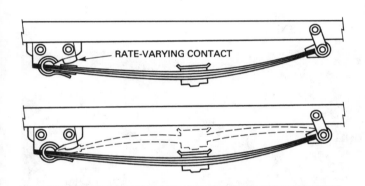

Figure 9–18 As the vehicle is loaded, the leaf spring contacts a section of the frame. This shortens the effective length of the spring, which makes it stiffer.

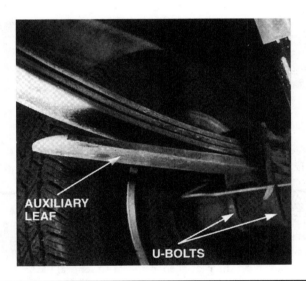

Figure 9–19 Many pickup trucks and sport utility vehicles (SUVs) use auxiliary leaf springs that contact the other leaves as the load is increased.

strong plastic leaf spring is the glass fibers running continuously from one end of the spring to the other and using 70 percent fiberglass with 30 percent epoxy composite. The single leaf composite spring helps isolate road noise and vibrations. It is more efficient than a multileaf spring because it eliminates the interleaf friction of the steel leaves and requires less space (see Figure 9–20).

Leaf spring rate increases when the thickness increases, and decreases as the length increases.

■ TORSION BARS

A torsion bar is a spring which is a long, *round*, hardened steel bar similar to a coil spring except for a *straight* bar (as shown in Figure 9–21).

One end is attached to the lower control arm of a front suspension and the other end to the frame. When the

wheels hit a bump, the bar twists and then untwists. Chrysler Corporation cars used torsion-bar front suspension both longitudinally and transversely (see Figure 9–22 on page 221).

Many manufacturers of pickup trucks currently use torsion-bar-type suspensions, especially on their four-wheel-drive models. Torsion bars allow room for the front drive axle and constant velocity joint and still provide for strong suspension.

As with all automotive springs, spring action is controlled by the shock absorbers. Unlike other types of springs, torsion bars are *adjustable* for correct ride height (see Figure 9–23 on page 221).

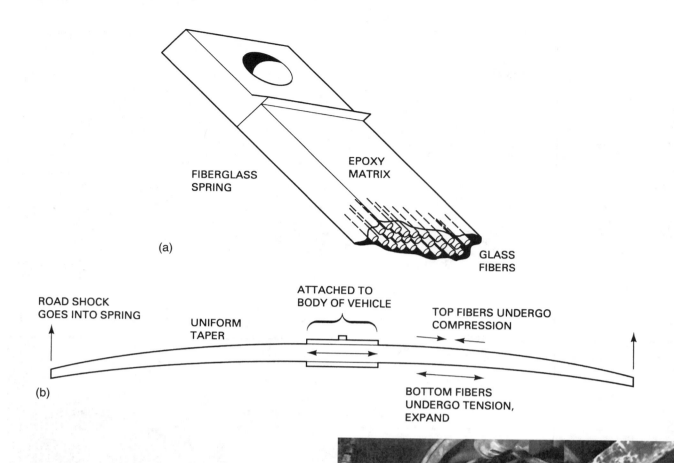

(a)

FIBERGLASS
SPRING

EPOXY
MATRIX

GLASS
FIBERS

ROAD SHOCK
GOES INTO SPRING

ATTACHED TO
BODY OF VEHICLE

TOP FIBERS UNDERGO
COMPRESSION

UNIFORM
TAPER

(b)

BOTTOM FIBERS
UNDERGO TENSION,
EXPAND

AUXILIARY
RUBBER SPRING

COMPOSITE
LEAF SPRING

(c)

Figure 9–20 (a) A fiberglass spring is composed of long fibers locked together in an epoxy (resin) matrix. (b) When the spring compresses, the bottom of the spring expands and the top compresses. (c) A composite transverse leaf spring used on the rear suspension of a General Motors front-wheel-drive car. The stiffness of the auxiliary rubber spring is used to fine-tune the rear suspension.

Figure 9–21 A torsion bar resists twisting and is used as a spring on some cars and many four-wheel-drive pickup trucks and sport utility vehicles. The larger the diameter, or the shorter the torsion bar, the stiffer the bar. (Courtesy of Ford Motor Company)

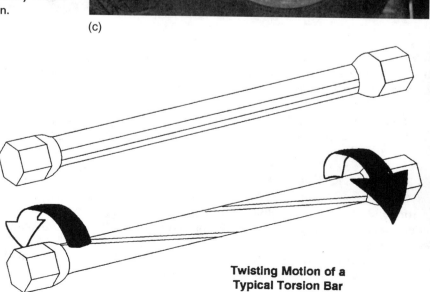

**Twisting Motion of a
Typical Torsion Bar**

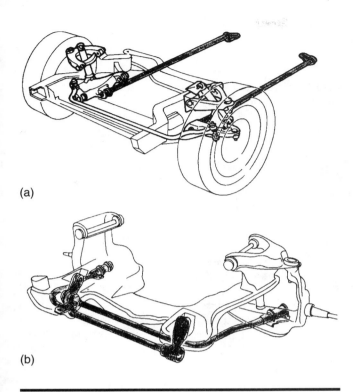

(a)

(b)

Figure 9–22 (a) Longitudinal torsion bars attach at the lower control arm at the front and an adjustable frame mount at the rear of the bar. (b) Transverse-mounted torsion bars attach to the lower control arm and the front of the frame. (Courtesy of Dana Corporation)

Torsion bars are usually used in front suspensions but also can be used in rear suspensions. Most torsion bars are labeled *left* or *right*, usually stamped into the end of the bars. The purpose of this designation is to make sure that the correct bar is installed on the original side of the vehicle. Torsion bars are manufactured without any built-in direction or preload. However, after being in a vehicle, the bar takes a set; reversing the side the torsion bar is used on causes the bar to be twisted in the opposite direction. Even though the bars are usually interchangeable, proper ride height can be accomplished even if the bars were installed on the side opposite from the original. But because the bar is being "worked" in the opposite direction, it can weaken and break. If a torsion bar breaks, the entire suspension collapses; this can cause severe vehicle damage, as well as a serious accident.

■ SUSPENSION PRINCIPLES

Suspensions use various links, arms, and joints to allow the wheels to move freely up and down; front suspensions also have to allow the front wheels to turn. All suspensions must provide for the following supports:

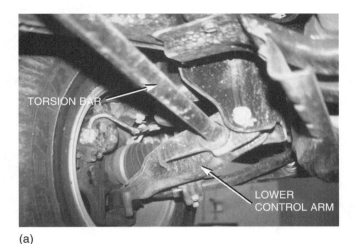

(a)

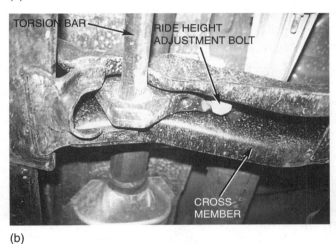

(b)

Figure 9–23 (a) The front end of the torsion bar attaches to the lower control arm on this pickup truck. (b) The other end of the torsion bar attaches to the crossmember through an adjustable socket. The suspension height can be adjusted by turning the adjusting bolt.

1. *Transverse (or side-to-side) wheel support.* As the wheels of the vehicle move up and down, the suspension must accommodate this movement and still keep the wheel from moving away from the vehicle or inward toward the center of the vehicle (see Figure 9–24). The control arm pivots on the vehicle frame. The wheels attach to a spindle that attaches to the ball joint at the end of the control arm. Transverse links are also called **lateral links.**
2. *Longitudinal (front-to-back) wheel support.* As the wheels of the vehicle move up and down, the suspension must allow for this movement and still keep the wheels from moving backward whenever a bump is hit. Note in Figure 9–24 how the separation of the pivot points, where the control arm meets the frame, provides support to prevent wheel movement front to back.

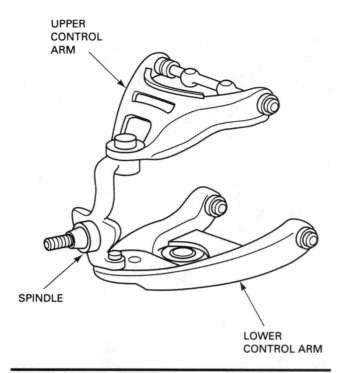

UPPER CONTROL ARM

SPINDLE

LOWER CONTROL ARM

Figure 9–24 The spindle supports the wheels and attaches to the control arm with ball and socket joints called *ball joints*. The control arm attaches to the frame of the vehicle through rubber bushings to help isolate noise and vibration between the road and the body. (Courtesy of Moog)

At least two suspension links or arms are required in order to provide for freedom of movement up or down, and to *prevent* any in-out or forward-back movement. Some suspension designs use an additional member (as shown in Figure 9–25) to control forward-back movement.

The design of the suspension and the location of the suspension mounting points on the frame or body are critical to the proper vehicle handling. Two very important design factors are called **anti-squat** and **anti-dive**.

1. *Anti-squat*. Anti-squat refers to the reaction of the body of a vehicle during acceleration. It is normal in most designs for the vehicle to squat down at the rear while accelerating. Most drivers are comfortable feeling this reaction, even on front-wheel-drive vehicles. Anti-squat refers to the degree to which this normal force is neutralized. If 100 percent anti-squat were designed into the suspension system, the vehicle would remain level while accelerating. Some Mercedes-Benz cars use over 100 percent anti-squat; the rear of the car actually *rises* while accelerating.

2. *Anti-dive*. Anti-dive refers to the force that causes the front of the vehicle to drop down while braking. Some front-nose dive feels normal to most drivers. If 100 percent anti-dive were

designed into a vehicle, it would remain perfectly level while braking.

The service technician cannot, and should not, attempt to change anti-squat or anti-dive characteristics built into the design of the vehicle. However, if the customer notices more squat or dive than normal, then the technician should carefully inspect all suspension components, especially those mounting points to the frame or body.

■ SOLID AXLES

Early cars and trucks used a solid (or *straight*) front axle to support the front wheels (see Figure 9–26).

A solid-axle front suspension is very strong and is still being used in the manufacture of medium and heavy trucks. The main disadvantage of solid-axle design is lack of ride quality. When one wheel hits a bump or dip in the road, the forces are transferred through the axle to the opposite wheel. Solid axles are currently used in the rear of most vehicles.

King Pins

At the end of many solid I-beam or tube axles are **king pins** that allow the front wheels to rotate for steering. King pins are hardened steel pins which attach the steering knuckle to the front axle, allowing the front wheels to move for steering. King pins usually have grease fittings to lubricate the king pin bushings. Failure to keep these bushings lubricated with chassis grease can cause wear and free play or can cause the pins to become galled (seized or frozen), resulting in hard steering and/or loud noise while turning (see Figure 9–27 on page 224).

■ TWIN I-BEAMS

A **twin I-beam** front suspension was used for over thirty years on Ford pickup trucks and vans, beginning in the mid-1960s. Strong steel twin beams that cross provide independent front suspension operation with the strength of a solid front axle. Early versions of the twin I-beam systems used king pins, while later models used ball joints to support the steering knuckle and spindle. Coil springs are usually used on twin I-beam suspensions, even though the original design and patent used leaf springs (see Figure 9–28 on page 224).

To control longitudinal (front-to-back) support, a **radius rod** is attached to each beam and is anchored to the frame of the truck using rubber bushings. These bushings allow the front axle to move up and down while still insulating road noise and vibration from the frame and body.

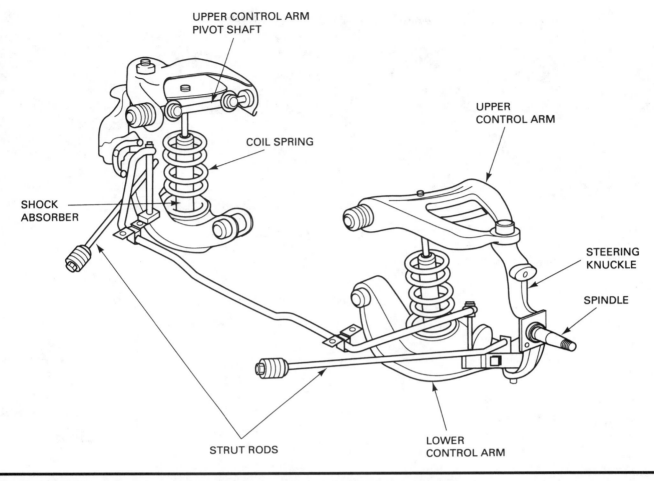

UPPER CONTROL ARM
PIVOT SHAFT

COIL SPRING

SHOCK
ABSORBER

UPPER
CONTROL ARM

STEERING
KNUCKLE

SPINDLE

STRUT RODS

LOWER
CONTROL ARM

Figure 9–25 The strut rods provide longitudinal support to the suspension to prevent forward or rearward movement of the control arms.

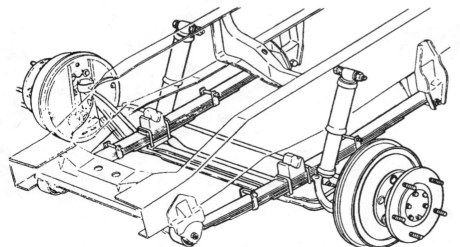

Figure 9–26 A solid-axle suspension with leaf springs. (Courtesy of Dana Corporation)

TECH TIP

Radius Rod Bushing Noise

When the radius rod bushing (see Figure 9–29) on a Ford truck or van deteriorates, the most common complaint from the driver is noise. Besides causing tire wear, worn or defective radius rod bushing deterioration can cause:

1. A clicking sound when braking (it sounds as if the brake caliper may be loose)
2. A clunking noise when hitting bumps

When the bushing deteriorates, the axles can move forward and backward with less control. Noise is the first sign that something is wrong. Without proper axle support, handling and cornering can also be affected. See Chapter 10 for additional suspension inspection and diagnostic procedures.

Figure 9–27 Typical king pin used with a solid axle. (Courtesy of Dana Corporation)

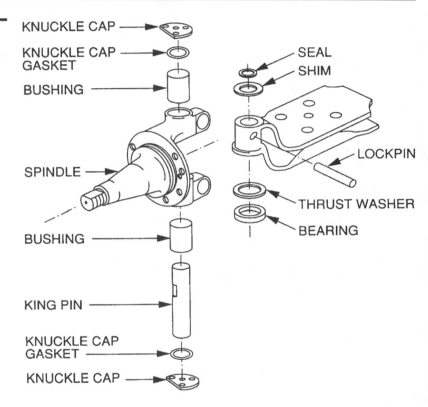

Figure 9–28 Twin I-beam front suspension. (Courtesy of Dana Corporation)

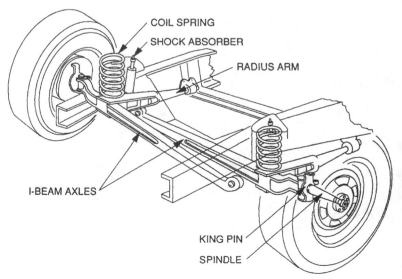

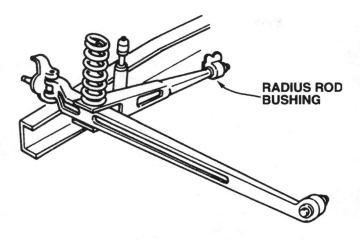

Figure 9–29 The rubber radius rod bushing absorbs road shocks and helps isolate road noise. (Courtesy of Moog)

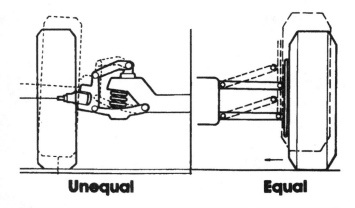

Figure 9–30 If the control arms were both the same length, the tire would move side to side as the wheels moved up and down as illustrated on the right. To keep the wheel moving straight up and down as the suspension moves, the upper control arm is shorter, as shown on the left. (Courtesy of Moog)

■ SHORT/LONG ARM SUSPENSIONS

Short/long arm suspension uses a short upper control arm and a larger lower control arm, and is usually referred to as the **SLA-type** suspension (see Figure 9–30).

SLA-type suspension can be used with either coil springs or torsion bars. Most vehicles use two A-shaped steel arms. One arm is positioned at the bottom of the "A," connected to the frame by rubber bushings. The other end of the arm, at the top of the "A," is connected to the steering knuckle by **ball joints** (see Figure 9–31). These **A-arms** are usually called **control arms** because they control the location of the front wheels and allow for up and down movement of the front wheels, plus turning of the front wheels for steering.

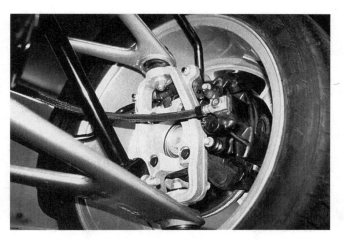

Figure 9–31 Note that both upper and lower control arms on this Chevrolet Corvette are constructed of aluminum alloy and attach to the knuckle through ball joints.

NOTE: SLA-type suspension is also called a *double wishbone* suspension because in some suspensions both control arms resemble the shape of a chicken or turkey wishbone.

The top control arm is called the **upper control arm.** The bottom control arm is called the **lower control arm.** The same terms also apply to the other parts of the control arms such as **lower ball joint, upper ball joint, upper control arm bushings,** and **lower control arm bushings.**

The upper control arm is shorter than the lower control arm. This permits the tires to remain as vertical as possible during suspension travel. Short/long arm suspensions are used in the front as well as the rear of many rear-wheel-drive and front-wheel-drive vehicles.

■ MACPHERSON STRUTS

The **MacPherson strut–type** suspension was patented in 1958 by Earle S. MacPherson, a vice president of engineering at Ford Motor Company. The original MacPherson strut–type suspension is now called a **modified strut type** (see Figure 9–32).

The most commonly used strut suspension combines the coil spring and the shock absorber into one structural suspension component, as shown in Figure 9–33.

A MacPherson strut suspension is lightweight and saves space in the vehicle because only one control arm is needed—the top of the strut simply attaches to the body of the vehicle. *The entire strut rotates when the front wheels are turned* (see Figure 9–34 on page 227).

The pivot points of a strut are at the lower ball joint and at a bearing assembly at the top of the strut. It is this upper bearing that supports the weight of the vehicle.

Figure 9–32 Modified strut suspension. The strut provides the structural support to the body with the coil spring acting directly on the lower control arm. (Courtesy of Dana Corporation)

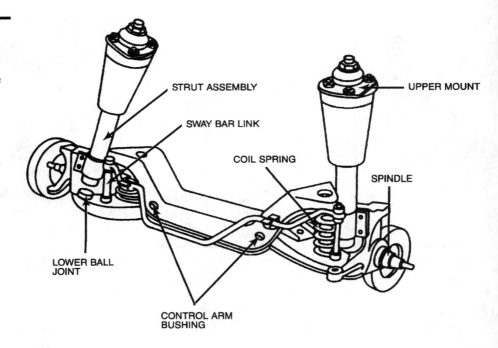

STRUT ASSEMBLY

UPPER MOUNT

SWAY BAR LINK

COIL SPRING

SPINDLE

LOWER BALL JOINT

CONTROL ARM BUSHING

Figure 9–33 Typical MacPherson strut–type front suspension on a front-wheel-drive vehicle. Note the drive axle shaft and CV joints.

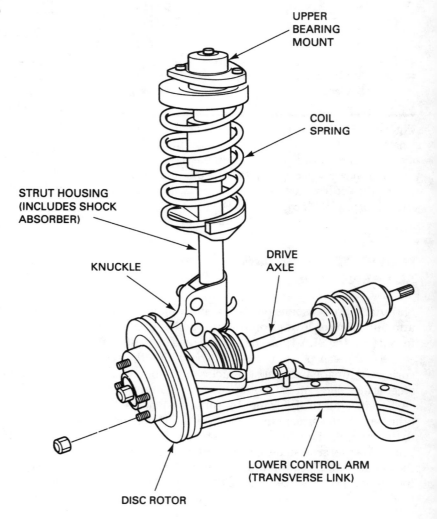

UPPER BEARING MOUNT

COIL SPRING

STRUT HOUSING (INCLUDES SHOCK ABSORBER)

DRIVE AXLE

KNUCKLE

LOWER CONTROL ARM (TRANSVERSE LINK)

DISC ROTOR

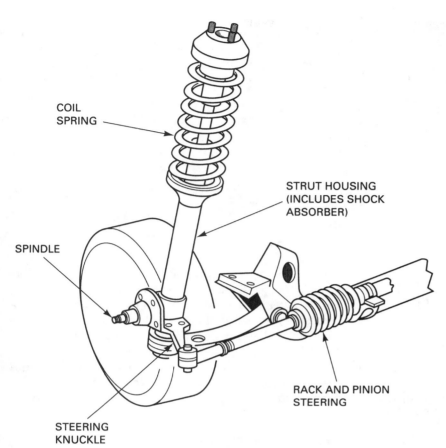

COIL
SPRING

STRUT HOUSING
(INCLUDES SHOCK
ABSORBER)

SPINDLE

RACK AND PINION
STEERING

STEERING
KNUCKLE

Figure 9–34 Typical MacPherson strut–type front suspension on a rear-wheel-drive vehicle. When the wheels are turned, the entire strut assembly and steering knuckle rotate on the upper strut mount and pivot on the lower ball joint at the end of the lower control arm.

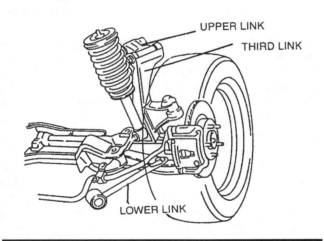

UPPER LINK

THIRD LINK

LOWER LINK

Figure 9–35 Multilink-type front suspension. Notice the third link that improves the geometry of the suspension to keep the wheels vertical during cornering and suspension movement. (Courtesy of Dana Corporation)

The lower ball joint is *not* load carrying. The simple MacPherson strut–type suspension is economical to manufacture and provides a reasonably smooth ride. The major disadvantage to a strut-type suspension is the height required for the strut assembly and its limited handling capability. Because the strut is part of the body, as the body leans into a corner, the strut itself leans and causes the tire to lean along with the strut. A strut-type front suspension is also called a **single arm front suspension** because it uses the top of the strut (shock) as the upper support and uses just one control arm at the bottom.

■ MULTILINK SUSPENSIONS

A suspension system similar to a MacPherson is a design that uses a third link to allow the wheels to remain vertical as the suspension moves up and down (see Figure 9–35).

Another variation of the MacPherson strut used in front-wheel-drive vehicles is called the *wishbone* because of the shape of the strut mount as it spans the drive axle shaft (see Figure 9–36).

■ BALL JOINTS

Ball joints are actually ball and socket joints, similar to the joints in a person's shoulder. Ball joints allow the front wheels to move up and down, as well as side to side (for steering).

A vehicle can be equipped with coil springs, mounted either above the upper control arm *or* on the lower control arm (see Figure 9–37).

Figure 9–36 A wishbone variation of the MacPherson strut. Notice the use of the wishbone-shaped strut mount that provides clearance for the drive axle shaft. (Courtesy of Dana Corporation)

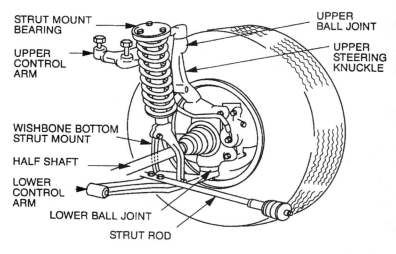

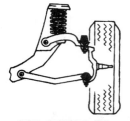

COIL SPRING OR TORSION BAR MOUNTED ON UPPER CONTROL ARM

COIL SPRING OR TORSION BAR MOUNTED ON LOWER CONTROL ARM

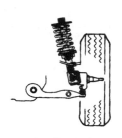
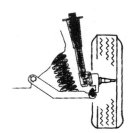

MACPHERSON STRUT

COIL SPRING MOUNTED ON LOWER CONTROL ARM WITH MODIFIED STRUT

Figure 9–37 Ball joints provide the freedom of movement necessary for steering and suspension movements. (Courtesy of Dana Corporation)

If the coil spring is attached to the top of the upper control arm, then the upper ball joint is carrying the weight of the vehicle and is called the **load-carrying** ball joint. The lower ball joint is called the **follower** ball joint (see Figure 9–38).

If the coil spring is attached to the lower control arm, then the lower ball joint is the load-carrying ball joint and the upper joint is the follower ball joint (see Figure 9–39).

If a torsion-bar-type spring is used, the lower ball joint is load-carrying because the torsion bar is attached

to the lower control arm on most vehicles that use torsion bars (see Figure 9–40 on page 230).

On vehicles equipped with a twin I-beam front suspension with ball joints, *both* ball joints are load-carrying and must, therefore, be replaced together if worn or defective (see Figure 9–41 on page 230).

MacPherson struts use a ball joint on the lower control arm. Since the weight of the vehicle is applied to the upper strut mount, the ball joint is non-load-carrying (see Figure 9–42 on page 230).

Ball Joint Design

There are two basic designs of ball joints: **compression loaded** and **tension loaded.** If the control arm rests on the steering knuckle, the ball joint is *compressed* into the control arm by the weight of the vehicle. If the knuckle rests on the control arm, the weight of the vehicle tends to pull the ball joint back into the control arm by *tension.* The type used is determined by the chassis design engineer, and the service technician cannot change the type of ball joint used for a particular application (see Figure 9–43 on page 231).

A specific amount of stud turning resistance is built into each ball joint to stabilize steering. A ball joint that does not support the weight of the vehicle and acts as a suspension pivot is often called a follower ball joint or a **friction** ball joint. The load-carrying (weight-carrying) ball joint is subjected to the greatest amount of wear and is the most frequently replaced (see Figure 9–44 on page 231).

■ STRUT RODS

Some vehicles are equipped with round steel rods which are attached between the lower control arm at one end and the frame of the vehicle with rubber bushings, called **strut rod** bushings, at the other end. The purpose of these strut rods is to provide forward/backward support to the control arms. Strut rods are used on vehicles equipped with MacPherson strut and many

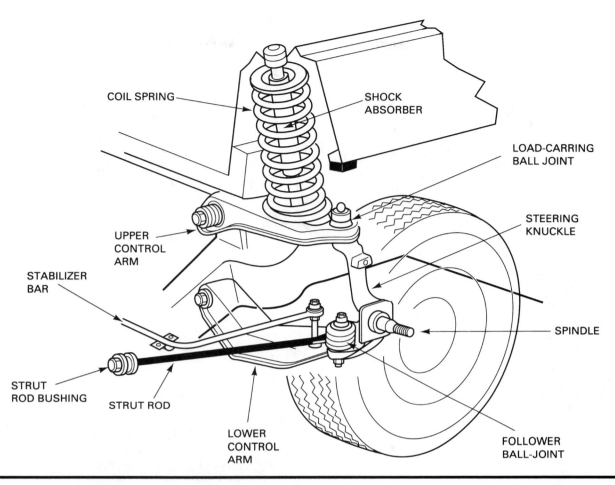

Figure 9–38 The upper ball joint is load-carrying in this type of suspension because the weight of the vehicle is applied through the spring, upper control arm, and ball joint to the wheel. The lower control arm is a lateral link, and the lower ball joint is called a *follower* ball joint.

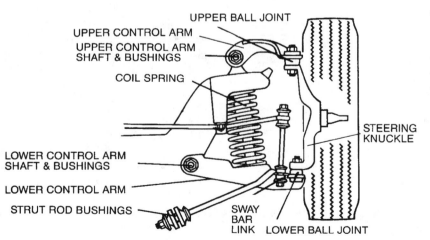

Figure 9–39 The lower ball joint is load-carrying in this type of suspension because the weight of the vehicle is applied through the spring, lower control arm, and ball joint to the wheel. (Courtesy of Dana Corporation)

short arm/long arm–type suspensions. The bushings are very important in maintaining proper wheel alignment while providing the necessary up-and-down movement of the control arms during suspension travel. Strut rods prevent lower control arm movement back and forth during braking (see Figure 9–45 on page 231).

Strut rods are also called **tension** or **compression rods** or simply **TC rods.** Tension rods attach in *front* of the wheels to the body or frame where the rod is being pulled in tension. Compression rods attach to the body or frame *behind* the wheels where the rod is being pushed or compressed. Some vehicle manufacturers call the

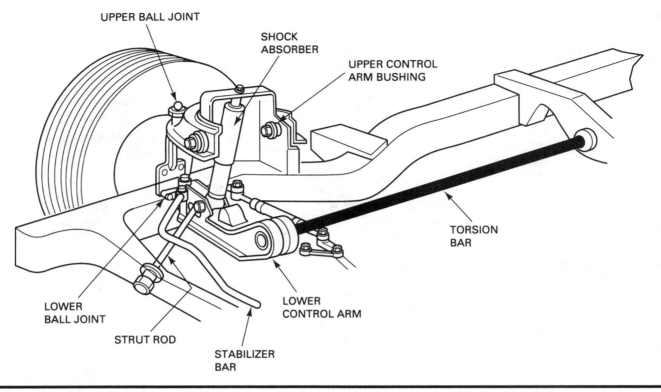

Figure 9–40 The lower ball joint is a load-carrying ball joint on this torsion-bar-type front suspension.

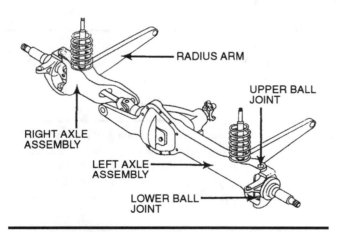

Figure 9–41 Both ball joints are load-carrying on twin I-beam-type suspension. This front suspension is on a four-wheel-drive Ford truck and is called a *twin traction beam suspension.* (Courtesy of Dana Corporation)

strut rod a **drag rod** because it was attached in front of the wheels, and therefore acted on the lower control arm as if to drag the wheels behind their attachment points.

The bushings are replaceable by removing a nut on the frame end of the strut rod (see Figure 9–46 on page 232). If a strut rod has a nut on *both* sides of the bushings, then the strut rod is used to adjust *caster.* See Chapter 11 for information on caster and other alignment angles.

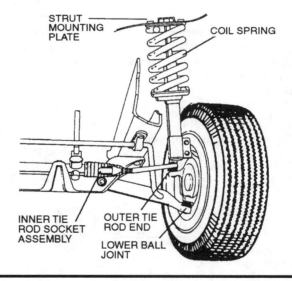

Figure 9–42 The ball joint used in a MacPherson-type suspension is not load-carrying. The vehicle weight is transferred through the spring and upper strut mount plate. (Courtesy of Dana Corporation)

■ STABILIZER BARS

Most cars and trucks are equipped with a stabilizer bar on the front suspension, which is a round, hardened steel bar (usually SAE 4560 or 4340 steel) attached to both lower control arms with bolts and

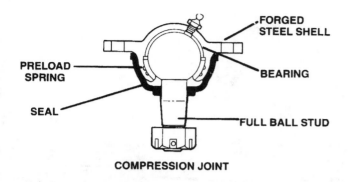

Figure 9–43 In this compression design ball joint, the wear surface is between the ball and the socket. (Courtesy of Dana Corporation)

rubber bushing washers called **stabilizer bar bushings** (see Figure 9–47).

A stabilizer bar is also called an **anti-sway bar (sway bar)** or **anti-roll bar (roll bar)**. A stabilizer bar operates by *twisting* the bar if one side of the vehicle moves up or down in relation to the other side, such as during cornering, hitting bumps, or driving over uneven road surfaces (see Figure 9–48).

The purpose of the stabilizer bar is to prevent excessive body roll while cornering and to add to stability while driving over rough road surfaces. The stabilizer bar is also used as a longitudinal (front/back) support to the lower control arm on many vehicles equipped with MacPherson struts. The effective force of a stabilizer

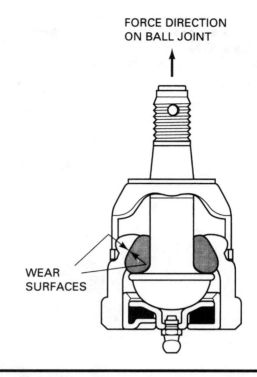

Figure 9–44 In this tension design ball joint, the wear surface is above the pivot ball and socket as shown.

bar is increased with the diameter of the bar. Therefore, optional suspensions often include larger diameter stabilizer bars and bushings.

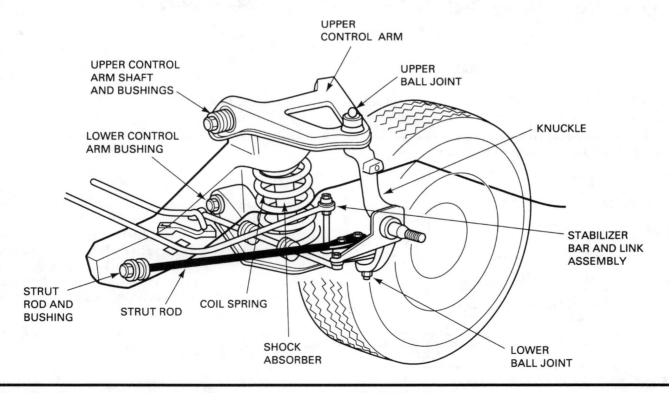

Figure 9–45 A strut rod is the longitudinal support to prevent front-to-back wheel movement.

Figure 9–46 Strut rod bushings insulate the steel bar from the vehicle frame or body. (Courtesy of Moog)

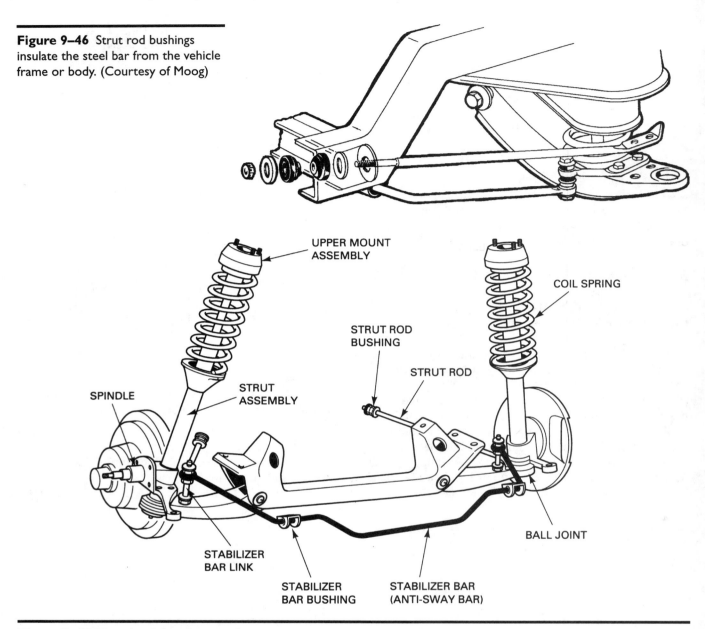

Figure 9–47 Typical stabilizer bar installation.

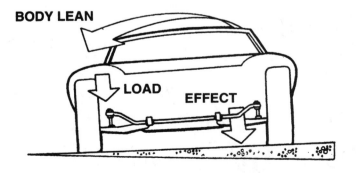

Figure 9–48 As the body of the vehicle leans, the stabilizer bar is twisted. The force exerted by the stabilizer bar counteracts the body lean. (Courtesy of Moog)

Stabilizer links connect the ends of the stabilizer bar to the lower control arm (see Figures 9–49 and 9–50). Careful inspection of the stabilizer bar links is important. Links are commonly found to be defective (cracked rubber washers or broken spacer bolts) because of the great amount of force that is transmitted through the links and the bushings. Defective links and/or bushings can cause unsafe vehicle handling and noise.

■ SHOCK ABSORBERS

Shock absorbers are used on all conventional suspension systems to dampen and control the motion of the vehicle's springs. Without shock absorbers (dampers),

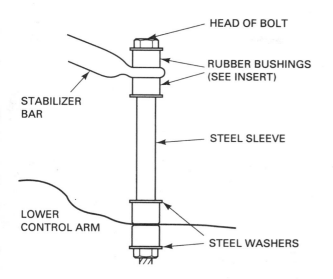

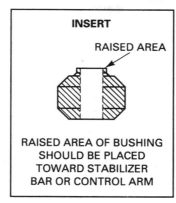

HEAD OF BOLT

RUBBER BUSHINGS
(SEE INSERT)

STABILIZER
BAR

STEEL SLEEVE

LOWER
CONTROL ARM

STEEL WASHERS

INSERT

RAISED AREA

RAISED AREA OF BUSHING
SHOULD BE PLACED
TOWARD STABILIZER
BAR OR CONTROL ARM

Figure 9–49 Stabilizer bar links are sold as a kit consisting of the long bolt with steel sleeve and rubber bushings. Steel washers are used on both sides of the rubber bushings as shown.

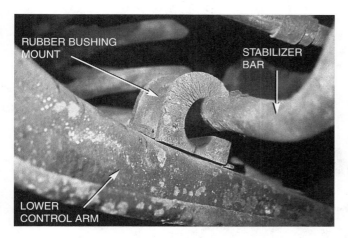

RUBBER BUSHING
MOUNT

STABILIZER
BAR

LOWER
CONTROL ARM

Figure 9–50 Notice how the stabilizer bar pulls down on the mounting bushing when the vehicle is hoisted off the ground, allowing the front suspension to drop down. These bushings are a common source of noise, especially when cold. Lubricating the bushings with paste silicone grease often cures the noise.

the vehicle would continue to bounce after hitting bumps (see Figure 9–51).

Struts are shock absorbers that are part of a MacPherson strut assembly. *The major purpose of any shock or strut is to control ride and handling.* Standard shock absorbers *do not* support the weight of a vehicle. *The springs support the weight of the vehicle; the shock absorbers control the actions and reactions of the springs.*

Most shock absorbers are *direct acting* because they are connected directly between the vehicle frame or body and the axles (see Figures 9–52 and 9–53).

As a wheel rolls over a bump, the wheel moves toward the body and compresses the spring(s) of the vehicle. As the spring compresses, it stores energy. The spring then releases this stored energy, causing the body of the vehicle to rise (rebound). See Figure 9–54 on page 235.

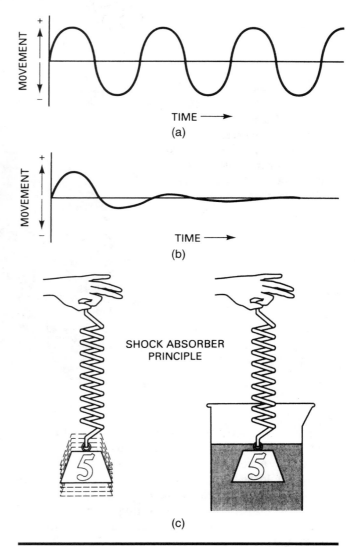

MOVEMENT

TIME
(a)

MOVEMENT

TIME
(b)

SHOCK ABSORBER
PRINCIPLE

(c)

Figure 9–51 (a) Movement of vehicle is supported by springs without a dampening device. (b) Spring action dampened with a shock absorber. (c) The function of any shock absorber is to dampen the movement or action of a spring similar to using a liquid to control the movement of a weight on a spring. (Courtesy of Ford Motor Company)

Figure 9–52 Shock absorbers work best when mounted as close to the spring as possible. Shock absorbers that are mounted straight up and down offer the most dampening.

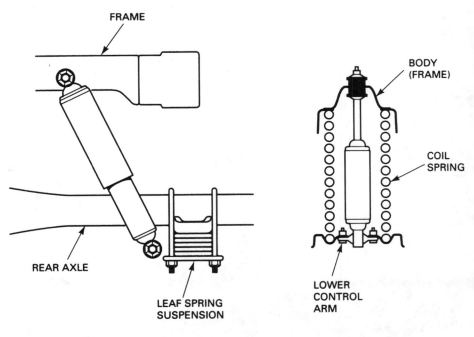

Figure 9–53 Some vehicles such as this Ford Mustang use four shock absorbers on the rear suspension. The vertical shock absorbers control vertical body movement and react with the rear coil springs. The longitudinal shocks help control the rear axle during acceleration and deceleration.

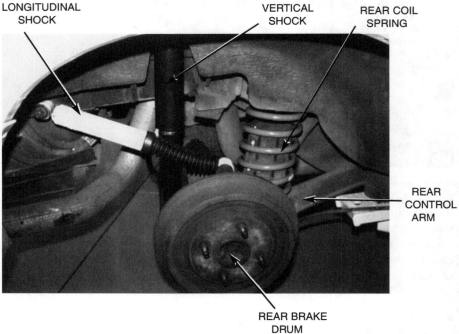

After the energy in the spring is used up, the body starts downward, causing the spring to compress. Without shock absorbers, the energy released from the spring would be very rapid and violent. The shock absorber helps dampen the rapid up-and-down movement of the vehicle springs by converting energy of movement into heat by forcing hydraulic fluid through small holes inside the shock absorber.

Shock Absorber Operation

The hydraulic shock absorber operates on the principle of fluid being forced through a small opening (orifice) (see Figures 9–55 and 9–56 on pages 235–236). Besides small openings, pressure relief valves are built into most shock absorbers to control vehicle ride under all operating conditions. The greater the pressure drop of the fluid inside the shock and the greater the amount of fluid moved through the orifice, the greater the amount of dampening; therefore, larger shock absorbers can provide better dampening than smaller units.

Gas-Charged Shocks

Most shock absorbers on new vehicles are gas charged. Pressurizing the oil inside the shock absorber helps smooth the ride over rough roads. This pressure helps prevent air pockets from forming in the shock absorber oil as it passes through the small passages in the shock. After the oil is forced through small passages, the pres-

REBOUND COMPRESSION

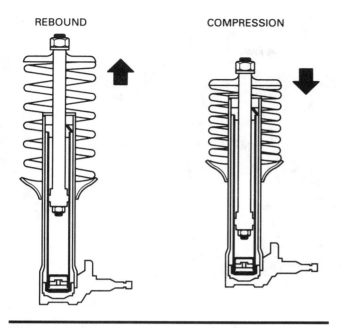

Figure 9–54 When a vehicle hits a bump in the road, the suspension moves upward. This is called *compression*. Rebound is when the spring (coil, torsion bar, or leaf) returns to its original position.

sure drops and the oil expands. As the oil expands, bubbles are created. The oil becomes foamy. This air-filled oil does not effectively provide dampening. The result of all of this aeration (air being mixed with the oil) is lack of dampening and a harsh ride.

The use of higher pressure radial tires and lighter vehicle weight has created the need for more effective shock absorbers. To meet this need, shock absorber design engineers use a pressurized gas that does not react chemically with the oil in the shock. If a substance does not react with any other substances, it is called *inert*. The gas most used is nitrogen, which is about 78 percent of our atmosphere. Typical gas-charged shocks are pressurized with 130–150 psi (900–1030 kPa) to aid in both handling and ride control. Some shocks use higher pressures, but the higher the pressure, the greater the possibility of leaks and the harsher the ride.

Some gas-charged shock absorbers use a single tube that contains two pistons that separate the high-pressure gas from the working fluid. Single tube shocks are also called **monotube** or **DeCarbon** after the French inventor of the principle and manufacturer of suspension components (see Figure 9–57).

■ BUMP STOPS

All suspension systems have a limit of travel. If the vehicle hits a large bump in the road, the wheels are forced upward toward the vehicle with tremendous force. This force is absorbed by the springs of the suspension system. If the bump is large enough, the suspension is compressed to its mechanical limit. Instead of allowing the metal components of the suspension to

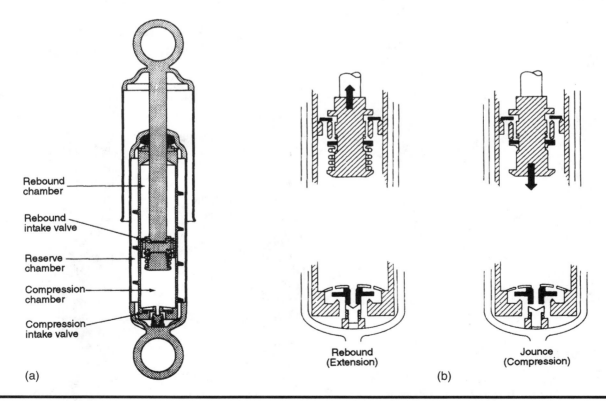

Rebound chamber

Rebound intake valve

Reserve chamber

Compression chamber

Compression intake valve

(a)

Rebound (Extension)

Jounce (Compression)

(b)

Figure 9–55 (a) A cutaway drawing of a typical double tube shock absorber. (b) Notice the position of the intake and compression valve during rebound (extension) and compression.

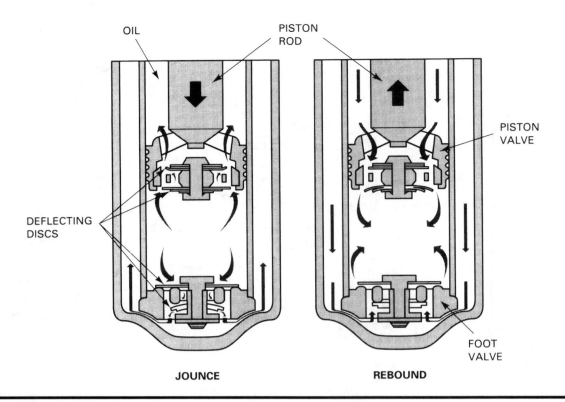

JOUNCE   **REBOUND**

Figure 9–56 Oil flow through a deflected disc–type piston valve. The deflecting disc can react rapidly to suspension movement. For example, if a large bump is hit at high speed, the disc can deflect completely and allow the suspension to reach its maximum jounce distance while maintaining a controlled rate of movement.

Figure 9–57 Gas-charged shock absorbers are manufactured with a double tube design similar to conventional shock absorbers and with a single or monotube design.

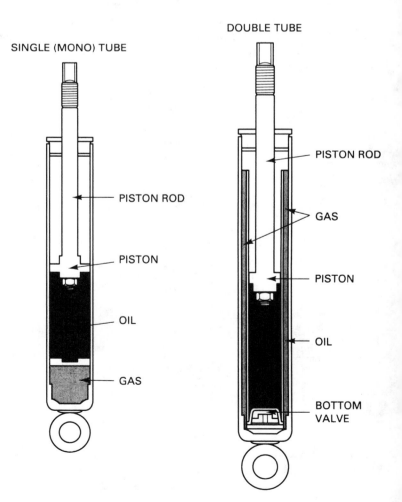

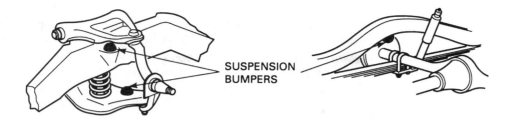

SUSPENSION
BUMPERS

Figure 9–58 Suspension bumpers are used on all suspension systems to prevent metal-to-metal contact between the suspension and the frame or body of the vehicle when the suspension "bottoms out" over large bumps or dips in the road. (Courtesy of Moog)

URETHANE
BUMPER

Figure 9–59 Typical urethane jounce bumper as used in a MacPherson strut. This stop is not visible until the strut is disassembled.

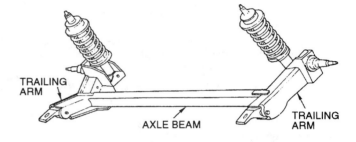

TRAILING
ARM

AXLE BEAM

TRAILING
ARM

Figure 9–60 A semi-independent axle beam rear suspension. It is called semi-independent because as a wheel hits a bump in the road, the trailing arm moves upward and twists the axle beam. This twisting of the axle beam helps isolate one side of the vehicle from the other. (Courtesy of Dana Corporation)

Most suspensions also use a rubber or foam stop to limit the downward travel of the suspension during rebound. (See Chapter 10 for additional information.) The rebound stop also prevents metal-to-metal contact of the suspension on the frame when the vehicle is on a body-contact-type hoist and the wheels are allowed to hang or droop down. Some stops are built into the shock absorber or strut.

hit the frame or body of the vehicle, a rubber or foam bumper is used to absorb and isolate the suspension from the frame or body (see Figure 9–58).

These bumpers are called **bump stops, suspension bumpers, strike-out bumpers,** or **jounce bumpers.** *Jounce* means jolt, or to cause to bounce or move up and down. Bumpers are made from rubber or microcellular urethane. Urethane is a high-strength material with good resistance to wear and tear as well as good chemical resistance to most fluids. Forming urethane foam with small, regular air cells makes the material ideal for jounce bumpers (see Figure 9–59).

Damaged suspension limiting bump stops can be caused by:

1. Sagging springs that result in lower than normal ride (trim) height
2. Worn or defective shock absorbers

■ REAR SUSPENSIONS
Solid-Axle Rear Suspensions

Solid- or straight-axle rear suspensions can use either coil or leaf springs (see Figures 9–60 and 9–61).

Leaf springs function as a load-carrying member and provide side-to-side support and stability. Coil springs, however, can function only as a load-carrying member and must depend on various other suspension members to provide side-to-side (lateral) as well as front-to-back rear axle support.

Longitudinal (back-and-forth) support is provided by rear **control arms,** also called rear **trailing arms** because the rear axle trails behind the control arm frame mounts. These rear control arms are usually angled to provide transverse (side-to-side) support as well as longitudinal support. Some manufacturers add an additional rear support member to ensure that the center of the body is kept directly over the center of the rear axle. These horizontal rear bars are called a **track rod** or a **panhard rod,** and are bolted to the rear axle at one end and the vehicle frame at the other end (see Figure 9–62).

A variation of a track rod is called a **Watt's link,** which uses two horizontal rods pivoting at the center of the rear axle (see Figures 9–63 and 9–64).

Live-Axle Rear Suspensions

A live rear axle is a rear axle that contains a drive axle shaft to provide engine power to the rear wheels. Besides providing up-and-down as well as forward-and-backward axle control, the suspension of a live rear axle must also control engine torque. As engine power is applied to the drive wheels, an opposite and equal torque is applied to the suspension. When the suspension system absorbs and controls the torque of the driving axle, the system is called a **Hotchkiss-type suspension.** Many live-axle rear suspensions mount the right (passenger side) shock absorbers in front of the axle and the left (driver's side) shocks behind the axle. This shock mounting arrangement is referred to as **staggered shock mounting;** this arrangement helps to control rear axle movement during acceleration and helps to prevent wheel hop. Figure 9–65 shows a Hotchkiss-type rear suspension where the leaf spring absorbs the engine torque.

Independent Rear Suspensions

Most newer front-wheel-drive vehicles use an independent rear suspension, often abbreviated **IRS.** An in-dependent rear suspension provides a smoother ride than a solid-axle suspension because each rear wheel can react to bumps and dips in the road without moving the entire rear axle. Most independent rear suspensions have lower unsprung weight. Many front-wheel-drive vehicles with IRS also provide for some alignment adjustments (see Figures 9–66 and 9–67 on pages 240–241).

This adjustability of the rear suspension allows for compensation to the alignment as the vehicle ages and the springs sag without the use of expensive rear shims. (See Chapter 12 for details on rear alignment.)

Independent rear suspension is also used on some rear-wheel-drive vehicles. To reduce unsprung weight, the differential is mounted to the body or frame (see Figure 9–68 on page 241).

Most rear-wheel-drive independent suspension systems use coil springs, yet the early Volkswagens and rear-engine Porsches used torsion bars. The constant velocity joints used on the rear of the rear-wheel-drive vehicles tend to last the life of the vehicle unless the protective boot is punctured. In a rear-wheel-drive vehicle, the drive axles and CV joints simply move up and down as the wheels roll over bumps and dips. (See Chapter 6 for CV joint service procedures.)

Air Shocks/Struts

Air-inflatable shock absorbers or struts are used in the rear of vehicles to provide proper vehicle ride height while carrying heavy loads. Many air shock/strut units are original equipment. They are often combined with a built-in air compressor and ride height sensor(s) to provide automatic ride height control.

Figure 9–61 A solid rear axle tube suspension with leaf springs.

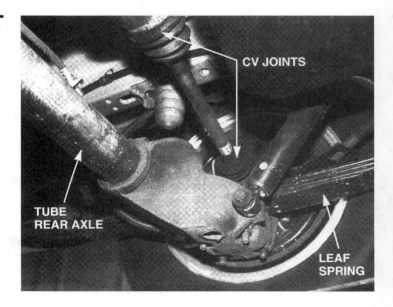

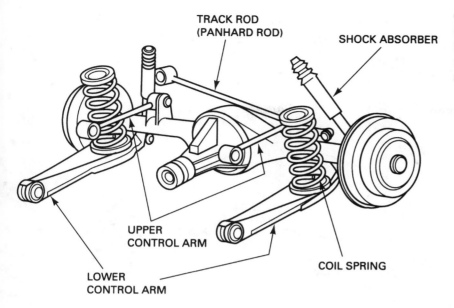

TRACK ROD
(PANHARD ROD)

SHOCK ABSORBER

UPPER
CONTROL ARM

LOWER
CONTROL ARM

COIL SPRING

Figure 9–62 The track rod (panhard rod) is used to keep the rear suspension centered under the vehicle and adds side-to-side stability. (Courtesy of Moog)

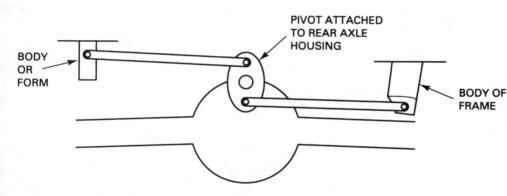

PIVOT ATTACHED
TO REAR AXLE
HOUSING

BODY
OR
FORM

BODY OF
FRAME

Figure 9–63 A Watt's link allows the suspension to travel up and down while preventing the axle from moving side-to-side.

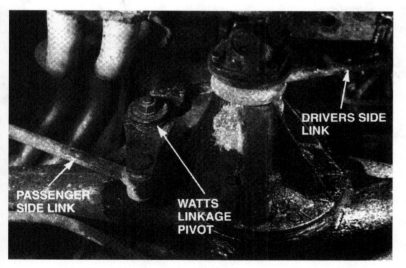

DRIVERS SIDE
LINK

PASSENGER
SIDE LINK

WATTS
LINKAGE
PIVOT

Figure 9–64 A photograph of a Watt's link used on an older Mazda RX-7.

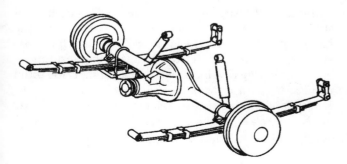

Figure 9–65 An example of a Hotchkiss-type suspension where the leaf springs absorb the engine torque as the vehicle is being accelerated. The front of the rear axle housing tends to rotate upward. This rotational force is transferred to the leaf springs and mounts through the U-bolts and spring perch. (Courtesy of Moog)

239

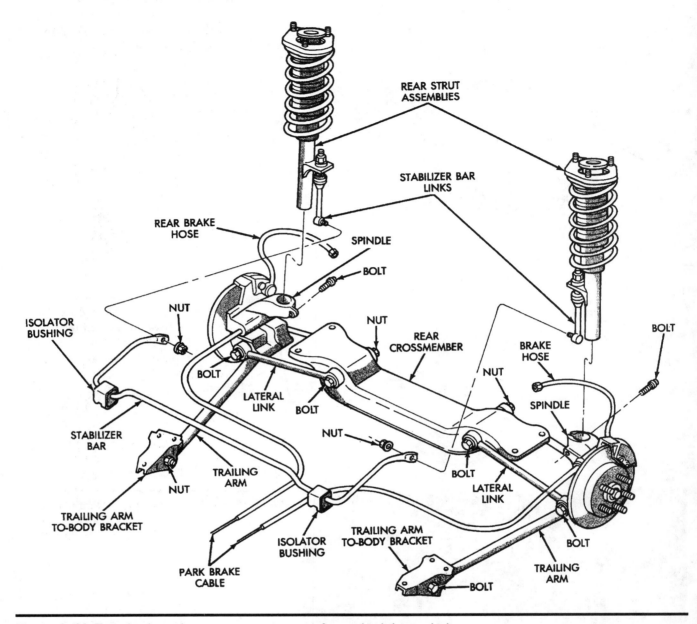

Figure 9–66 Typical independent rear suspension on a front-wheel-drive vehicle. (Courtesy of Chrysler Corporation)

Air-inflatable shocks are standard shock absorbers with an air chamber with a rubber bag built into the dust cover (top) of the shock (see Figure 9–69).

Air pressure is used to inflate the bag, which raises the installed height of the shock. As the shock increases in height, the rear of the vehicle is raised. Typical maximum air pressure in air shocks ranges from 90 to 150 psi (620 to 1030 kPa). As the air pressure increases in the air-inflatable reservoir of the shock, the stiffness of the suspension increases. This additional stiffness is due to the shock taking weight from the spring, and therefore, the air in the air shock becomes an air spring. Now, with two springs to support the vehicle, the spring rate increases and a harsher ride often results. *It is important that the load capacity of the vehicle not be exceeded or serious damage can occur to the vehicle's springs, axles, bearings, and shock support mounts.*

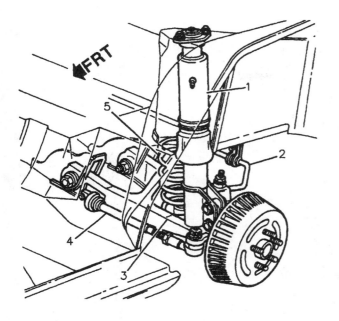

1 STRUT
2 STABILIZER SHAFT
3 CONTROL ARM
4 SUSPENSION ADJUSTMENT LINK
5 COIL SPRING

Figure 9–67 Many independent rear suspensions use an adjustment link that looks like the tie rods used on front suspension. (Courtesy of Oldsmobile)

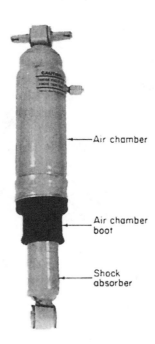

Figure 9–69 Typical air shock.

Air chamber

Air chamber boot

Shock absorber

Electronically Controlled Suspensions

Many vehicle manufacturers offer some type of electronically controlled suspensions. Most use conventional springs or air springs. Few production vehicles use true active or reactive suspensions.

Passive electronically controlled suspension systems are more common and use conventional or airbag springs to support the weight of the vehicle instead of hydraulic cylinders used on active systems. Most nonactive electronically controlled suspension systems can be grouped into four categories:

Figure 9–68 The differential is attached to the frame of the vehicle in this example of an IRS. (Courtesy of Moog)

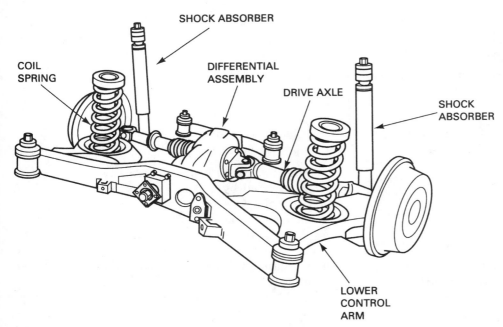

COIL SPRING

SHOCK ABSORBER

DIFFERENTIAL ASSEMBLY

DRIVE AXLE

SHOCK ABSORBER

LOWER CONTROL ARM

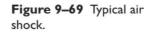

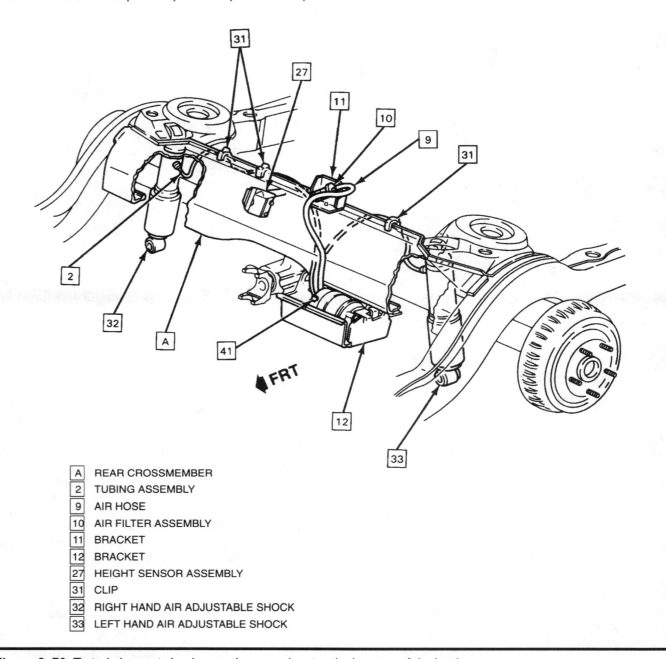

A	REAR CROSSMEMBER
2	TUBING ASSEMBLY
9	AIR HOSE
10	AIR FILTER ASSEMBLY
11	BRACKET
12	BRACKET
27	HEIGHT SENSOR ASSEMBLY
31	CLIP
32	RIGHT HAND AIR ADJUSTABLE SHOCK
33	LEFT HAND AIR ADJUSTABLE SHOCK

Figure 9–70 Typical electronic level-control system showing the location of the height sensor and the compressor assembly. (Courtesy of Oldsmobile)

Type I. Electronically Controlled Rear Air-Inflatable Shock Absorbers
The main purpose of this system is to maintain controlled rear ride (trim) height under all vehicle load conditions. Some vehicles are equipped with rear air shocks that can be controlled electronically to adjust the ride height of the vehicle regardless of vehicle load. A typical electronic level-control system includes the following components:

1. Air-adjustable rear shocks (or struts, in some cases)
2. Small air compressor (mounted under the hood or under the vehicle at the rear)
3. Electronic height sensor
4. Air dryer
5. Exhaust solenoid
6. Relay, wiring, and tubing

The compressor is usually a small single-piston air pump powered by a 12-volt permanent-magnet (PM) electric motor (see Figure 9–70).

An air dryer is usually attached to the pump to remove moisture from the air before it is sent to the shocks and through the dryer (to dry the chemical dryer) during the release of air from the shocks (see Figure 9–71).

The height sensor operates the compressor or exhaust solenoid, based on the height of the rear of the vehicle. Some systems operate only when the ignition

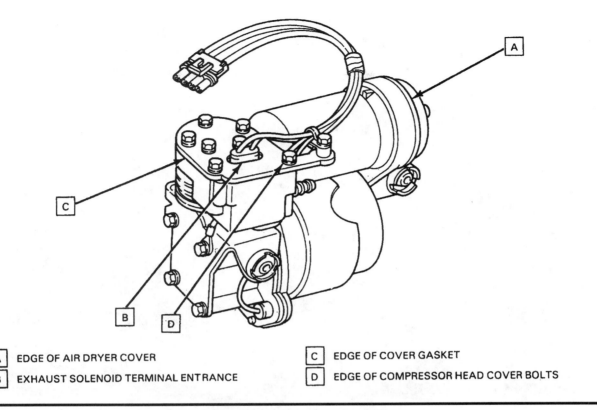

| A | EDGE OF AIR DRYER COVER | C | EDGE OF COVER GASKET |
| B | EXHAUST SOLENOID TERMINAL ENTRANCE | D | EDGE OF COMPRESSOR HEAD COVER BOLTS |

Figure 9–71 All air compressors used for suspension height control use a dryer to remove moisture from the air before it is used to pressurize the air chambers of the shock absorbers, struts, or air springs. (Courtesy of Oldsmobile)

switch is on, while other systems operate any time because the compressor is wired to a voltage source that is "hot" at all times. To avoid unnecessary ride height increase or decrease due to variations in the road surface, the system operates only after a time delay of about fifteen seconds. Most compressors are also equipped with a timer circuit that limits compressor "on" time to about three minutes, to prevent compressor damage if the air system has a leak.

Type 2. Four-Wheel-Ride Height Control

This system uses air-inflatable springs to support vehicle weight with the inflation pressure varied by an electronic controller. The main purpose of this type of system is to maintain the same ride height, both front and rear as well as side to side, under all driving conditions (see Figures 9–72, 9–73, and 9–74 on pages 244–245).

Type 3. Computer-Controlled Shock Absorbers

This system is used with conventional metal or fiberglass composite springs. The main purpose of this type of system is to permit a smoother ride over rough road surfaces yet permit a stiffer ride at higher speeds, or to stiffen the shocks during braking, accelerating, or cornering.

CAUTION: Many vehicles equipped with air suspension, such as many Lincoln automobiles, have a trunk-mounted on/off switch. Before servicing, hoisting, jacking, or towing, this switch must be in the "off" position.

Type 4. Active Suspension Systems

These systems were first developed by Lotus of England. In fact, the term **active suspension** is a registered trademark of Lotus. An active suspension system uses hydraulic cylinders at each wheel to keep the tires on the road and the body of the vehicle level under all driving conditions. The purpose of an active suspension system is to provide a smooth ride and superior handling in one vehicle. With conventional suspension systems, a smooth ride requires soft springs and suspension bushings to allow the suspension to absorb road bumps. A soft suspension, however, allows the vehicle to lean while cornering. It tends to "float" over bumps and dips in the road. For improved handling, a stiff, less compliant suspension is required. Stiffer springs and bushings are often used on performance vehicles or vehicles with sport, high-performance or touring suspension. This stiffer suspension allows less of the road impact to be absorbed by the suspension; more

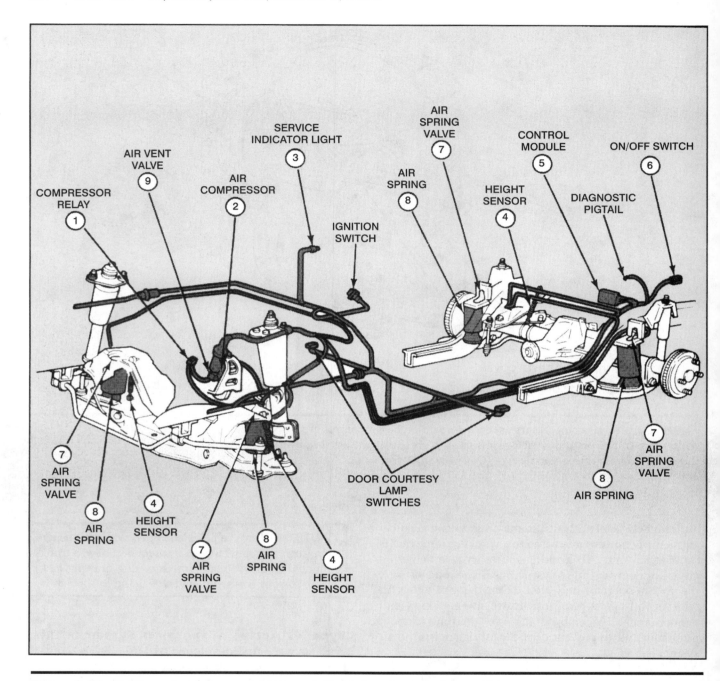

Figure 9–72 Air suspension components. Note on/off switch at the rear of the vehicle. This switch *must* be turned "off" before hoisting the vehicle. (Courtesy of Ford Motor Company)

of the motion of the road surface is transmitted through the suspension and into the passenger compartment. The active suspension system can provide the smooth ride of a soft suspension and still provide the control and handling of a stiff suspension. The typical active suspension system consists of the following components:

Controller. The on-board computer (controller) uses information from sensors to activate hydraulically operated cylinders located at each wheel.

Sensors. Each wheel has a height sensor and an acceleration sensor. Other sensors used by the controller include vehicle speed sensor, steering wheel movement sensor, and brake switch sensor.

Hydraulic pump and actuators. An engine-driven hydraulic pump provides the power necessary to raise and lower the vehicle and maintain proper vehicle height. These high-speed actuators must be capable of raising or lowering the vehicle in as little as 3 milliseconds (3/1000 of one second).

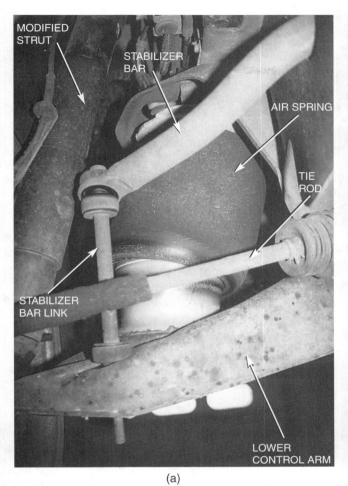

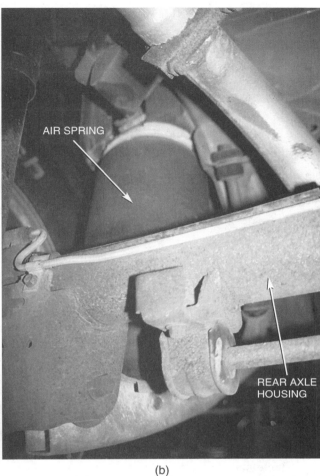

(a)   (b)

Figure 9–73 (a) The front suspension of a Lincoln with an air spring suspension. (b) The rear suspension of the same Lincoln.

Figure 9–74 Always check in the trunk for a cutoff switch for a vehicle equipped with an air suspension before hoisting or towing the vehicle.

■ SUMMARY

1. The lighter the wheel/tire combination, the lower the "unsprung weight" and the better the ride and handling of the vehicle.

2. All springs—including the coil, leaf, and torsion bar types—share Hooke's Law, which states that the force exerted by the spring is directly proportional to the amount the spring is deflected.

3. All springs are similar to torsion bars. As the torsion bar becomes longer or smaller in diameter, the easier it is to twist. If a coil spring is cut, the remaining spring is shorter, yet stiffer.

4. Ball joints attach to control arms and allow the front wheels to move up and down, as well as turn.

5. Suspension designs include a straight or solid-axle, two-control-arm type called an SLA or a MacPherson strut.

6. All shock absorbers dampen the motion of the suspension to control ride and handling.

7. Active (or reactive) suspension systems use sensors and a hydraulic pump to maintain ride height under all vehicle maneuvers.

■ REVIEW QUESTIONS

1. List the types of suspensions and name their component parts.
2. Explain Hooke's Law.
3. Describe the purpose and function of a stabilizer bar.
4. Explain the difference between a MacPherson strut and a modified strut–type suspension.

■ ASE CERTIFICATION-TYPE QUESTIONS

1. The spring rate of a spring is measured in units of
   a. lb. per inch
   b. lb. ft.
   c. psi
   d. in.-lb.

2. Two technicians are discussing torsion bars. Technician A says that many torsion bars are adjustable to allow for ride height adjustment. Technician B says that torsion bars are usually marked left and right and should not be switched side to side. Which technician is correct?
   a. Technician A only
   b. Technician B only
   c. Both Technician A and B
   d. Neither Technician A nor B

3. A vehicle makes a loud noise while traveling over bumpy sections of road. Technician A says that worn or deteriorated control arm bushings could be the cause. Technician B says that worn or deteriorated strut rod bushings could be the cause. Which technician is correct?
   a. Technician A only
   b. Technician B only
   c. Both Technician A and B
   d. Neither Technician A nor B

4. Two technicians are discussing MacPherson struts. Technician A says that the entire strut assembly rotates when the front wheels are turned. Technician B says a typical MacPherson strut suspension system uses only one control arm and one ball joint per side. Which technician is correct?
   a. Technician A only
   b. Technician B only
   c. Both Technician A and B
   d. Neither Technician A nor B

5. Technician A says that replacement regular shock absorbers will raise the rear of a vehicle that is sagging down. Technician B says that replacement springs will be required to restore the proper ride height. Which technician is correct?
   a. Technician A only
   b. Technician B only
   c. Both Technician A and B
   d. Neither Technician A nor B

6. A vehicle used for 24-hour-a-day security was found to have damaged suspension-limiting rubber jounce bumpers. Technician A says sagging springs could be the cause. Technician B says defective or worn shock absorbers could be the cause. Which technician is correct?
   a. Technician A only
   b. Technician B only
   c. Both Technician A and B
   d. Neither Technician A nor B

7. The part of many rear suspension systems that controls side-to-side movement is called the
   a. Rear control arm
   b. Track rod or panhard rod
   c. Stabilizer bar
   d. Trailing arm

8. Two technicians are discussing air shocks. Technician A says that air is forced through small holes to dampen the ride. Technician B says that air shocks are conventional hydraulic shock absorbers with an airbag to control vehicle ride height. Which technician is correct?
   a. Technician A only
   b. Technician B only
   c. Both Technician A and B
   d. Neither Technician A nor B

9. The owner of a pickup truck wants to cut the coil springs to lower the vehicle. Technician A says that the ride will be harsher than normal if the springs are cut. Technician B says that the springs could be damaged especially if a cutting torch is used to cut the springs. Which technician is correct?
   a. Technician A only
   b. Technician B only
   c. Both Technician A and B
   d. Neither Technician A nor B

10. A MacPherson strut is a structural part of the vehicle.
    a. True
    b. False

Suspension System Diagnosis and Service

Objectives: After studying Chapter 10, the reader should be able to:

1. Explain how to perform a road test, a dry park test, a visual inspection, and a bounce test.
2. Discuss the procedures for testing load-carrying and follower-type ball joints.
3. Describe ball joint replacement procedures.
4. List the steps required to replace control arm and stabilizer bar bushings.
5. Explain routine service procedures of the suspension system.

Suspension systems are designed and manufactured to provide years of trouble-free service with a minimum amount of maintenance. In fact, the suspension system is often "invisible" or "transparent" to the driver because the vehicle rides and handles as expected. It is when the driver notices that the vehicle is not riding or handling as it did, or should, that a technician is asked to repair, align, or fix the problem.

The smart technician should always road-test any vehicle before and after servicing (see Tech Tip for details). *The purpose of any diagnosis is to eliminate known good components.* (See the suspension diagnostic chart, Figure 10–1, for a list of components that can cause the problem or customer complaint.)

■ ROAD TEST DIAGNOSIS

If possible, perform a road test of the vehicle with the owner of the vehicle. It is also helpful to have the owner drive the vehicle. While driving, try to determine when and where the noise or problem occurs such as:

1. In cold or warm weather
2. With cold or warm engine/vehicle
3. While turning, left only, right only

A proper road test for any suspension system problem should include:

1. *Drive beside parked vehicles.* Any noise generated by the vehicle suspension or tires is reflected off solid objects, such as a row of parked vehicles along a street. For best results, drive with the windows down and drive close to the parked vehicles or a retaining wall on the left side. Repeat the drive for the right side. Defective wheel bearings or power steering pumps usually make noise and can be heard during this test.
2. *Drive into driveways.* Suspension problems often occur when turning at the same time the suspension hits a bump. This action is best repeated by driving slowly into a driveway with a curb. The curb causes the suspension to compress while the wheels are turned (see the Diagnostic Story titled "The Rock-Hard Problem" for an example). Defective stabilizer bar bushings, control arm bushings, and ball joints will usually make noise during this test procedure.

SUSPENSION PROBLEM DIAGNOSIS CHART

CHECK	PROBLEM					
	Noise	Instability	Pull to One Side	Excessive Steering Play	Hard Steering	Shimmy
Tires/Wheels	Road/tire noise	Low/uneven air pressure; radials mixed with bias-belted ply tires	Low/uneven air pressure; mismatched tire sizes	Low/uneven air pressure	Low/uneven air pressure	Wheel out of balance/uneven tire wear/ over worn tires; radials mixed with belted bias ply tires
Shock Dampners (Struts/ Absorbers)	Loose/worn mounts/ bushings	Loose/worn mounts/ bushings; worn/ damaged struts/ shock absorbers	Loose/worn mounts/ bushings		Loose/worn mounts/ bushings on strut assemblies	Worn/damaged struts/shock absorbers
Strut Rods	Loose/worn mounts/ bushings	Loose/worn mounts/ bushings	Loose/worn mounts/ bushings			Loose/worn mounts/ bushings
Springs	Worn/damaged	Worn/damaged	Worn/damaged, especially rear		Worn/damaged	
Control Arms	Steering knuckle contacting control arm stop; worn/damaged mounts/bushings	Worn/damaged mounts/ bushings	Worn/damaged mounts/ bushings		Worn/damaged mounts/ bushings	Worn/damaged mounts/ bushings
Steering System	Component wear/damage	Component wear/damage	Component wear/damage	Component wear/damage	Component wear/damage	Component wear/damage
Alignment		Front and rear, especially caster	Front, camber and caster	Front	Front, especially caster	Front, especially caster
Wheel Bearings	On turns/speed changes: front-wheel bearings	Loose/worn (front and rear)	Loose/worn (front and rear)	Loose/worn (front and rear)		Loose/worn (front and rear)
Brake System			On braking		On braking	
Other	Clunk on speed changes: trans-axle: click on turn: CV joints; ball joint lubrication				Ball joint lubrication	Loose/worn friction ball joints

CAUTION: More than one factor may be the cause of a problem. Be sure to inspect all suspension components, and repair all parts that are worn or damaged. Failure to do so may allow the problem to reoccur and cause premature failure of other suspension components.

Figure 10–1 This chart helps to illustrate that a particular problem could have many possible causes. (Courtesy of Dana Corporation)

3. *Drive in reverse while turning.* This technique is usually used to find possible defective outer CV joints used on the drive axle shaft of front-wheel-drive vehicles (see Chapter 6 for additional information). This technique also forces the suspension system to work in reverse of normal. Any excessive clearances in the suspension system are reversed and often make noise or cause a vibration during the test. Besides defective CV joints, this test can often detect worn control arm bushings, ball joints, stabilizer bar bushings, or links. This test can also detect defective or worn steering system components such as an idler arm, tie rod end, or center link.

4. *Drive over a bumpy road.* Worn or defective suspension (and steering) components can cause the vehicle to bounce or dart side to side while traveling over bumps and dips in the road. Worn or defective ball joints, control arm bushings, stabilizer bar bushings, stabilizer bar links, or worn shock absorbers can be the cause.

Once the problem has been confirmed, then a further inspection can be performed in the service bay.

■ DRY PARK TEST (SUSPENSION)

A dry park test is described in Chapter 8 as a method to use to find loose or defective steering components. This same test can also be used to help locate worn or defective suspension components. To review, the dry park test is performed by having an assistant move the steering wheel side to side while feeling and observing for any free play in the steering or suspension. For best results, the vehicle should be on a level floor or on a drive-on-type hoist with the front wheels pointing straight ahead. In this suspension analysis using the dry park test, the technician should observe the following for any noticeable play or unusual noise:

1. *Front wheel bearings.* Loose or defective wheel bearings are often overlooked as a possible cause of poor handling or darting vehicle performance. (See Chapter 4 for wheel bearing inspection and service procedures.)
2. *Control arm bushing wear or movement.* Check for any abnormal movement in upper control arm bushings or lower control arm bushings.
3. *Ball joint movement.* Check for any noticeable play or noise from both load-carrying and follower-type ball joints on both sides of the vehicle.

NOTE: The dry park test (and many other chassis system tests) rely on the experience of the technician to be able to judge normal wear from abnormal wear. It is extremely important that all beginning technicians work closely with an experienced technician to gain this knowledge.

■ VISUAL INSPECTION

All suspension components should be carefully inspected for signs of wear or damage. A thorough visual inspection should include checking all of the following:

Road Test—Before and After

Many times technicians will start to work on a vehicle based on the description of the problem by the driver or owner. A typical conversation was overheard where the vehicle owner complained that the vehicle handled "funny," especially when turning. The owner wanted a wheel alignment, and the technician and shop owner wanted the business. The vehicle was aligned, but the problem was still present. The real problem was a defective tire. The service technician should have road-tested the vehicle *before* any service work was done to confirm the problem and try to determine its cause. Every technician should test-drive the vehicle *after* any service work is performed to confirm that the service work was performed correctly and that the customer complaint has been resolved. This is especially true for any service work involving the steering, suspension, or braking systems.

Figure 10–2 A leaking shock absorber. This shock absorber definitely needs to be replaced.

1. Shock absorbers (see Figure 10–2)
2. Springs (see Figure 10–3)
3. Stabilizer bar links

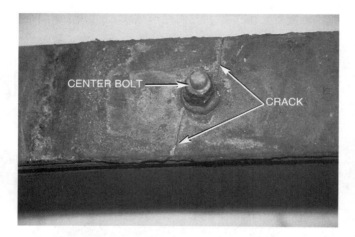

Figure 10–3 The center bolt is used to hold the leaves of a leaf spring together. However, the hole for the center bolt also weakens the leaf spring. The crack shown is what a technician discovered when the leaf spring was removed during the diagnosis of a sagging rear suspension.

4. Stabilizer bar bushings
5. Upper and lower shock absorber mounting points
6. Bump stops
7. Body-to-chassis mounts
8. Engine and transmission (transaxle) mounts
9. Suspension arm bushings (see Figure 10–4)

Alignment equipment and procedures can often determine if a suspension component is bent or has been moved out of position. A careful *visual* inspection can often reveal suspected damaged components by observing scrape marks or rusty sections that could indicate contact with the road or another object. (Figure 10–5 shows an example of under-vehicle damage.)

While an assistant bounces the vehicle up and down, check to see if there is any free play in any of the suspension components (see Figure 10–6).

■ BALL JOINTS

Diagnosis and Inspection

The life of ball joints depends on driving conditions, vehicle weight, and lubrication. Even with proper care and lubrication, the load-carrying (weight-carrying) ball joints wear more than the follower ball joints. Ball joints should be replaced in pairs, both lower or both upper, to ensure the best handling.

Defective or worn ball joints can cause looseness in the suspension and these common driver complaints:

1. Loud popping or squeaking whenever driving over curbs, such as into a driveway
2. Shimmy-type vibration felt in the steering wheel

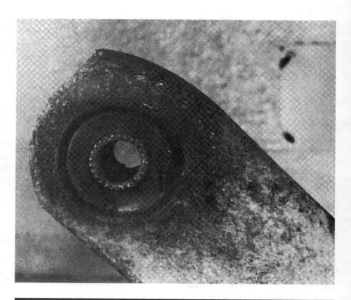

Figure 10–4 Worn control arm bushing.

3. Vehicle wander or a tendency not to track straight
4. Excessive free play in the steering wheel

A ball joint inspection should also be performed when an alignment is performed or as part of any other comprehensive vehicle inspection.

Many load-carrying ball joints have wear indicators with a raised area around the grease fitting (see Figures 10–7 and 10–8 on page 252).

Always check wear-indicator-type ball joints with the wheels of the vehicle on the ground. If the raised area around the grease fitting is flush or recessed with the surrounding area, the ball joint is worn more than 0.050″ and must be replaced.

HINT: Most ball joints must be replaced if the joint has more than 0.050″ axial (up-and-down) movement. To help visualize this distance, consider that the thickness of an American nickel coin is about 0.060″. It is helpful to know that maximum wear should be less than the thickness of a nickel. There are dial indicators (gauges) available that screw into the grease fitting hole of the ball joint that can accurately measure the wear (see Figure 10–9 on page 252).

To perform a proper ball joint inspection, the force of the vehicle's springs *must* be *unloaded* from the ball joint. If this force is not relieved, the force of the spring pushes the ball and socket joint tightly together and any wear due to movement will not be detected.

HINT: The location of the load-carrying ball joint is closest to the seat of the spring or torsion bar.

Figure 10–5 Sagging springs due to overloading caused this vehicle to scrape the road. All four springs had to be replaced to restore the proper ride height.

Figure 10–6 (a) This upper control arm bushing did not look worn or defective until the vehicle was pushed down. (b) As the vehicle was pushed down, the upper control arm moved off center indicating a defective bushing.

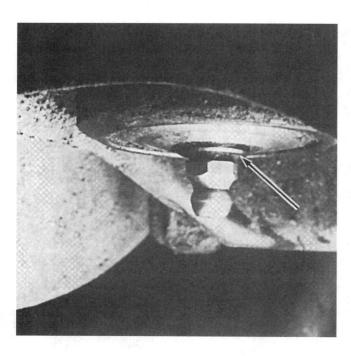

Figure 10–7 Grease fitting projecting down from surrounding area of ball joint. The ball joint should be replaced when the area around the grease fitting is flush or recessed.

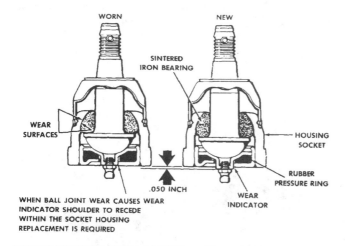

WORN | NEW

SINTERED IRON BEARING

WEAR SURFACES

HOUSING SOCKET

.050 INCH

RUBBER PRESSURE RING

WEAR INDICATOR

WHEN BALL JOINT WEAR CAUSES WEAR INDICATOR SHOULDER TO RECEDE WITHIN THE SOCKET HOUSING REPLACEMENT IS REQUIRED

Figure 10–8 Indicator ball joints should be checked with the weight of the vehicle on the ground. (Courtesy of Moog)

If the coil spring or torsion bar is attached to the *lower* control arm, the *lower* ball joint is the load-carrying ball joint (see Figure 10–10). This includes vehicles equipped with modified MacPherson strut–type suspension (see Figure 10–11).

1. Place the jack under the lower control arm as close to the ball joint as possible and raise the wheels

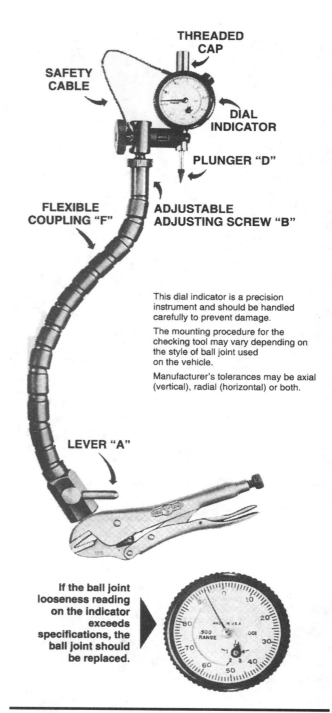

THREADED CAP

SAFETY CABLE

DIAL INDICATOR

PLUNGER "D"

FLEXIBLE COUPLING "F"

ADJUSTABLE ADJUSTING SCREW "B"

This dial indicator is a precision instrument and should be handled carefully to prevent damage.

The mounting procedure for the checking tool may vary depending on the style of ball joint used on the vehicle.

Manufacturer's tolerances may be axial (vertical), radial (horizontal) or both.

LEVER "A"

If the ball joint looseness reading on the indicator exceeds specifications, the ball joint should be replaced.

Figure 10–9 Typical dial indicator used to measure suspension component movement. The locking pliers attach the gauge to a stationary part of the vehicle and the flexible coupling allows the dial indicator to be positioned at any angle. (Courtesy of Moog)

approximately 1–2″ (2.5–5 cm) off the ground to unload the lower ball joint.

2. Using a pry bar under the tire, lift the wheel up. If there is excessive vertical movement in the ball joint itself, it must be replaced. Most manufacturers specify a maximum vertical play of approximately

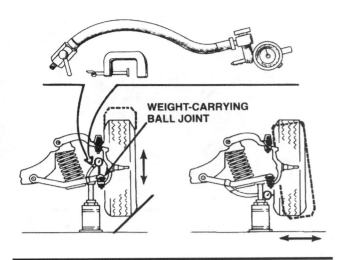

Figure 10–10 If the spring is attached to the lower control arm as in this SLA suspension, the jack should be placed under the lower control arm as shown. A dial indicator should be used to measure the amount of free play in the ball joints. Be sure that the looseness being measured is not due to normal wheel bearing end play. (Courtesy of Dana Corporation)

TECH TIP ✔

The Chrysler Minivan Ball Joint Test

Minivans manufactured by Chrysler Corporation, including the Dodge Caravan and the Plymouth Voyager, are easily checked for worn ball joints. Simply grasp the grease fitting and attempt to move it with your fingers. If the grease fitting moves, the ball joint is worn and requires replacement. This simple test should be done with the weight of the vehicle on the floor—in other words, there is no need to raise or support the vehicle to do this simple test.

0.050″ (1.3 mm), or the thickness of a nickel. Always check the manufacturers' specifications for exact maximum allowable movement. Vertical movement of a ball joint is often called axial play because the looseness is in the same axis as the ball joint stud. To check for *lateral* play, grip the tire at the top and bottom and move your hands in opposite directions.

NOTE: Be sure that the cause of any free play or looseness is not due to the wheel bearing. Closely observe the exact source of the free-play movement.

If the coil spring is attached to the *upper* control arm, the *upper* ball joint is the load-carrying ball joint (see Figure 10–12).

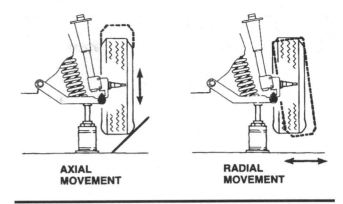

Figure 10–11 The jack should be placed under the lower control arm of this modified MacPherson-type suspension. (Courtesy of Dana Corporation)

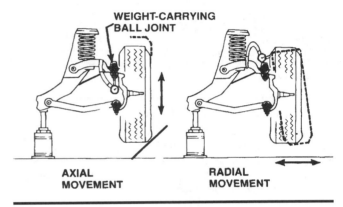

Figure 10–12 If the spring is attached to the upper control arm, the jack should be placed under the frame to check for ball joint wear. (Courtesy of Dana Corporation)

1. Place a block of wood (2″ × 4″) between the upper control arm and the frame, or use a special tool designed for this purpose (see Figure 10–13).
2. Place the jack under the vehicle's *frame* and raise the wheel approximately 1″ to 2″ off the ground. (The wood block or special tool keeps the weight of the vehicle off the upper ball joint.)
3. Using a pry bar under the tire, lift the wheel. If there is excessive vertical movement in the ball joint itself, it must be replaced. Always check the manufacturers' specifications for exact maximum allowable movement.

Follower ball joints (friction ball joints) should also be inspected while testing load-carrying ball joints. Grasp the tire at the top and the bottom and attempt to shake the wheel assembly while looking directly at the follower ball joint. Generally, there should be no lateral or axial movement at all in the follower ball joint. However, some manufacturers specify a maximum of 0.250″ (1/4″ or 0.6 cm) measured at the top of the tire. Always check the manufacturers' specifications before condemning a ball joint because of excessive play.

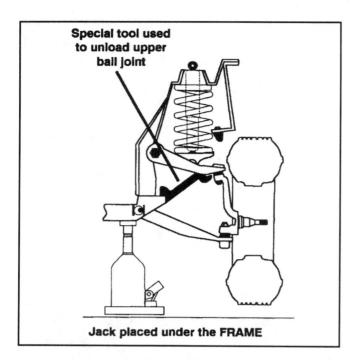

Figure 10–13 A special tool or a block of wood should be inserted between the frame and the upper control arm before lifting the vehicle off the ground. This tool stops the force of the spring against the upper ball joint so that a true test can be performed on the condition of the ball joint. (Courtesy of Ford Motor Company)

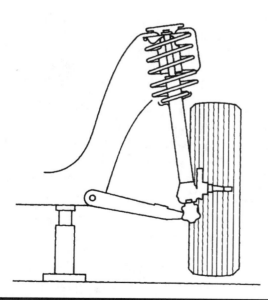

Figure 10–14 The jacking point is under the frame for checking the play of a lower ball joint used with a MacPherson strut.

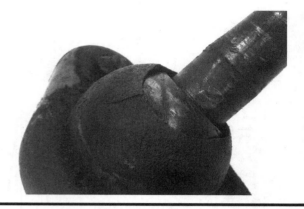

Figure 10–15 Failed ball joints that are cracked.

MacPherson strut–equipped vehicles do not have a load-carrying ball joint. The weight of the vehicle is carried through the upper strut mount and bearing assembly (see Figure 10–14).

After checking axial play in a ball joint, grasp the tire from the side at the top and bottom and alternately push and pull. *Any* lateral movement at the ball joint should generally be considered a good reason to replace the ball joints (see Figure 10–15).

Ball Joint Removal

Take care to avoid damaging grease seals when separating ball joints from their mounts. *The preferred method to separate tapered parts is to use a puller-type tool that applies pressure to the tapered joint as the bolt is tightened on the puller.* (The same applies to steering components such as tie rod ends, as discussed in Chapter 8.) See Figure 10–16.

> **CAUTION:** Using tapered "pickle forks" should be avoided, unless the part is to be replaced, because they often damage the grease seal of the part being separated.

Sometimes the shock of a hammer can be used to separate the ball joint from the steering knuckle. For best results, another hammer should be used as a backup while striking the joint to be separated on the side with a heavy hammer.

Some ball joint studs have a slot or groove where a **pinch bolt** is used to hold the ball joint to the steering knuckle (see Figure 10–17). Use penetrating oil in the steering knuckle groove and rotate the knuckle several times. Do not use a hammer on the pinch bolt because this can cause damage to the bolt and the ball

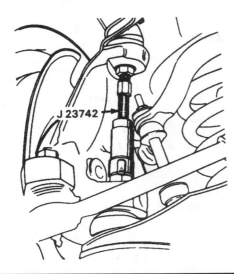

Figure 10–16 Taper breaker tool being used to separate the upper ball joint from the steering knuckle. (Courtesy of Oldsmobile)

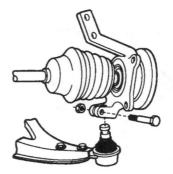

Figure 10–17 A pinch bolt attaches the steering knuckle to the ball joint. Remove the pinch bolt by turning the nut, not the bolt. (Courtesy of Moog)

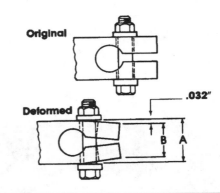

Figure 10–18 If the pinch bolt is overtightened, the steering knuckle can be deformed. A deformed knuckle can cause the pinch bolt to break and the ball joint could become separated from the steering knuckle. (Courtesy of Moog)

| **Frequently Asked Question** | **???** |

What Is the Difference between a Low-Friction Ball Joint and a Steel-on-Steel Ball Joint?

Before the late 1980s, most ball joints were constructed with a steel ball that rubbed on a steel socket. This design created friction and provided for a tight high-friction joint until wear caused looseness in the joint.

Newer designs use a polished steel ball which is installed in a hard plastic polymer resulting in a low-friction joint assembly. Because of the difference in friction characteristics, the vehicle may handle differently than originally designed if the incorrect style ball joints are installed. Most component manufacturers indicate that low-friction ball joints in a vehicle originally equipped with steel-on-steel high-friction ball joints are usually acceptable, but high-friction replacement ball joints should be avoided on a vehicle originally equipped with low-friction ball joints.

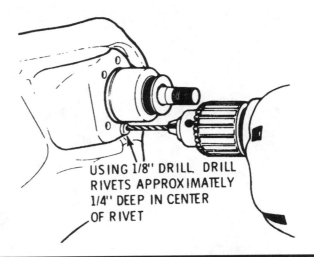

Figure 10–19 By drilling into the rivet, the holding force is released. (Courtesy of Oldsmobile)

joint. Do not widen the slot in the steering knuckle. Once separated, check the shape of the steering knuckle (see Figure 10–18). If the pinch bolt has been deformed by overtightening, the steering knuckle should be replaced.

When removing ball joints that are riveted in place, always cut off or drill rivet heads before separating the ball joint from the spindle. This provides a more solid base to assist in removing rivets. *The preferred method to remove rivets from ball joints is to center punch and drill out the center of the rivet before using a drill or an air-powered chisel to remove the rivet heads. Be careful not to drill or chisel into the control arms (see Figures 10–19 through 10–22).*

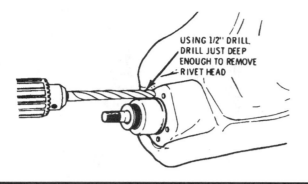

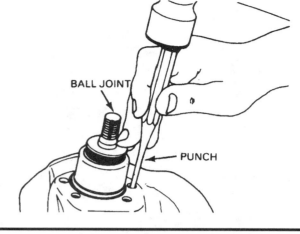

Figure 10–20 The head of the rivet can be removed by using a larger diameter drill bit as shown. (Courtesy of Oldsmobile)

Figure 10–21 Using a punch and a hammer to remove the rivet after drilling down through the center and removing the head of the rivet. (Courtesy of Oldsmobile)

(a)

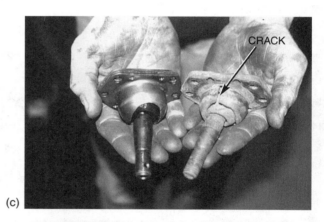

(c)

(b)

(d)

Figure 10–22 (a) A safety stand is used to support the weight of the vehicle by placing it under the lower control arm while replacing the upper ball joint. (b) The old ball joint has been removed. (c) The new replacement ball joint is on the left and the old ball joint is on the right. Note the crack right down the middle of the old ball joint—it is a miracle that the vehicle made it into the shop. (d) The replacement ball joint is bolted in place using the hardened steel bolts and nuts supplied with the replacement ball joint.

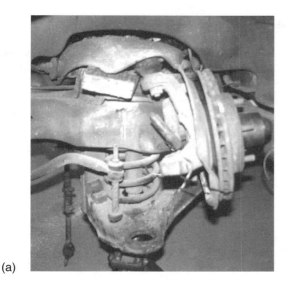

(a)

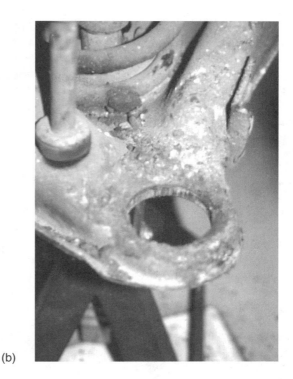

(b)

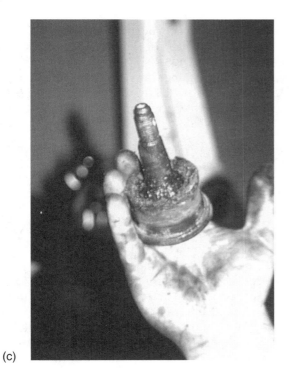

(c)

(d)

Figure 10–23 (a) A block of wood was used to hold the upper control arm, and a safety stand is used to support the vehicle under the lower control arm during the replacement of the lower ball joint. (b) A photo of the lower control arm after the old press-in-type ball joint was removed. (c) The old press-in ball joint was replaced because the wear indicator was recessed (instead of extended) in the housing indicating that the joint was worn beyond serviceable limits. (d) A large C-clamp press is being used with adapters and sleeves to install the replacement ball joint.

Press-in-type ball joints are removed and installed using a special C-clamp-type tool (see Figure 10–23).

NOTE: Many replacement press-in-type ball joints are slightly larger in diameter—about 0.050″ (1.3 mm)—than the original ball joint to provide the same press fit. If the ball joints have been replaced before, then the control arm must be replaced.

Avoid using heat to remove suspension or steering components. Many chassis parts use rubber and plastic that can be damaged if heated with a torch. *If heat is used to remove a part, it must be replaced.* For best results, try soaking with a penetrating oil and use the proper tools and procedures as specified by the manufacturer.

Many vehicles are equipped with nonreplaceable ball joints, and the entire control arm must be replaced if the ball joint is worn or defective.

> **CAUTION:** Always follow manufacturers' recommended installation instructions whenever replacing any suspension or other chassis component part. Tie rod ends and ball joints use a taper to provide the attachment to other components. Whenever a nut is used to tighten a tapered part, it is important not to back off (loosen) the nut after tightening. As the nut is being tightened, the taper is being pulled into the taper of the adjoining part. The specified torque on the nut ensures that the two pieces of the taper are properly joined. If the cotter key does not line up with the hole in the tapered stud when the nut has been properly torqued, tighten it more to line up a hole—never loosen the nut.

■ KING PIN DIAGNOSIS AND SERVICE

King pins are usually used on trucks, sport utility vehicles, and other heavy-duty vehicles (see Figure 10–24).

King pins are designed to rotate inside **king pin bushings** with a clearance between them of approximately 0.001″ to 0.003″ (0.025 to 0.075 mm). As wear occurs, this clearance distance increases and the king pin becomes loose. Looseness in the king pin causes looseness in the wheels and steering. *King pins can also gall or seize due to lack of lubrication resulting in very hard steering.*

Diagnostic Story

The Rattle Story

A customer complained that a rattle was heard every time the vehicle hit a bump. The noise sounded as if it came from the rear. All parts of the exhaust system and suspension system were checked. Everything seemed okay until the vehicle was raised with a frame-type hoist instead of a drive-on type. Then, whenever the right rear wheel was lifted, the noise occurred. The problem was a worn (elongated) shock absorber mounting hole. A washer with the proper size hole was welded over the worn lower frame mount and the shock absorber was bolted back into place.

After supporting the vehicle safely off the ground, inspect for looseness by positioning a dial gauge (indicator) on the extreme inside bottom of the edge of the wheel and rock the tire. If the dial indicator registers more than 1/4″ (6 mm), replace the king pin and/or king pin bushings.

To remove a typical king pin, follow these basic steps:

Step 1 Remove the tire, brake drum, and backing plate or caliper.

Step 2 Remove the lock pin. The lock pin is usually tapered with a threaded end for the nut. Drive the lock pin out with a drift (punch).

Step 3 Remove the grease caps and drive the king pin from the steering knuckle and axle with a hammer and a brass punch or an hydraulic press, if necessary (see Figure 10–25).

To replace the king pin bushings and/or king pin, refer to the manufacturer's procedure and specifications (see Figure 10–26).

Some bronze bushings must be sized by reaming or honing to provide from 0.001″ to 0.003″ (0.025 to 0.075 mm) clearance between the king pin and the bushing.

■ SHOCK ABSORBERS AND STRUTS

Diagnosis

Shock absorber life depends on how and where the vehicle is driven. Original equipment (OE) shock absorbers are carefully matched to the vehicle springs and bushings to provide the best ride comfort and control. As the control arm bushings and ball joints age, the energy built up in the springs of the vehicle is controlled less by the friction of these joints and bushings, requiring more control from the shock absorbers. Shock absorber action is also reduced as the seals inside wear. Replacement shock absorbers may be required when any or all of the following symptoms appear:

1. *Ride harshness.* As the effectiveness of the shock absorber decreases, the rapid forces of the springs are not as dampened or controlled. Worn shocks can cause ride harshness and yet not cause the vehicle to bounce after hitting a bump.
2. *Frequent bottoming out on rough roads.* Shock absorbers provide a controlled movement of the axle whenever the vehicle hits a bump or dip in the road. As fluid is lost or wear occurs, the shock absorber becomes weaker and cannot resist the forces acting on the axle. The worn shock absorber can allow the spring to compress enough so that the axle contacts the jounce bumper (bump stop) on the body or frame of the vehicle.

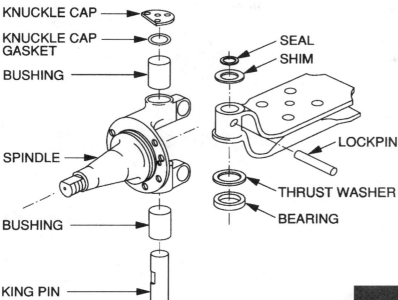

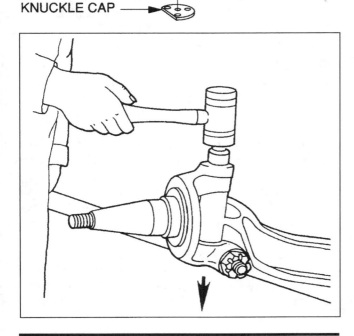

Figure 10-25 Driving a king pin out with a hammer. (Courtesy of Dana Corporation)

Figure 10-24 Typical king pin assembly. (Courtesy of Dana Corporation)

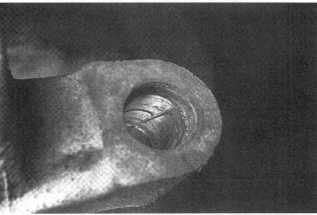

Figure 10-26 This galled king pin bushing has to be replaced.

NOTE: Frequent bottoming out is also a symptom of reduced ride height due to sagging springs. Before replacing the shock absorbers, always check for proper ride height as specified in the vehicle service manual or any alignment specification booklet available from suppliers or companies of alignment or chassis parts and equipment. See Appendix 7 for names and addresses of chassis and alignment equipment manufacturers.

TECH TIP

The Shock Stud Trick

Front shock absorbers used on many rear-wheel-drive vehicles are often difficult to remove because the attaching nut is rusted to the upper shock stub. A common trick is to use a deep-well 9/16" socket and a long extension and simply bend the shock stud until it breaks off. At first you would think that this method causes harm, and it does ruin the shock absorber—but the shock absorber is not going to be reused and will be discarded anyway.

The usual procedure followed by many technicians is to simply take a minute or two to break off the upper shock stud, then hoist the vehicle to allow access to the lower two shock bolts and the shock can easily be removed. To install the replacement shock absorber, attach the lower bolts and lower the vehicle and attach the upper rubber bushings and retaining nut.

3. *Extended vehicle movement after driving on dips or a rise in the road.* The most common shock absorber test is the bounce test. Push down on the body of the vehicle and let go; the vehicle should return to its normal ride height and stop. Worn shock absorbers can cause poor driver control due to excessive up and down suspension movements. If the vehicle continues to bounce two or three times, then the shocks or struts are worn and must be replaced.

4. *Cuppy-type tire wear.* Defective shock absorbers can cause **cuppy**-type tire wear. This type of tire wear is caused by the tire's bouncing up and down as it rotates.

5. *Leaking hydraulic oil.* When a shock or strut leaks oil externally, this indicates a defective seal. The shock absorber or strut cannot function correctly when low on oil inside.

Shock absorbers should be replaced in pairs. Both front or both rear shocks should be replaced together to provide the best handling and control.

> **NOTE:** Shock absorbers do not affect ride height except where special air shocks or coil overload carrying shocks are used. If a vehicle is sagging on one side or in the front or the rear, the springs should be checked and replaced if necessary.

Shock absorbers are filled with fluid and sealed during production. They are not refillable; and if worn, damaged, or leaking, they must be replaced. A slight amount of fluid may bleed by the rod seal in cold weather and deposit a light film on the upper area of the shock absorber. This condition will not hurt the operation of the shock and should be considered normal. If noisy when driving, always check the tightness and condition of all shock absorber mounts.

Replacement shock absorbers and/or struts should match the original equipment unit in physical size.

Front Shock Replacement

Front shock absorbers provide ride control and are usually attached to the lower control arm by bolts and nuts. The upper portion of the shock usually extends through the spring housing and is attached to the frame of the vehicle with rubber grommets to help insulate noise, vibration and harshness.

Most front shock absorbers can be replaced either with the vehicle still on the ground or while on a hoist. The front suspension of most vehicles allows the removal of the shocks without the need to support the

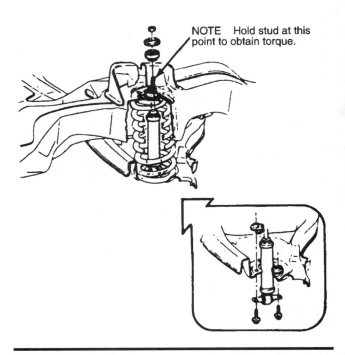

Figure 10–27 Most shock absorbers used on the front suspension can be removed from underneath the vehicle after removing the attaching bolts or nuts. (Courtesy of Oldsmobile)

downward travel of the lower control arm. The downward travel limit is stopped by a rubber stop or by the physical limit of the suspension arms.

Special sockets and other tools are available to remove the nut from the shocks that use a single mounting stem. The removal of the lower mounting of most front shock absorbers usually involves the removal of two bolts. After removal of the attaching hardware, the shock absorber is simply pulled through the front control arm (see Figure 10–27).

Reverse the removal procedure to install replacement shock absorbers. Always consult the manufacturer's recommended procedures and fastener tightening torque. See Figure 10–28 for an example of the steps necessary to change the shock absorber on the front of a Ford Mustang equipped with a modified MacPherson strut–type front suspension.

Rear Shock Replacement

Before removing the rear shock absorbers, the rear axle must be supported to prevent stretching the rear brake flexible hose. Shocks are attached to the frame or body of the vehicle at the top and to a bracket on the rear axle housing at the bottom. Often the top of the rear shock absorber is fastened *inside* the vehicle. Consult the manufacturer's installation instructions for exact procedures and fastener torque values (see Figure 10–29 on page 262).

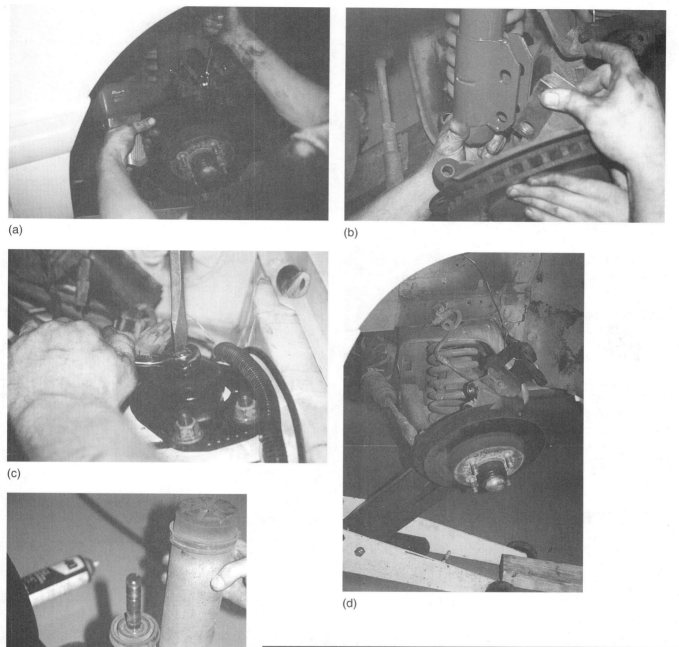

Figure 10–28 (a) After supporting the vehicle and the lower control arm, the attaching bolts that retain the shock absorber from the front knuckle can be removed. (b) After the bolts are removed, the shock is then separated from the knuckle. In this case, the shock does not fall downward because it is still retained at the upper mount. (c) After the bottom of the shock has been disconnected, the upper retainer is removed and the shock is removed from underneath the vehicle. (d) With the shock removed, note how the lower control arm is supported by a hydraulic jack. The jack makes it easy to raise or lower the lower control arm and steering knuckle while installing the replacement shock. (e) In this situation, the dust boot from the old shock has to be removed and installed on the new shock before installing the replacement shock into the vehicle.

Figure 10–29 Whenever replacing shock absorbers (especially rear shock absorbers) be sure to support the rear axle. The rear axle assembly could drop and cause damage to the vehicle and possible personal injury if the axle is not supported and the shock absorber attaching bolt (or nut) is removed.

Air Shock Installation Air-adjustable shock absorbers are a popular replacement for conventional rear shock absorbers. Air shocks can be used to level the vehicle while towing a trailer or when heavily loaded. When the trailer or load is removed, air can be released from the air shocks to return the vehicle height to normal.

Figure 10–30 To prevent melting the plastic air shock lines, the lines should be clipped well away from the exhaust system.

Most replacement air shocks are directional and are labeled *left* and *right*. This ensures that the plastic air hose line exits the shock absorber toward the center or rear of the vehicle and is kept away from the wheels.

The plastic air shock line attaches to the shock absorber with an O-ring or brass ferrule and nut. An air leak can result if this O-ring or ferrule is not installed according to the manufacturer's recommendations. Route the plastic air line along the body, keeping it away from the exhaust and any other body parts where the line could be damaged (see Figure 10–30). Attach the line to both shocks to a junction at a convenient location for adding or releasing air.

> **NOTE:** Chrysler Corporation does not recommend the use of high-pressure gas shocks or air-adjustable shocks on the rear of certain models of minivans. The high-pressure gas inside the shocks may raise the ride height of the rear of the vehicle. When the rear height is elevated, the brake proportioning valve that measures rear suspension height could reduce the maximum braking force to the rear brakes. Always consult the vehicle manufacturer or shock absorber manufacturer's printed application guides to be assured of purchasing the correct shock absorber or strut.

■ MACPHERSON STRUT REPLACEMENT

On most vehicles equipped with MacPherson strut suspensions, strut replacement involves the following steps.

> **CAUTION:** Always follow the manufacturer's recommended methods and procedures whenever replacing a MacPherson strut assembly or component.

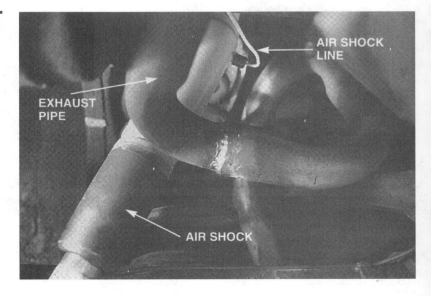

Figure 10–31 Using correction fluid to mark just one strut mounting stud. This helps ensure that the assembly is correctly positioned back in the vehicle.

1. Hoist the vehicle, remove the wheels, and mark the attaching bolts/nuts as shown in Figure 10–31.
2. Remove the upper strut mounting bolts except for one to hold the strut until ready to remove the strut assembly (see Figure 10–32).
3. Remove the brake caliper or brake hose from the strut housing (see Figures 10–33 and 10–34).
4. After removing all lower attaching bolts, remove the final upper strut bolt and remove the strut assembly from the vehicle (see Figure 10–35). Place the strut assembly into a strut spring compressor fixture as shown in Figures 10–36 and 10–37 on page 265. Manual spring compressors can also be used (see Figures 10–38 and 10–39 on pages 265–266).
5. Compress the coil spring enough to relieve the tension on the strut rod nut. Remove the strut rod nut as shown in Figure 10–40 on page 266.
6. After removing the strut rod nut, remove the upper strut bearing assembly and the spring (see Figures 10–41, 10–42, and 10–43 on pages 266–267).

NOTE: The bearing assembly should be carefully inspected and replaced if necessary. Some automotive experts recommend replacing the bearing assembly whenever the strut is replaced.

7. Many MacPherson struts are replaced as an entire unit assembly. Other struts can have the cartridge inside the housing replaced. The cartridge is installed after removing the **gland nut** at the top of the strut tube and removing the original strut rod, valves, and hydraulic oil. Always replace the cartridge assembly per the manufacturer's recommended procedure (see Figures 10–44 through 10–50 on pages 267–270).
8. Reinstall the strut in the vehicle.

CAUTION: Many GM front-wheel-drive vehicles use a MacPherson strut with a large-diameter spring seat area. GM chassis engineers call this a *cow catcher* design. If the coil spring breaks, the extra large cow catcher will prevent the end of the broken spring from moving outward where it could puncture a tire and possibly cause an accident. Always use a replacement strut that has this feature (see Figure 10–51 on page 270).

Modified MacPherson Strut Service

A modified MacPherson strut assembly is a type of front suspension that does not use the coil spring around the strut but is mounted on the lower control arm. The strut

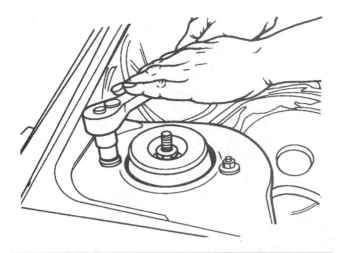

Figure 10–32 Removing the upper strut mounting bolts. Some experts recommend leaving one of the upper strut mount nuts loosely attached to prevent the strut from falling when the lower attaching bolts are removed. (Courtesy of Dana Corporation)

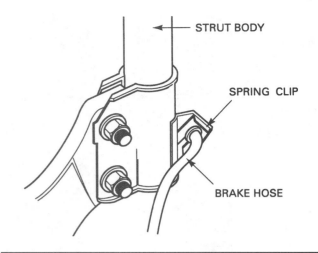

Figure 10–33 A brake hydraulic hose is often attached to the strut housing. Sometimes all that is required to separate the line from the strut is to remove a spring clip.

Figure 10–34 A commonly used procedure of cutting through the strut clamp eliminates the need to disconnect the brake hose and open the hydraulic system. This operation saves time because the technician does not have to bleed the air out of the brake system after installing new struts.

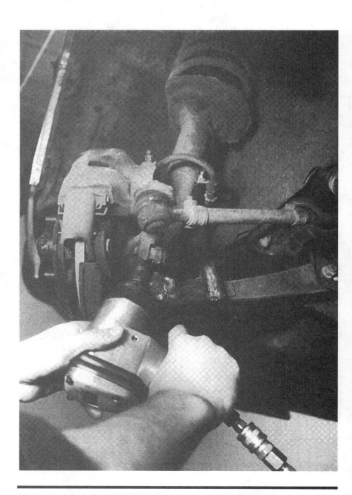

Figure 10–35 Removing the bolts that attach the strut to the knuckle.

Figure 10–36 Typical spring compressor designed for General Motors MacPherson strut assemblies. The bottom of the strut is attached to the fixture with pins and secured to the top with the top bearing nuts. The unit is held to a workbench with a conventional holding fixture that can also hold transmissions or transaxles.

unit provides the upper pivot point of the suspension. It mounts to the body at the top and the lower control arm at the bottom. The usual procedure for the replacement of a strut cartridge or assembly in a modified MacPherson strut system includes the following steps:

1. Raise the vehicle safely and support the lower control arms with a floor jack or safety stand.
2. Remove the front brake caliper.
3. Unbolt the strut from the spindle.
4. Remove the upper shaft nut (remove the strut assembly from the vehicle). See Figure 10–52 on page 270.

> **NOTE:** On most modified MacPherson strut–equipped vehicles, it is not necessary to remove the upper strut mounting bolts on the body of the vehicle. Often a rivet is used to maintain proper wheel alignment; this rivet should *not* be removed.

5. Install the replacement unit in the reverse order, being certain that the jounce bumper (stop bumper) is properly located and that all fasteners are tightened to the manufacturer's recommended specifications.

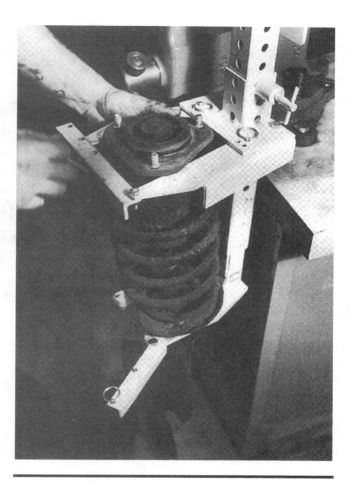

Figure 10–37 A universal-type MacPherson strut spring compressor. Note where the compressor is mounted—under the bearing and on top of the spring. This spring compressor is clamped in a vise attached to a workbench.

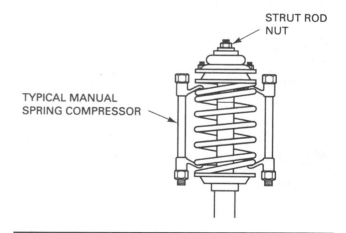

Figure 10–38 Two individual spring compressors like the one shown can also be used to compress the spring on a MacPherson strut before removing the strut retaining nut.

Figure 10–39 These individual spring compressors have a rubberized coating on the jaws that touch the spring. This is important because the use of metal jaws can remove the rust preventative coating on the spring. If this coating is removed, even in a small area, rust can start and can weaken or cause the spring to break.

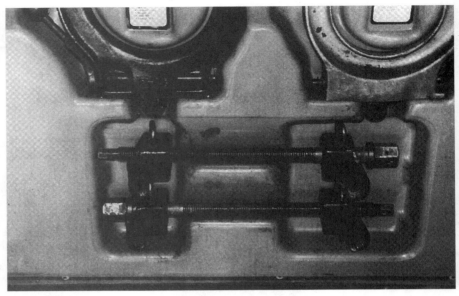

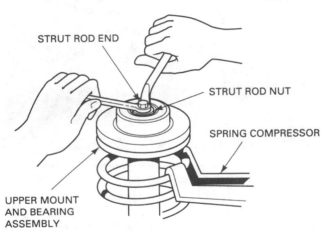

Figure 10–40 Removing the strut rod nut. The strut shaft is being helped with one wrench while the nut is being removed with the other wrench. Notice that the spring is compressed before the nut is removed.

■ STABILIZER BAR LINK AND BUSHINGS

Diagnosis

Stabilizer bars twist whenever a vehicle turns a corner or whenever one side of the vehicle rises or lowers. The more the body of the vehicle leans, the more the bar twists, and the more the bar counteracts the roll of the body. A great deal of force is transferred from the stabilizer bar to the body or frame of the vehicle through the stabilizer links at the control arm and the stabilizer bar mounting bushings on the body or frame. *The most common symptom of defective stabilizer bar links or bushings is noise while turning, especially over curbs.* When driving up and over a curb at an angle, one wheel is pushed upward. Since the stabilizer bar con-

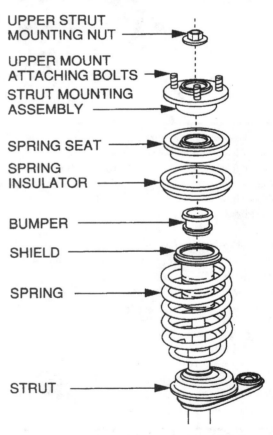

Figure 10–41 Typical MacPherson strut showing the various components. (Courtesy of Dana Corporation)

nects both wheels, the bar tends to resist this motion by twisting. If one or both links are broken, a loud knock or clanking sound is heard as the bar contacts the control arm. Any sound heard when driving should be investigated and the cause determined by a thorough visual inspection.

Figure 10–42 It is often a good idea to mark the location of the spring, spring seat, and upper strut mount to make sure it is reassembled correctly. This assembly was marked with correction fluid ("White Out") because it dries quickly and is easy to use.

Figure 10–44 After the spring has been lifted off the strut, the gland nut can be removed. While a special wrench designed for this purpose should be used, a pipe wrench can also get the job done.

Figure 10–43 After removing the strut end nut, the spring compressor can be expanded. Note how much longer the spring is now that it is allowed to expand to its normal length.

Replacement

Stabilizer links are usually purchased as a kit consisting of replacement rubber bushings, retainers, a long bolt, and spacer with a nut and lock nut. Most manufacturers recommend that stabilizer links be replaced in pairs and two kits purchased so that the links on both the left and the right can be replaced at the same time (see Figure 10–53 on page 270).

Stabilizer links are replaced simply by unbolting the old and installing replacement parts. Special precautions are usually not necessary, but as always, consult a service manual for the exact recommended procedure and fastener torque.

If the mounting bushings (also called *isolator bushings*) on the body or frame are worn or defective, a loud squeaking or knocking sound is usually heard (see Figure 10–54 on page 270).

Stabilizer bar bushings often fail because the rubber has deteriorated from engine oil or other fluid leaks. Before replacing defective bushings, make certain that all other fluid leaks are stopped. Stabilizer bar

Figure 10–45 The gland nut being removed from the strut housing.

Figure 10–46 A little jerk on the strut rod and the shock components are lifted up and out of the strut tube.

bushing replacement involves the removal (unbolting) of the bushing retainers that surround the bushing and attach to the vehicle body or frame. The bushings are usually split so that they can be easily removed and replaced on the stabilizer bar and slide into proper position. Reinstall bushing retainer bolts and tighten to specified torque.

■ STRUT ROD BUSHINGS

Diagnosis

Strut rods are used on the front suspension of many front-wheel-drive and rear-wheel-drive vehicles. As with any rubber suspension component, strut rod bushings can deteriorate and crack. When strut rod bushings fail, the lower control arm can move forward and backward during braking or when hitting bumps in the road (see Figure 10–55 on page 271).

Since the lower control position is important to vehicle control, when the bushing fails, a pulling or drifting of the vehicle often occurs. *A common symptom of a defective strut rod bushing is noise and a pull toward one side while braking.*

Replacement

To replace a strut rod bushing, the nut on the end of the strut rod has to be removed (see Figure 10–56 on page 271). As the nut is being removed, the lower control arm is likely to move forward or backward, depending on the location of the strut rod frame mount. After removing the strut rod nut, remove the one or two fasteners that retain the strut rod to the lower control arm. The strut rod can now be removed from the vehicle and the replacement rubber bushings can be installed. Install the strut rod in the reverse order of installation. Most bushings use a serrated spacer to maintain the proper force on the rubber bushings; this prevents the technician from compressing the rubber bushing when the strut rod nut is tightened (see Figure 10–57 on page 271).

> **NOTE:** Many vehicles use the strut rod as an adjustable suspension component for alignment. If the end of the strut rod has a nut on both sides of the frame mount, then the strut rod is the adjustable component for changing the caster angle. See Chapters 12 and 13 for additional information on alignment angles and procedures.

Figure 10–47 After pulling out the strut shaft (in the drain pan), remove the strut and pour out all of the hydraulic fluid that was in the shock. Dispose of this waste oil in the proper method.

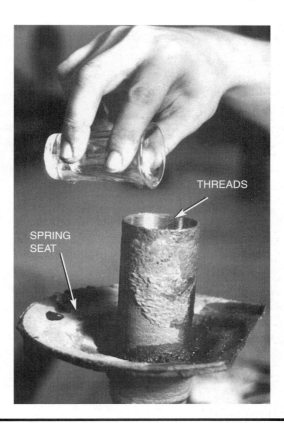

Figure 10–48 A measured amount of engine oil should be poured into the empty strut housing before installing a sealed strut replacement cartridge. The purpose of the oil is to transfer heat from the insert to the strut housing. The technician is using a shot glass to measure the amount of oil. Most manufacturers recommend 1 or 2 oz (28 to 56 cc) of oil.

■ REAR COIL SPRINGS

Replacement

Coil springs in the rear are easily replaced on both front-wheel-drive and rear-wheel-drive vehicles. The procedure includes the following steps:

1. Raise the vehicle safely on a hoist.
2. Remove both rear wheels.
3. Support the rear axle assembly with tall safety stands.
4. Remove the lower shock absorber mounting bolts/nut and disconnect the shock absorber from the rear axle assembly.
5. Slowly lower the rear axle assembly by either lowering the height of the adjustable safety stands or raising the height of the vehicle on the hoist.
6. Lower the rear axle just enough to remove the coil springs (see Figure 10–58 on page 272).

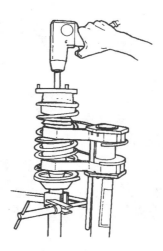

Figure 10–49 After installing the replacement strut cartridge, reinstall the spring and upper bearing assembly after compressing the spring. Notice that the strut is being held in a strut vise. (Courtesy of Dana Corporation)

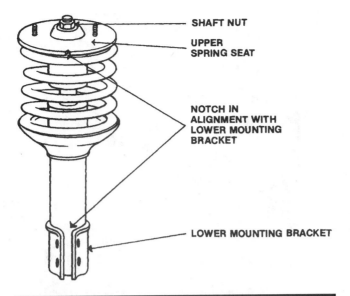

Figure 10–50 Before final assembly, make sure the marks you made are aligned. Some struts are manufactured with marks to ensure proper reassembly. (Courtesy of Moog)

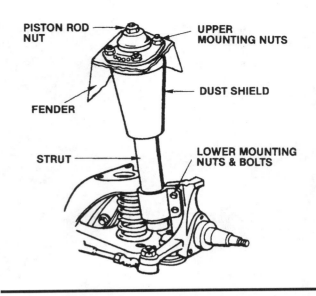

Figure 10–52 The strut on a modified MacPherson strut assembly can be replaced by removing the upper mounting nuts. (Courtesy of Moog)

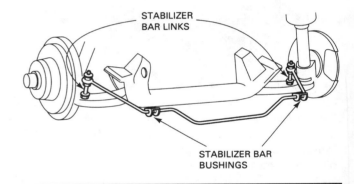

Figure 10–53 Stabilizer bar links should be replaced as a pair. (Courtesy of Moog)

Figure 10–51 A strut assembly on a front-wheel-drive Buick Regal. Note that the drive axle shaft has to be removed to replace this strut assembly. This strut assembly is being replaced because the ball joint was excessively worn. The joint is not available as a separate part and is sold only as part of the entire strut assembly.

Figure 10–54 Stabilizer bar bushing clamp unbolted from the frame.

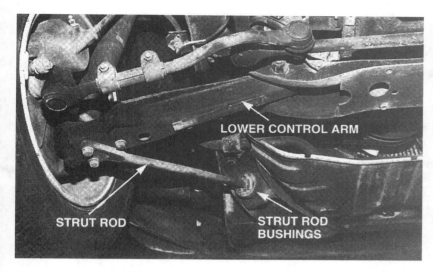

LOWER CONTROL ARM

STRUT ROD

STRUT ROD BUSHINGS

Figure 10–55 Notice that the strut rod prevents the control arm from moving forward or rearward as the vehicle accelerates, brakes, and turns.

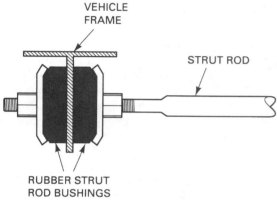

VEHICLE FRAME

STRUT ROD

RUBBER STRUT ROD BUSHINGS

Figure 10–56 Typical strut rod bushing with rubber on both sides of the frame to help isolate noise, vibration, and harshness from being transferred to the passengers. (Courtesy of Ford Motor Company)

SERRATED SPACERS

Figure 10–57 Parts of a strut rod bushing. Note that the serrated spacers are used between the two rubber bushings to maintain a fixed distance between the bushings. This bushing helps prevent the nut from being tightened too much, thereby compressing the bushings too much.

Diagnostic Story

The Rock-Hard Problem

The owner of a six-month-old full-size pickup truck complained that occasionally when the truck was driven up into a driveway, a loud grinding sound was heard. Several service technicians worked on the truck, trying to find the cause for the noise. After the left front shock absorber was replaced, the noise did not occur for two weeks, then started again. Finally, the service manager told the technician to replace anything and everything in the front suspension in an attempt to solve the customer's intermittent problem. Five minutes later, a technician handed the service manager a small, deformed rock. This technician had taken a few minutes to *carefully* inspect the entire front suspension. Around the bottom coil spring seat, the technician found the rock. Apparently when the truck made a turn over a bump, the rock was forced between the coils of the coil spring, making a very loud grinding noise. But the rock did not always get between the coils. Therefore, the problem occurred only once in a while. The technician handed the rock to the very happy customer.

CAUTION: The shock absorber is usually the only component that limits the downward movement of the rear axle to allow removal of the rear coil springs. Some vehicles may be equipped with rear suspension height sensors for the adjustable suspension system or an adjustable rear proportioning valve for the rear brakes. Some vehicles also require that the rear stabilizer bar or track rod be disconnected or removed before lowering the rear axle assembly. Always consult a service manual for the exact procedure and torque specifications for the vehicle you are servicing.

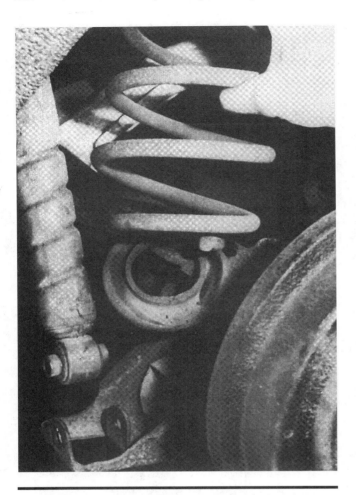

Figure 10–58 Coil spring being removed from a rear suspension.

■ FRONT COIL SPRINGS

Diagnosis

Coil springs should be replaced in pairs if the vehicle ride height is lower than specifications. Sagging springs can cause the tires to slide laterally (side to side) across the pavement, causing excessive tire wear (see Figure 10–59). Sagging springs can also cause the vehicle to bottom out against the suspension bumpers when traveling over normal bumps and dips in the road.

If a vehicle is overloaded, the springs of the vehicle can *take a set* and not recover to the proper ride height. This commonly occurs with all types of vehicles whenever a heavy load is carried, even on a short trip.

It is normal for the rear of a vehicle to sag when a heavy load is carried, but it can permanently damage the spring by exceeding the yield point of the steel spring material.

Replacement

The only solution recommended by the vehicle manufacturer is to replace the damaged springs in pairs (both

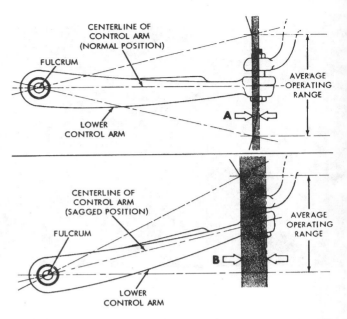

Figure 10–59 Notice that if the front coil springs are sagging, the resulting angle of the lower control arm causes the wheels to move from side to side as the suspension moves up and down. Note the difference between the distance at "A" with good springs and the distance at "B" with sagging springs. (Courtesy of Moog)

front and/or both rear, or all four). Several aftermarket alternatives include:

Helper or auxiliary left springs. These helper springs are usually designed to increase the load-carrying capacity of leaf springs or to restore the original ride height to sagging springs.

Spring inserts for coil springs. Hard rubber or metal spaces are *not recommended* because they create concentrated pressure points on the coils that can cause the spring to break. Hard rubber stabilizers that bridge the gap between two coils without raising the ride height stiffen the ride by locking two coils together. With two coils connected, the spring rate increases and the spring becomes stiffer.

Air shocks or airbag devices. Air shocks or airbags are generally used to restore or increase the load-carrying capacity to the rear of the vehicle. While these devices do allow the ride height to be maintained, the extra load is still being supported by the tires and rear axle bearings. Caution should be used not to overload the basic chassis or power train of the vehicle.

> **CAUTION:** Do not carry a load in any vehicle that exceeds its rated gross vehicle weight capacity as indicated on the door sticker (see the Diagnostic Story, "It's Not Far, It Can Take It" on page 276).

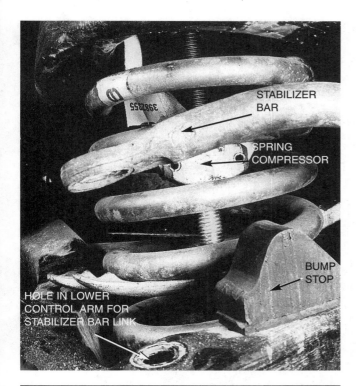

Figure 10–60 Spring compressing tool in place to hold the spring as the ball joint is separated. Note that the stabilizer bar links have been removed to allow the lower control arm to move downward enough to remove the coil spring.

There are two designs of vehicles that use coil springs. The most commonly used design places the coil spring between the vehicle frame and the lower control arm. An older design, used on many Ford products, places the coil spring between the upper control arm and the body.

Both front suspension designs require that the front shock absorbers be removed in order to replace the coil springs. This first step in the coil spring removal procedure can be performed on the ground or on any type of hoist. After removing the shock absorber, use a coil spring compressor and install it through the center of the coil spring (see Figure 10–60). Hook the arms of the coil spring compressor over the rungs of the coil spring and rotate the adjusting nut on the spring compressor with a wrench to shorten the spring. Coil spring clips can also be used to retain the coil spring (see Figure 10–61).

CAUTION: When compressed, all springs contain a great deal of stored energy. If a compressed spring were to become disconnected from its spring compressor or clips, it could be projected outward with enough force to cause injury or death.

After the coil spring is retained, the control arm can be separated from the steering knuckle and the coil spring removed as shown in Figures 10–62 and 10–63.

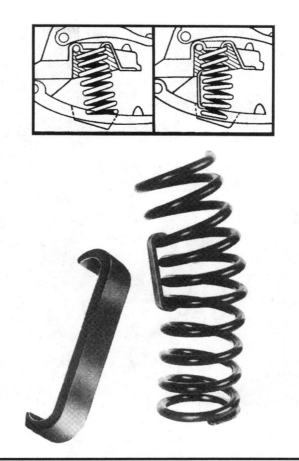

Figure 10–61 Clamp-type tools are used with adapters to remove and install the control arm bushings. (Courtesy of Moog)

Replacement springs should be compressed and installed using the reverse procedure.

NOTE: Many automotive experts recommend that new coil insulators be installed every time the coil springs are replaced.

Make sure that the spring is positioned correctly in the control arm. Most control arms use two holes for the purpose of coil spring seating. The end of the spring should cover one hole completely and partially cover the second hole (see Figure 10–64).

■ STEERING KNUCKLES

Diagnosis

Most steering knuckles are constructed of cast or forged cast iron. The steering knuckle usually incorporates the wheel spindle and steering arm. The steering knuckle/steering arm can become bent if the vehicle is in an accident or hits a curb sideways. Often this type of damage is not apparent until vehicle handling or excessive

Figure 10–62 The steering knuckle has been disconnected from the ball joints. The lower control arm and coil spring are being held up by a floor jack.

tire wear is noticed. Unless a thorough inspection is performed during a wheel alignment, a bent steering knuckle is often overlooked. A quick and easy test is shown in Figure 10–65. (For additional measurements and testing, see Chapter 12.)

Replacement

If the steering knuckle is bent or damaged, it must be replaced. It should *not* be bent back into shape or repaired. To replace the steering knuckle, both ball joints have to be disconnected from the knuckle and the brake components removed. Be sure to support the control arm and spring properly during the procedure. See a factory service manual for the exact procedure and fastener torque for the vehicle you are servicing.

■ TORSION BARS

Adjustment

Most torsion bar suspensions are designed with an adjustable bolt to permit the tension on the torsion bar to be increased or decreased to change the ride height. Unequal side-to-side ride height can be corrected by adjusting (turning) the torsion bar tension bolt (see Figure 10–66).

Figure 10–63 The spring is being held with a spring compressor as the lower control arm is being pushed down to release the spring.

Spring to be installed with tape at lowest position. Bottom of spring is coiled helical, and the top is coiled flat with a gripper notch near end of wire.

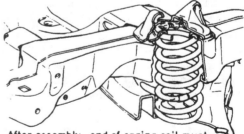

After assembly, end of spring coil must cover all or part of one inspection drain hole. The other hole must be partly exposed or completely uncovered.

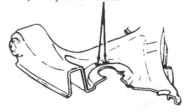

Figure 10–64 Note that the holes in the spring seat of the lower control arms are actually used to help ensure that the spring is properly installed. (Courtesy of Oldsmobile)

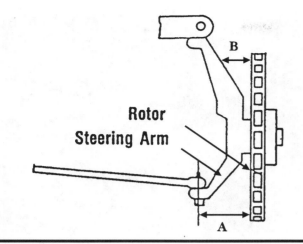

Figure 10–65 The distance between the rotor and the steering knuckle should be the same on both sides of the vehicle. Dimension "A" checks if the steering arm is bent, and dimension "B" checks if the spindle is bent.

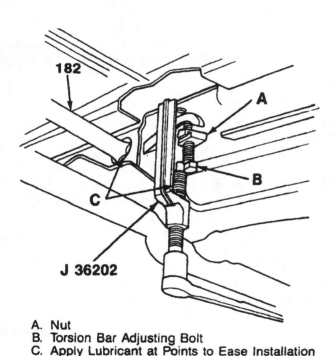

A. Nut
B. Torsion Bar Adjusting Bolt
C. Apply Lubricant at Points to Ease Installation
182. Torsion Bar

Figure 10–66 By rotating the adjusting bolt, the vehicle can be raised or lowered. (Courtesy of Oldsmobile)

Torsion bar adjustment should be made if the difference in ride height from one side to another exceeds 1/8″ (0.125″ or 3.2 mm). If the ride height difference side to side is greater than 1/8″, the vehicle will tend to wander or be unstable, with constant steering wheel movements required to maintain straight-ahead direction.

Figure 10–67 Using a large C-clamp-type tool to remove and install upper control arm bushings.

■ CONTROL ARM BUSHINGS

Diagnosis

Defective control arm bushings are a common source of vehicle handling and suspension noise problems. Most suspension control arm bushings are constructed of three parts: an inner metal sleeve, the rubber bushing itself, and an outer steel sleeve. (Some vehicles use a two-piece bushing that does not use an outer sleeve.)

Replacement

To remove an old bushing from a control arm, the control arm must first be separated from the suspension and/or frame of the vehicle. Several methods can be used to remove the bushing from the control arm, but all methods apply force to the *outer* sleeve. While an air chisel is frequently used to force the steel sleeve out of the suspension member, a puller tool (such as shown in Figure 10–67) is most often recommended by vehicle manufacturers. The puller can be used to remove the old and install the replacement bushing without harming the control arm or the new bushing. All bushings should be tightened with the vehicle on the ground and the wheels in a straight-ahead position; this prevents the rubber bushing from exerting a pulling force on the suspension.

The upper control arm bushings can be replaced in most vehicles that use a short/long-arm-type suspension by following just four easy steps:

> **NOTE:** Some replacement control arm bushings have a higher *durometer* or hardness rating than the original. Urethane bushings are often used in sporty or race-type vehicles and deflect less than standard replacement bushings. These harder bushings also transfer more road noise, vibration, and harshness to the body.

Diagnostic Story

It's Not Far—It Can Take It

An automotive instructor needed to transport several V-8 engines just a couple of miles. A truck was not available, so the instructor carried the three engines in a station wagon. The rear of the station wagon sagged under the load. After the engines were unloaded, the rear of the station wagon remained lower than normal. The steel of the coil spring had exceeded its yield point and did not return to its original position. The rear coil springs were ruined and "took a set" due to the excessive load. The rear coil springs had to be replaced to restore the proper ride height.

> **NOTE:** Leaf springs too can be easily overloaded and take a set or break! Overloading *any* vehicle can also damage the wheel bearings.

Never carry a load in any vehicle that exceeds its design capacity.

Step 1 Raise the vehicle and support the lower control arm with a safety stand or floor jack.

Step 2 Disconnect the upper control arm from the frame by removing the frame-attaching nuts or bolts.

Step 3 Using the upper ball joint as a pivot, rotate the upper control arm outward into the wheel well area. With the control arm accessible, it is much easier to remove and replace the upper control arm pivot shaft and rubber bushings.

Step 4 After replacing the bushings, simply rotate the upper control arm back into location and reattach the upper control arm pivot shaft to the vehicle frame.

> **NOTE:** An alignment should always be performed after making any suspension-related repairs.

■ REAR LEAF SPRINGS

Replacement

Rear leaf springs often need replacement due to one of these common causes:

1. **Individual leaves of a leaf spring often crack, then break.** When a leaf spring breaks, the load-carrying capacity of the vehicle decreases and it often sags on the side with the broken spring. Metal fatigue, corrosion, and overloading are three of the most common causes of leaf spring breakage (see Figure 10–68).

Figure 10–68 The leaves of this leaf spring are probably broken because the rear of the vehicle is sagging and the leaves are separated.

> **NOTE:** When one rear spring on one side sags, the opposite front end of the vehicle tends to *rise*. For example, if the right rear spring breaks or sags, the left front of the vehicle tends to rise higher than the right front. This unequal vehicle height can make the vehicle difficult to handle, especially around corners or curves.

2. **If the center bolt breaks,** the individual leaves can move and the rear axle is no longer held in the correct location. When one side of the rear axle is behind the other side, the vehicle will *dog track*. Dog tracking refers to the sideways angle of the vehicle while traveling straight. It is commonly caused by the rear axle's steering the vehicle toward one side while the driver controls the direction of the vehicle with the front wheels (see Chapter 12 for additional information).

> **NOTE:** *Leaf springs should be replaced in pairs.*

To replace leaf springs in the rear of a rear-wheel-drive vehicle:

1. Raise the vehicle safely on a hoist.
2. Support the rear axle with safety stands.
3. Remove the rear shackle bolts and forward mounting bolt or mounting bracket.
4. Remove the U-bolts.
5. Being careful of any nearby brake line, remove the spring (see Figure 10–69).
6. Install the new spring, being careful to position the center bolt correctly into the hole on the axle pedestal.

Special tools are often required when removing a transverse leaf spring (as shown in Figure 10–70).

Figure 10–69 Removing the spring shackle while the spring is supported with a jack.

■ TROUBLESHOOTING ELECTRONIC LEVELING SYSTEMS

The first step with any troubleshooting procedure is to check for normal operation. Some leveling systems require that the ignition key be on (run), while other systems operate all the time. Begin troubleshooting by placing approximately 300 lb. (135 kg) on the rear of the vehicle. If the compressor does not operate, check to see if the sensor is connected to a rear suspension member and that the electrical connections are not corroded.

Also check the condition of the compressor ground wire. It must be tight and free of rust and corrosion where it attaches to the vehicle body. If the compressor still does not run, check to see if 12 volts are available at the power lead to the compressor. If necessary, use a jumper wire directly from the positive (+) of the battery to the power lead of the compressor. If the compressor does not operate, it must be replaced.

If the ride height compressor runs excessively, check the air compressor, the air lines, and the air shocks (or struts) with soapy water for air leaks. Most air shocks or air struts are not repairable and must be replaced. Most electronic level-control systems provide some adjustments of the rear ride height by adjusting the linkage between the height sensor and the rear suspension (see Figure 10–71).

Figure 10–70 A special tool used to straighten a transverse composite leaf spring so that it can be removed from the vehicle.

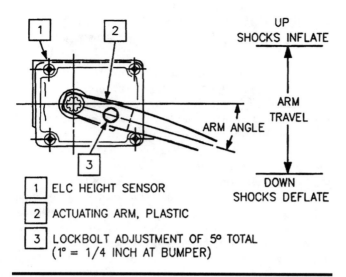

UP
SHOCKS INFLATE

ARM
TRAVEL

ARM ANGLE

DOWN
SHOCKS DEFLATE

1	ELC HEIGHT SENSOR
2	ACTUATING ARM, PLASTIC
3	LOCKBOLT ADJUSTMENT OF 5° TOTAL (1° = 1/4 INCH AT BUMPER)

Figure 10–71 Most electronic level-control sensors can be adjusted such as this General Motors unit. (Courtesy of Oldsmobile)

■ SUMMARY

1. A thorough road test of a suspension problem should include driving beside parked vehicles and into driveways in an attempt to determine when and where the noise occurs.

2. A dry park test should be performed to help isolate defective or worn suspension components.

3. Ball joints must be unloaded before testing. The ball joints used on vehicles with a MacPherson strut suspension are *not* load-carrying. Wear-indicator ball joints are observed with the wheels on the ground.

4. Always use a taper-braker puller or two hammers to loosen tapered parts to remove them. Never use heat unless you are replacing the part; heat from a torch can damage rubber and plastic parts.

5. When installing a tapered part, always tighten the attaching nut to specifications. Never loosen the nut to install a cotter key. If the cotter key will not line up with a hole in the tapered part, tighten the nut more until the cotter key hole lines up with the nut and stud.

6. Defective shock absorbers can cause ride harshness as well as frequent bottoming out on rough roads.

7. While many rear axles must be supported when changing rear shock absorbers, most front shocks can be replaced without having to limit front suspension downward travel.

8. Always follow manufacturers' recommended procedures whenever replacing springs or MacPherson struts. Never remove the strut end nut until the coil spring is compressed and the spring force is removed from the upper bearing assembly.

PHOTO SEQUENCE Strut Replacement

PS10–1 After removing the wheel and tire, hoist the vehicle safely to a good working height. Use correction fluid or other marking material to mark one stud of the MacPherson strut unit. This helps ensure that the unit will be properly reinstalled into the vehicle.

PS10–2 Remove the upper strut mounting fasteners.

PS10–3 Remove the two large lower retaining bolts that attach the bottom of the strut to the steering knuckle.

PS10–4 Disconnect the brake hose from the strut housing and install a CV joint boot protector before separating the strut from the steering knuckle.

PS10–5 Remove the strut assembly from underneath the vehicle.

PS10–6 Attach the strut assembly to the MacPherson strut spring compressor fixture.

Strut Replacement—continued

PS10–7 Line up the three upper bearing bolts into the appropriate fixture holes in this special General Motors MacPherson strut compressor fixture.

PS10–8 Tighten the screw thread on the fixture to compress the coil spring on the strut.

PS10–9 After the coil spring has been compressed, remove the strut rod retaining nut.

PS10–10 After the nut has been removed, turn the screw on the fixture to remove the force on the coil spring.

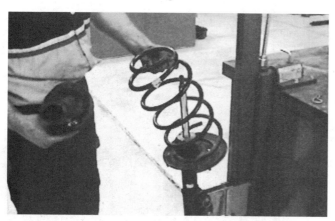

PS10–11 With the fixture extended almost all the way, the upper bearing assembly, jounce bumper, and coil spring can be removed from the assembly.

PS10–12 During reassembly, note the proper location of the end of the spring on the strut housing.

Strut Replacement—continued

PS10–13 Reassembly of the strut assembly requires that the strut rod be aligned with the upper bearing and the coil spring compressed enough to allow the strut retaining nut to be installed.

PS10–14 Reinstall the strut retaining nut.

PS10–15 Remove the assembled strut assembly from the MacPherson strut compressor fixture.

PS10–16 Insert the top of the strut back into the vehicle being sure that the marked stud is reinstalled into the correct hole. Reinstall the lower MacPherson strut retaining bolts and nuts, and torque to factory specifications.

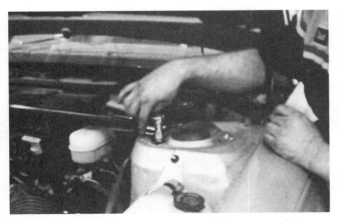

PS10–17 Install the upper strut retaining nuts, and torque to factory specifications.

PS10–18 Install the wheel/tire assembly and align the vehicle before returning the vehicle to the customer.

■ REVIEW QUESTIONS

1. Describe how to perform a proper road test for the diagnosis of suspension-related problems.

2. List four symptoms of worn or defective shock absorbers.

3. Explain the procedure for replacing front shock absorbers on an SLA-type suspension vehicle.

4. Describe the testing procedure for ball joints.

5. Describe the general procedure to remove and replace tapered suspension components correctly.

■ ASE CERTIFICATION-TYPE QUESTIONS

1. Unusual noise during a test drive can be caused by
   a. Defective wheel bearings or stabilizer bar links
   b. Defective or worn control arm bushings or ball joints
   c. Worn or defective CV joints
   d. All of the above

2. Two technicians are discussing ball joint inspection. Technician A says that the vehicle should be on the ground with the ball joints *loaded*, then checked for free play. Technician B says that the ball joints should be *unloaded* before checking for free play. Which technician is correct?
   a. Technician A only
   b. Technician B only
   c. Both Technician A and B
   d. Neither Technician A nor B

3. Most manufacturers specify a maximum axial play for ball joints of about
   a. 0.003″ (0.076 mm)
   b. 0.010″ (0.25 mm)
   c. 0.030″ (0.76 mm)
   d. 0.050″ (1.27 mm)

4. The preferred method to separate tapered chassis parts is to use
   a. A pickle fork tool
   b. A torch to heat the joint until it separates
   c. Two hammers to shock and deform the taper or a puller tool
   d. A drill to drill out the tapered part

5. A light film of oil is observed on the upper area of a shock absorber. Technician A says that this condition should be considered normal. Technician B says that a rod seal may bleed fluid during cold weather, causing the oil film. Which technician is correct?
   a. Technician A only
   b. Technician B only
   c. Both Technician A and B
   d. Neither Technician A nor B

6. Before the strut cartridge can be removed from a typical MacPherson strut assembly, which operation is necessary to prevent possible personal injury?
   a. The brake caliper and/or brake hose should be removed from the strut housing.
   b. The coil spring should be compressed.
   c. The upper strut mounting bolts should be removed.
   d. The lower attaching bolts should be removed.

7. What should the technician do when replacing stabilizer bar links?
   a. The stabilizer bar should be removed from the vehicle before replacing the links.
   b. The links can be replaced individually, yet the manufacturer often recommends that the links at both ends be replaced together.
   c. The stabilizer bar must be compressed using a special tool before removing or installing stabilizer bar links.
   d. Both b and c are correct.

8. A noise and a pull toward one side during braking is a common symptom of a worn or defective
   a. Shock absorber
   b. Strut rod bushing
   c. Stabilizer bar link
   d. Track rod bushing

9. To help prevent vehicle wandering on a vehicle with torsion bars, the ride height (trim height) should be within _____ side to side.
   a. 0.003″ (0.076 mm)
   b. 0.050″ (1.27 mm)
   c. 0.100″ (2.5 mm)
   d. 0.125″ (3.2 mm)

10. A vehicle equipped with a coil spring front suspension and a leaf spring rear suspension "dog tracks" while driving on a straight, level road. Technician A says that a broken center bolt could be the cause. Technician B says defective upper control arm bushings on the front could be the cause. Which technician is correct?
    a. Technician A only
    b. Technician B only
    c. Both Technician A and B
    d. Neither Technician A nor B

Wheel Alignment Principles

A wheel alignment is the adjustment of the suspension and steering to ensure proper vehicle handling with minimum tire wear. When a vehicle is new, the alignment angles are set at the factory. After many miles and/or months of driving, the alignment angles can change slightly. The change in alignment angles may result from one or more of the following:

1. Wear of the steering and the suspension components
2. Bent or damaged steering and suspension parts
3. Sagging springs, which can change the ride height of the vehicle and, therefore, the alignment angles

By adjusting the suspension and steering components, proper alignment angles can be restored. An alignment includes checking and adjusting, if necessary, both front and rear wheels.

■ ALIGNMENT-RELATED PROBLEMS

Most alignment diagnosis is symptom-based. This means that the problem with the alignment is determined from symptoms such as excessive tire wear or a pull to one side of the road. The definitions of alignment symptom terms used in this book include:

Pull

A pull is generally defined as a definite tug on the steering wheel toward the left or the right while driving straight on a level road (see Figure 11–1). Bent, damaged, or worn suspension and/or steering components can cause this problem, as well as a tire problem.

Lead or Drift

A lead or drift is a mild pull that does not cause a force on the steering wheel that the driver must counteract. A lead or drift is observed by momentarily removing your hands from the steering wheel while driving on a straight, level road. When the vehicle moves toward one side or the other, this is called a *lead* or a *drift*.

CAUTION: When test-driving a vehicle for a lead or a drift, make sure that the road is free of traffic and that your hands remain close to the steering wheel. Your hands should be held away from the steering wheel for just a second or two—just long enough to check for a lead or drift condition.

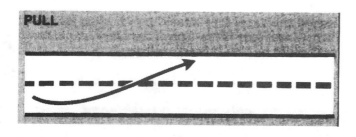

Figure 11–1 A pull is usually defined as a tug on the steering wheel toward one side or the other. (Courtesy of Ford Motor Company)

Road Crown Effects

Most roads are constructed with a slight angle to permit water to drain from the road surface. On a two-lane road, the center of the road is often higher than the berms, resulting in a "crown" (see Figure 11–2).

On a four-lane expressway (freeway), the crown is often *between the two sets* of lanes. Because of this slight angle to the road, some vehicles may lead or drift away from the road crown. In other words, it may be perfectly normal for a vehicle to lead toward the right while being driven in the slow lane and toward the left while being driven in the fast (or inside) lane of a typical divided highway.

Wander

A wander is a condition during which constant steering wheel corrections are necessary to maintain a straight-ahead direction on a straight, level road (see Figure 11–3).

Worn suspension and/or steering components are the most likely cause of this condition. Incorrect or unequal alignment angles such as caster and toe, as well as defective tire(s), can also cause this condition.

Stiff Steering or Slow Return to Center

Hard-to-steer problems are commonly caused by leaks, either low tire pressure (due to the leak of air) and/or

lack of proper power steering (due to the leak of power steering fluid). Other causes include excessive positive caster on the front wheels or binding steering linkage.

Tramp or Shimmy Vibration

Tramp is a vertical-type (up-and-down) vibration usually caused by out-of-balance or defective tires or wheels (see Chapter 3). Shimmy is a back-and-forth vibration that can be caused by an out-of-balance tire or defective wheel (see Chapter 3) or by an alignment problem.

> **NOTE:** Wheel alignment will not correct a tramp-type vibration.

■ CAMBER

Camber *is the inward or outward tilt of the wheels from true vertical as viewed from the front or rear of the vehicle* (see Figure 11–4).

1. If the top of the tire is tilted out, then camber is positive (+), as shown in Figure 11–5.
2. If the top of the tire is tilted in, then camber is negative (−), as shown in Figure 11–6.
3. Camber is zero degrees (0°) if the tilt of the wheel is true vertical, as shown in Figure 11–7.
4. Camber is measured in degrees or fractions of degrees.
5. *Camber can cause tire wear if not correct.*
   a. *Excessive positive camber* causes scuffing and wear on the outside edge of the tire, as shown in Figure 11–8 on page 286.
   b. *Excessive negative camber* causes scuffing and wear on the inside edge of the tire, as shown in Figure 11–9 on page 286.
6. Camber can cause pull if it is unequal side to side. **The vehicle will pull toward the side with the most camber.** A difference of more than 1/2 degree from one side to the other will cause the vehicle to pull (see Figures 11–10, 11–11, and 11–12 on pages 286–287).

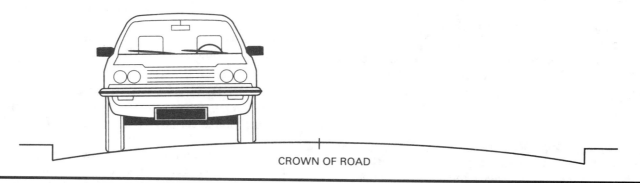

CROWN OF ROAD

Figure 11–2 The crown of the road refers to the angle or slope of the roadway needed to drain water off the pavement.

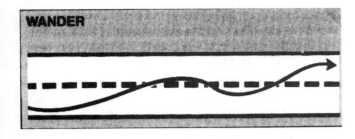

Figure 11–3 Wander is an unstable condition requiring constant driver corrections. (Courtesy of Ford Motor Company)

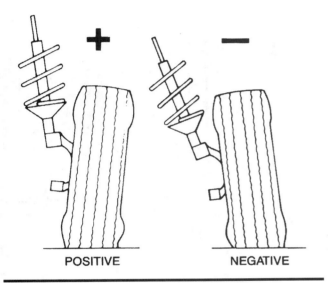

POSITIVE   NEGATIVE

Figure 11–4 Positive and negative camber. (Courtesy of Dana Corporation)

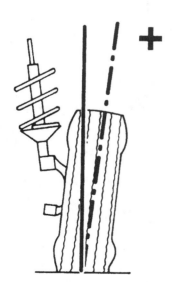

Figure 11–5 Positive camber. The solid vertical line represents true vertical, and the dotted line represents the angle of the tire. (Courtesy of Hunter Engineering Company)

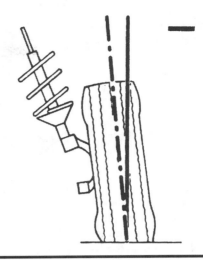

Figure 11–6 Negative camber. The solid vertical line represents true vertical, and the dotted line represents the angle of the tire. (Courtesy of Hunter Engineering Company)

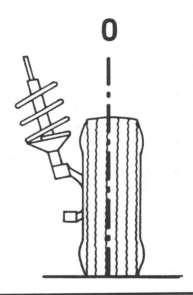

Figure 11–7 Zero camber. Note that the angle of the tire is true vertical. (Courtesy of Hunter Engineering Company)

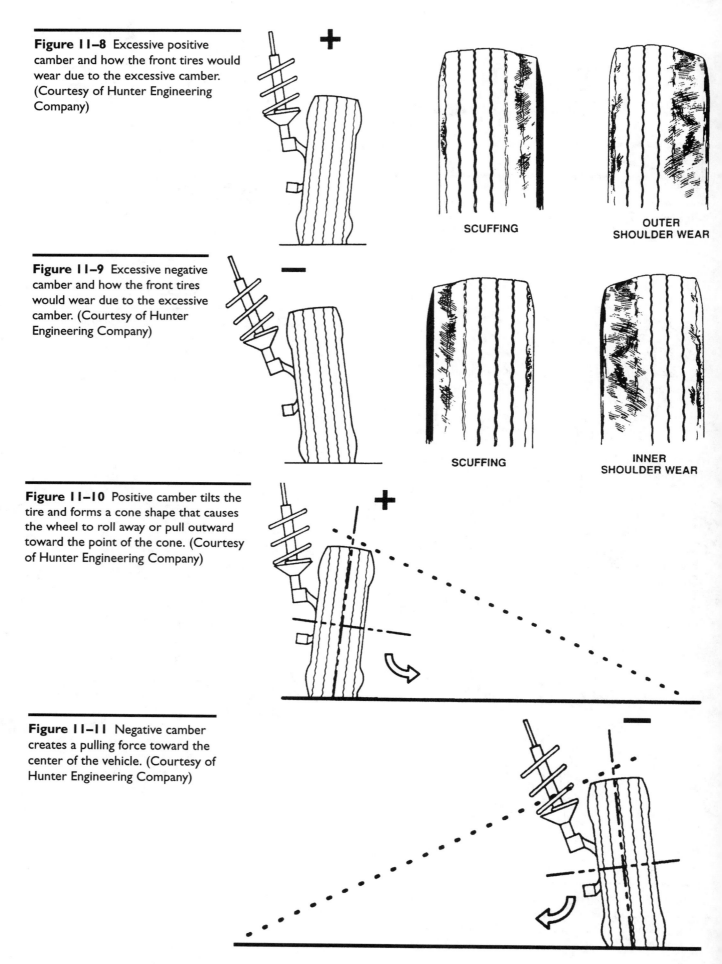

Figure 11–8 Excessive positive camber and how the front tires would wear due to the excessive camber. (Courtesy of Hunter Engineering Company)

+

SCUFFING

OUTER SHOULDER WEAR

Figure 11–9 Excessive negative camber and how the front tires would wear due to the excessive camber. (Courtesy of Hunter Engineering Company)

−

SCUFFING

INNER SHOULDER WEAR

Figure 11–10 Positive camber tilts the tire and forms a cone shape that causes the wheel to roll away or pull outward toward the point of the cone. (Courtesy of Hunter Engineering Company)

+

Figure 11–11 Negative camber creates a pulling force toward the center of the vehicle. (Courtesy of Hunter Engineering Company)

−

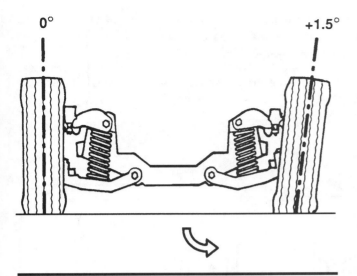

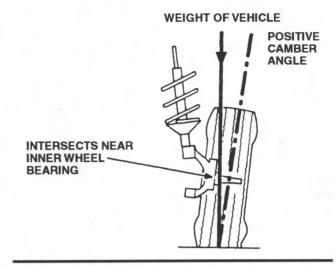

Figure 11–12 If camber angles are different from one side to the other, the vehicle will pull toward the side with the most camber. (Courtesy of Hunter Engineering Company)

Figure 11–13 Positive camber applies the vehicle weight toward the larger inner wheel bearing. This is desirable because the larger inner bearing is designed to carry more vehicle weight than the smaller outer bearing. (Courtesy of Hunter Engineering Company)

7. Incorrect camber can cause excessive wear on wheel bearings, as shown in Figures 11–13 and 11–14. Many vehicle manufacturers specify positive camber so that the vehicle's weight is applied to the larger inner wheel bearing and spindle. As the vehicle is loaded or when the springs sag, camber usually decreases. If camber is kept positive, then the running camber is kept near zero degrees for best tire life.

NOTE: Many front-wheel-drive vehicles that use sealed wheel bearings often specify negative camber.

8. Camber is *not* adjustable on many vehicles.

9. If camber is adjustable, the change is made by moving the upper or the lower control arm or strut assembly by means of one of the following methods:
   a. Shims
   b. Eccentric cams
   c. Slots
   See Chapter 12 for camber adjusting methods and procedures.
10. Camber should be equal on both sides; however, if camber cannot be adjusted exactly equal, make certain that there is more camber on the front of the left side to help compensate for the road crown (1/2 degree maximum difference).

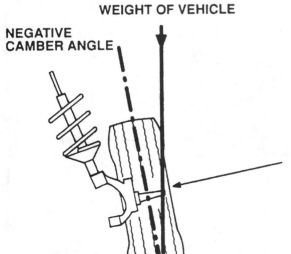

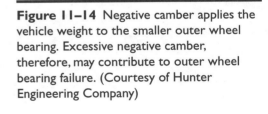

Figure 11–14 Negative camber applies the vehicle weight to the smaller outer wheel bearing. Excessive negative camber, therefore, may contribute to outer wheel bearing failure. (Courtesy of Hunter Engineering Company)

■ TOE

Toe *is the difference in distance between the front and rear of the tires.* As viewed from the top of the vehicle (a bird's-eye view), zero toe means that both wheels on the same axle are parallel, as shown in Figure 11–15.

Toe is also described as a comparison of horizontal lines drawn through both wheels on the same axle, as shown in Figure 11–16.

If the front of the tires is closer than the rear of the same tires, then the toe is called *toe-in* or positive (+) toe (see Figure 11–17).

If the front of the tires is farther apart than the rear of the same tires, then the wheels are *toed-out*, or have negative (−) toe (see Figure 11–18).

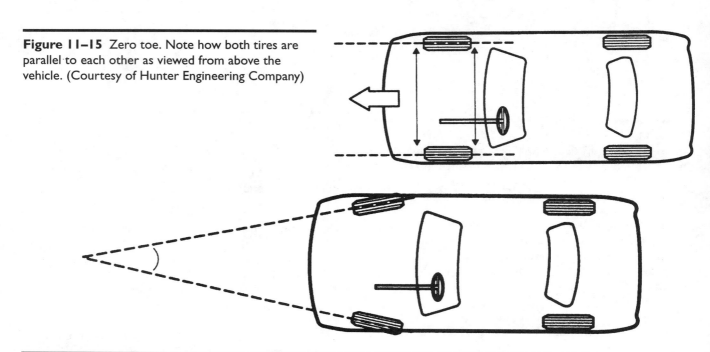

Figure 11–15 Zero toe. Note how both tires are parallel to each other as viewed from above the vehicle. (Courtesy of Hunter Engineering Company)

Figure 11–16 Total toe is often expressed as an angle. Because both front wheels are tied together through the tie rods and center link, the toe angle is always equally split between the two front wheels when the vehicle moves forward. (Courtesy of Hunter Engineering Company)

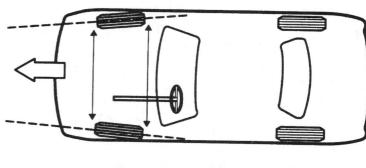

Figure 11–17 Toe-in; also called positive (+) toe. (Courtesy of Hunter Engineering Company)

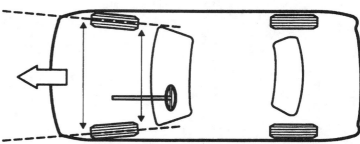

Figure 11–18 Toe-out; also called negative (−) toe. (Courtesy of Hunter Engineering Company)

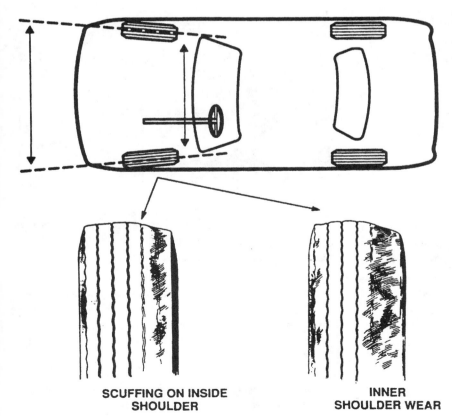

Figure 11–19 This tire is just one month old! It was new and installed on the front of a vehicle that had about 1/4″ (6 mm) of toe out. By the time the customer returned to the tire store for an alignment, the tire was completely bald on the inside. Note the almost new tread on the outside.

The purpose of the correct toe setting is to provide maximum stability with a minimum of tire wear when the vehicle is being driven.

1. Toe is measured in fractions of degrees or in fractions of an inch (usually 1/16s), millimeters (mm), or decimals of an inch (such as 0.06″).
2. *Incorrect toe is the major cause of excessive tire wear!* (See Figure 11–19.)

NOTE: If the toe is improper by just 1/8″ (3 mm) the resulting tire wear is equivalent to dragging a tire sideways 28 feet (8.5 m) for every mile (1.6 km) traveled.

Toe causes camber-type wear on one side of the tire, if not correct (see Figures 11–20 and 11–21). Feather-edge wear is also common especially if the vehicle is equipped with nonradial tires (see Figure 11–22).

3. *Incorrect front toe does not cause a pull condition.* Incorrect toe on the front wheels is split equally as the vehicle is driven because the forces acting on the tires are exerted through the tie rod and steering linkage to both wheels.
4. *Incorrect (or unequal) rear toe can cause tire wear* (see Figures 11–23, 11–24, and 11–25 on page 291). If the toe of the rear wheels is not equal, the steering wheel will not be straight and will pull toward the side with the most toe-in (see Chapter 12 for details).

Figure 11–20 Excessive toe-out and the type of wear that can occur to the inside of the left front tire. (Courtesy of Hunter Engineering Company)

SCUFFING ON INSIDE SHOULDER

INNER SHOULDER WEAR

Figure 11–21 Excessive toe-in and the type of wear that can occur to the outside of the left front tire. (Courtesy of Hunter Engineering Company)

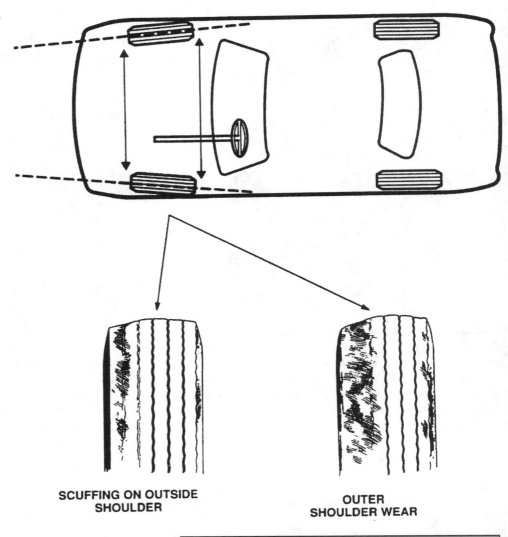

SCUFFING ON OUTSIDE SHOULDER

OUTER SHOULDER WEAR

Figure 11–22 Feather-edge wear pattern caused by excessive toe-in or toe-out. (Courtesy of Hunter Engineering Company)

5. Front toe adjustment must be made correctly by adjusting the tie rod sleeves (see Figure 11–26).
6. Most vehicle manufacturers specify a slight amount of toe-in to compensate for the natural tendency of the front wheels to spread apart (become toed-out) due to centrifugal force of the rolling wheels acting on the steering linkage.

Frequently Asked Question **???**

Why Doesn't Unequal Front Toe on the Front Wheels Cause the Vehicle to Pull?

Each wheel could have individual toe, but as the vehicle is being driven, the forces on the tires tend to split the toe, causing the steering wheel to cock at an angle as the front wheels both track the same. If the toe is different on the rear of the vehicle, the rear will be "steered" similar to a rudder on a boat because the rear wheels are not tied together as are the front wheels.

NOTE: Some manufacturers of front-wheel-drive vehicles specify a toe-out setting to compensate for the toe-in forces created by the engine drive forces on the front wheels.

7. Normal wear to the tie rod ends and other steering linkage parts usually causes toe-out.

Figure 11–23 Rear toe-in (+). The rear toe (unlike the front toe) can be different for each wheel while the vehicle is moving forward because the rear wheels are not tied together as they are in the front. (Courtesy of Hunter Engineering Company)

Figure 11–24 Incorrect toe can cause the tire to run sideways as it rolls resulting in a diagonal wipe. (Courtesy of Hunter Engineering Company)

Figure 11–25 Diagonal wear such as shown here is usually caused by incorrect toe on the rear of a front-wheel-drive vehicle.

TECH TIP

**Smooth In, Toed-In;
Smooth Out, Toed-Out**

Whenever the toe setting is not zero, a rubbing action occurs that causes a feather-edge-type wear (see Figure 11–27). A quick, easy method to determine if incorrect toe could be causing problems is simply to rub your hand across the tread of the tire. If it feels smoother moving your hand toward the center of the vehicle than when you move your hand toward the outside, then the cause is excessive toe-in. The opposite effect is caused by toe-out. This may be felt on all types of tires, including radial ply tires where the wear may not be seen as feather edged. Just remember this simple saying: "Smooth in, toed-in; smooth out, toed-out."

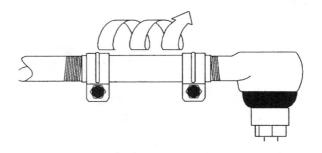

Figure 11–26 Toe on the front of most vehicles is adjusted by turning the tire rod sleeve as shown. (Courtesy of Ford Motor Company)

Feathered or Sawtooth
Tire wear pattern

Sharp edges point in the direction
of the toe problem
(IN - Toe In / OUT Toe Out)

Figure 11–27 Although the feathered or sawtooth tire tread wear pattern may not be noticeable to the eye, this wear can usually be felt by rubbing your hand across the tread of the tire. (Courtesy of Ford Motor Company)

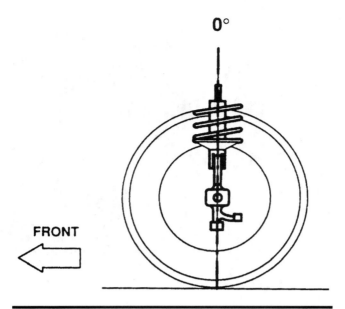

Figure 11–28 Zero caster. (Courtesy of Hunter Engineering Company)

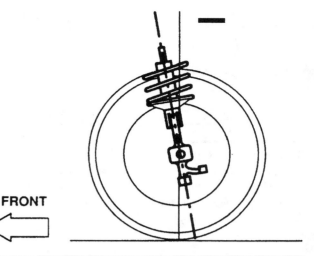

Figure 11–30 Negative (−) caster is seldom specified on today's vehicles because it tends to make the vehicle unstable at highway speeds. Negative caster was specified on some older vehicles not equipped with power steering to help reduce the steering effort. (Courtesy of Hunter Engineering Company)

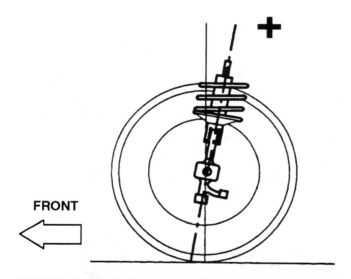

Figure 11–29 Positive (+) caster. (Courtesy of Hunter Engineering Company)

Excessive front toe-out will cause wandering (lack of directional stability), especially during braking. Incorrectly set toe will cause an uncentered steering wheel. If toe is unequal in the *rear*, the vehicle will pull toward the side with the most toe-in.

■ CASTER

Caster *is the forward or rearward tilt of the steering axis in reference to a vertical line as viewed from the side of the vehicle.* The steering axis is defined as the line drawn through the upper and lower steering pivot

points. On an SLA suspension system, the upper pivot is the upper ball joint and the lower pivot is the lower ball joint. On a MacPherson strut system, the upper pivot is the center of the upper bearing mount and the lower pivot point is the lower ball joint. Zero caster means that the steering axis is straight up and down, also called zero degrees or perfectly vertical, as shown in Figure 11–28.

1. Positive (+) caster is present when the upper suspension pivot point is behind the lower pivot point (ball joint) as viewed from the side (see Figure 11–29).
2. Negative (−) caster is present when the upper suspension pivot point is ahead of the lower pivot point (ball joint) as viewed from the side (see Figure 11–30).
3. Caster is measured in degrees or fractions of degrees.
4. Caster is not a tire-wearing angle, but positive caster does cause changes in camber during a turn (see Figure 11–31). This condition is called **camber roll** (see the Tech Tip, "Caster Angle Tire Wear").
5. Caster is a stability angle.
   a. If caster is excessively positive, the vehicle steering will be very stable (will tend to go straight with little steering wheel correction needed). This degree of caster helps with steering wheel **returnability** after a turn (see Figure 11–32).
   b. If the caster is positive, steering effort will increase with increasing positive caster. Greater road shocks will be felt by the driver when driving over rough road surfaces (see

Figure 11–31 As the spindle rotates, it lifts the weight of the vehicle due to the angle of the steering axis. (Courtesy of Hunter Engineering Company)

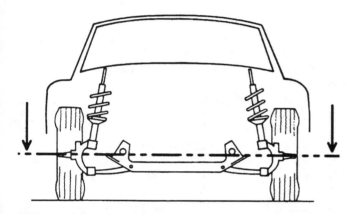

Figure 11–32 Vehicle weight tends to lower the spindle, which returns the steering to the straight-ahead position. (Courtesy of Hunter Engineering Company)

Figure 11–33). Vehicles with as many as 11 degrees positive caster usually use a steering dampener to control possible shimmy at high speeds and to dampen the snapback of the spindle after a turn (see Figure 11–34).

c. If caster is negative, or excessively unequal, the vehicle will not be as stable and will tend to

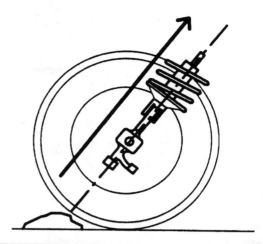

Figure 11–33 High caster provides a road shock path to the vehicle. (Courtesy of Hunter Engineering Company)

wander (constant steering wheel movement will be required to maintain straight-ahead direction). If a vehicle is heavily loaded in the rear, caster increases (as shown in Figure 11–35).

6. Caster can cause pull if unequal; **the vehicle will pull toward the side with the least caster.** However, the pulling force of unequal caster is only about one-fourth the pulling force of camber. It would require a difference of caster of one full degree to equal the pulling force of only 1/4 degree difference of camber.

7. Caster is *not* adjustable on many vehicles.

8. If caster is adjustable, it is changed by moving either the lower or the upper pivot point forward or backward by means of:
   a. Shims
   b. Eccentric cams
   c. Slots
   d. Strut rods

9. Caster should be equal on both sides; however, if caster cannot be adjusted to be exactly equal, make certain that there is more caster on the right side (maximum 1/2 degree difference) to help compensate for the crown of the road.

STEERING DAMPER

Figure 11–34 Steering damper is used on many pickup trucks, sport utility vehicles (SUVs), and many luxury vehicles designed with a high-positive caster setting. The damper helps prevent steering wheel kickback when the front tires hit a bump or hole in the road and also helps reduce steering wheel shimmy that may result from the high-caster setting. (Courtesy of Hunter Engineering Company)

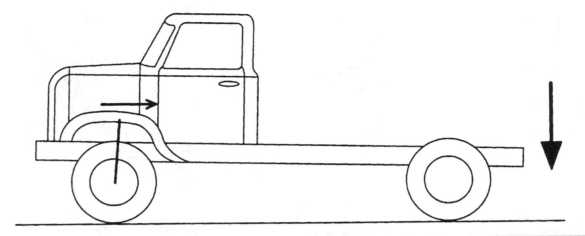

Figure 11–35 As the load increases in the rear of a vehicle, the top steering axis pivot point moves rearward increasing positive (+) caster. (Courtesy of Hunter Engineering Company)

TECH TIP ✔

Caster Angle Tire Wear

The caster angle is generally considered to be a *non*-tire-wearing angle. However, excessive or unequal caster can *indirectly* cause tire wear. When the front wheels are turned on a vehicle with a lot of positive caster, they become angled. This is called *camber roll*. (Caster angle is a measurement of the difference in camber angle from when the wheel is turned inward compared to when the wheel is turned outward.) Most vehicle manufacturers have positive caster designed into the suspension system. This positive caster increases the directional stability.

However, if the vehicle is used exclusively in city driving, positive caster can cause tire wear to the outside shoulders of both front tires (as seen in Figure 11–36).

NOTE: Caster is only measured on the front turning wheels of the vehicle. Although some caster is built into the rear suspension of many vehicles, rear caster is not measured as part of a four-wheel alignment.

■ STEERING AXIS INCLINATION (SAI)

The steering axis is the angle formed between true vertical and an imaginary line drawn between the upper and lower pivot points of the spindle (see Figure 11–37). Steering axis inclination (SAI) is the inward tilt of the steering axis. SAI is also known as king pin inclination (KPI) and is the imaginary line drawn through the king pin as viewed from the front. SAI is also called ball joint inclination (BJI), if SLA-type suspension is used, or MacPherson strut inclination (MSI).

The purpose of SAI is to provide an upper suspension pivot location that causes the spindle to travel in an arc when turning, which tends to raise the vehicle, as shown in Figure 11–38.

Vehicle weight tends to keep the front wheels in a straight-ahead position when driving, thereby increasing vehicle stability, directional control, and steering wheel returnability. The greater the SAI, the more stable the vehicle. It also helps center the steering wheel after making a turn and reduces the need for excessive positive caster. The SAI/KPI angle of all vehicles ranges between 2 degrees and 16 degrees. Front-wheel-drive vehicles usually have greater than 9 degrees SAI (typically 12°–16°) for directional stability, whereas rear-wheel-drive vehicles usually have less than 8 degrees of SAI. The steering axis inclination angle and the camber angle together are called the *included angle*.

■ INCLUDED ANGLE

The included angle is the SAI added to the camber reading of the front wheels only. *The included angle is determined by the design of the steering knuckle, or strut construction* (see Figures 11–39 and 11–40 on page 296).

Included angle is an important angle to measure for diagnosis of vehicle handling or tire wear problems. For example, if the cradle is out of location due to previous service work or an accident, knowing SAI, camber, and the included angle can help in determining what needs to be done to correct the problem (as shown in Figure 11–41 on page 296). (See Chapter 12 for complete alignment angle diagnosis and correction.)

If the included angles are equal side to side, but the camber is unequal on both sides, then the SAI must be unequal. For best handling, the included angle should be within 1/2 degree of the SAI of the other side of the vehicle.

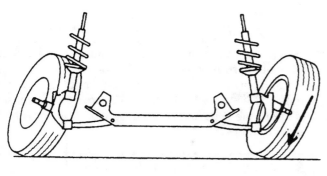

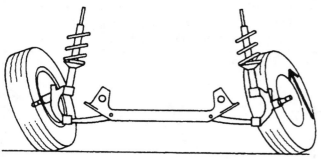

OUTSIDE TURN
SPINDLE MOVES DOWN

INSIDE TURN
SPINDLE MOVES UP

Figure 11–36 Note how the front tire becomes tilted as the vehicle turns a corner with positive caster. The higher the caster angle, the more the front tires tilt causing camber-type tire wear. (Courtesy of Hunter Engineering Company)

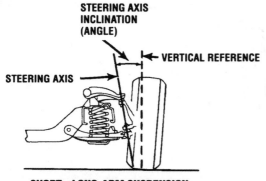

SHORT - LONG-ARM SUSPENSION

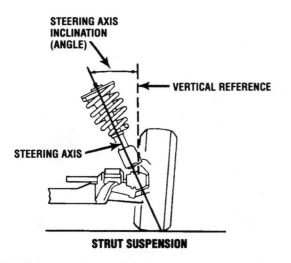

STRUT SUSPENSION

Figure 11–37 The top illustration shows that the steering axis inclination angle is determined by drawing a line through the center of the upper and lower ball joints. This represents the pivot points of the front wheels when the steering wheel is rotated during cornering. The lower illustration shows that the steering axis inclination angle is determined by drawing a line through the axis of the upper strut bearing mount assembly and the lower ball joint. (Courtesy of Oldsmobile)

■ SCRUB RADIUS

Scrub radius refers to the *distance* between the line through the steering axis and the centerline of the wheel at the contact point with the road surface (see Figure 11–42 on page 297).

Scrub radius is *not* adjustable and cannot be measured. Scrub radius can be zero, positive, or negative. **Zero** scrub radius means that the line through the steering axis intersects the centerline of the tire at the road surface. **Positive** scrub radius means that the line intersects the centerline of the tire below the road surface. **Negative** scrub radius means that the line intersects the centerline of the tire above the road surface.

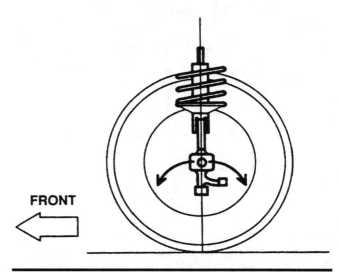

FRONT

Figure 11–38 The SAI causes the spindle to travel in an arc when the wheels are turned. The weight of the vehicle is therefore used to help straighten the front tires after a turn and to help give directional stability. (Courtesy of Hunter Engineering Company)

Scrub radius is also called **steering offset** by some vehicle manufacturers. If a wheel is permitted to roll rather than pivot, then steering will be more difficult because a tire can pivot more easily than it can roll while turning the front wheels. If the point of intersection is inside the centerline of the tire and below the road surface, this creates a toe-out force on the front wheels. Negative scrub radius is required on front-wheel-drive vehicles to provide good steering stability during braking (see Figures 11–43 and 11–44 on page 298).

Scrub radius is designed into each vehicle to provide acceptable handling and steering control under most conditions. Scrub radius also causes resistance to rolling of the front wheels to exert force on the steering linkage. This tends to dampen the effect of minor movements of the front wheels. Negative scrub radius causes the tire to toe in during acceleration, braking, or traveling over bumps.

> **NOTE:** It is this tendency to toe in caused by the negative scrub radius and engine torque that requires many front-wheel-drive vehicles to specify a toe-out setting for the front-drive wheels.

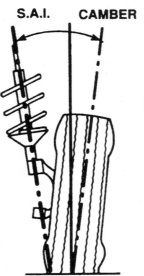

S.A.I. + CAMBER = INCLUDED ANGLE

Figure 11–39 Included angle on a MacPherson strut–type suspension. (Courtesy of Hunter Engineering Company)

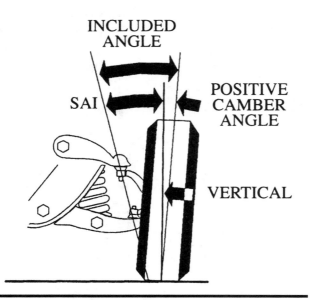

Figure 11–40 Included angle on a SLA-type suspension. The included angle is the SAI angle and the camber angle added together. If the camber angle is negative (−) (tire tilted inward at the top), the camber is subtracted from the SAI angle to determine the included angle. (Courtesy of Ford Motor Company)

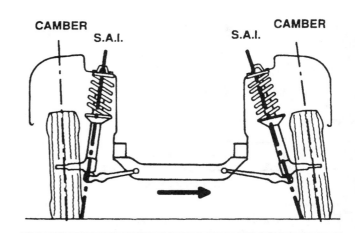

Figure 11–41 Cradle placement. If the cradle is not replaced in the exact position after removal for a transmission or clutch replacement, the SAI, camber, and included angle will not be equal side-to-side. (Courtesy of Hunter Engineering Company)

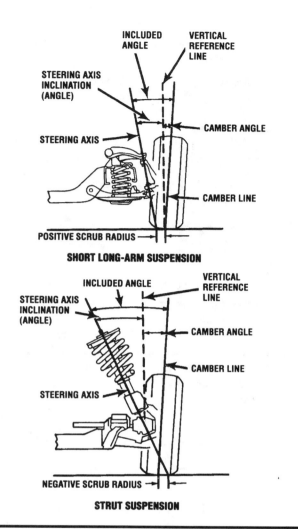

Figure 11–42 A positive scrub radius is usually built into most SLA front suspensions, and a negative scrub radius is usually built into most MacPherson strut–type front suspensions. (Courtesy of Oldsmobile)

Zero scrub radius is acceptable; positive scrub radius is less desirable because it causes the wheel to toe out during acceleration, braking, or traveling over bumps and causes instability. Positive scrub radius is commonly used on rear-wheel-drive vehicles and requires a toe-in setting to help compensate for the tendency to toe out.

A bent spindle can cause a change in the scrub radius and could cause hard steering, wander, or pull.

Also, changing tire or wheel sizes can affect the centerline location of the wheel or the height of the tire assembly and will change the scrub radius, which can negatively affect the steering control. When larger diameter tires and positive offset wheels are installed, the scrub radius becomes positive and the wheels tend to toe out causing wander, poor handling, and tire wear.

■ TURNING RADIUS (TOE-OUT ON TURNS)

Whenever a vehicle turns a corner, the inside wheel has to turn at a sharper angle than the outside wheel because the inside wheel has a shorter distance to travel (see Figure 11–45).

Turning radius is also called **toe-out on turns,** abbreviated **TOT** or **TOOT,** and is determined by the angle of the steering knuckle arms. **Turning radius is a nonadjustable angle.** The turning radius can and should be measured as part of an alignment to check if the steering arms are bent or damaged. Symptoms of out-of-specification turning angle include:

1. Tire squeal noise during normal cornering, even at low speeds
2. Scuffed tire wear

The proper angle of the steering arms is where imaginary lines drawn from the steering arms should intersect exactly at the center of the rear axle (see Figure 11–46 on page 299). This angle is called the Ackerman effect (named for its promoter, an English publisher, Rudolph Ackerman, circa 1898).

■ SETBACK

Setback is the angle formed by a line drawn perpendicular (at 90 degrees) to the front axles (see Figure 11–47 on page 299).

Setback is a nonadjustable measurement, even though it may be corrected. Positive setback means the right front wheel is set back farther than the left; negative setback means the left front wheel is set back farther than the right.

Setback can be measured with a four-wheel alignment machine or can be determined by measuring the wheel base on both sides of the vehicle.

> **NOTE:** The wheel base of any vehicle is the distance between the center of the front wheel and the center of the rear wheel on the same side. The wheel base should be within 1/8″ (3 mm) side to side.

The causes of setback include:

1. Cradle placement not correct on a front-wheel-drive vehicle. This can be caused by incorrectly installing the cradle after a transmission, clutch or engine replacement, or service (see Figure 11–48 on page 299).
2. An accident that affected the frame or cradle of the vehicle and was unnoticed or not repaired.

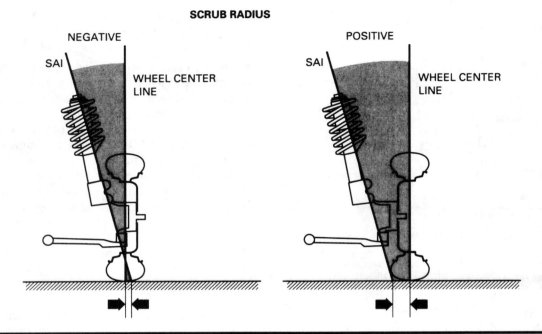

Figure 11–43 With negative scrub radius, the imaginary line through the steering axis inclination (SAI) intersects the road outside of the centerline of the tire. With positive scrub radius, the SAI line intersects the road inside the centerline of the tires.

Figure 11–44 With a positive scrub radius, the pivot point, marked with a + mark, is inside the centerline of the tire and will cause the wheel to turn toward the outside especially during braking. Zero scrub radius does not create any force on the tires and is not usually used on vehicles because it does not create an opposing force on the tires, which in turn makes the vehicle more susceptible to minor bumps and dips in the road. Negative scrub radius, as is used with most front-wheel-drive vehicles, generates an inward force on the tires.

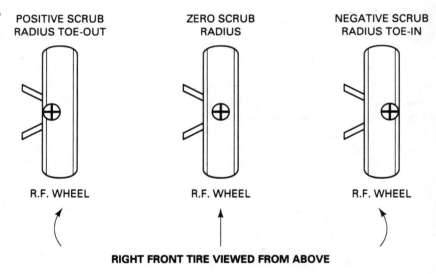

Figure 11–45 To provide handling, the inside wheel has to turn at a greater turning radius than the outside wheel.

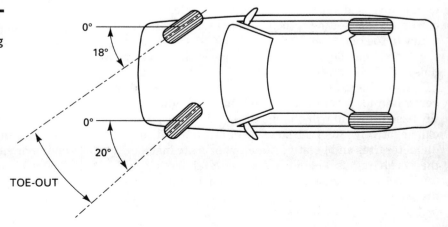

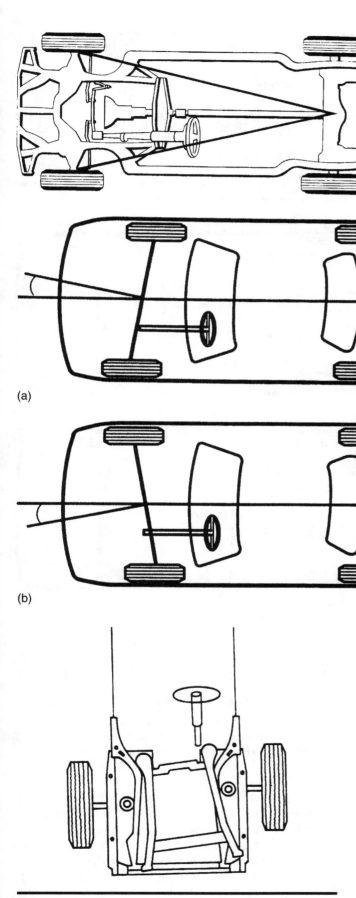

Figure 11–46 The proper toe-out on turns is achieved by angling the steering arms. (Courtesy of Hunter Engineering Company)

Figure 11–47 (a) Positive setback. (Courtesy of Hunter Engineering Company) (b) Negative setback. (Courtesy of Hunter Engineering Company)

(a)

(b)

Figure 11–48 Cradle placement affects setback. (Courtesy of Hunter Engineering Company)

Most vehicle manufacturers do not specify a minimum setback specification. However, a reading of 0.50 degree or 0.5″ (13 mm) or less of setback is generally considered to be acceptable.

■ THRUST ANGLE

Thrust angle is the angle of the rear wheels as determined by the total rear toe. If both rear wheels have zero toe, then the thrust angle is the same as the geometric centerline of the vehicle. The total of the rear toe setting determines the **thrust line,** or the direction the rear wheels are pointed (see Figure 11–49).

On vehicles with an independent rear suspension, if both wheels do not have equal toe, the vehicle will pull in the direction of the side with the most toe-in. (See Chapter 12 for details on performing a thrust line alignment.)

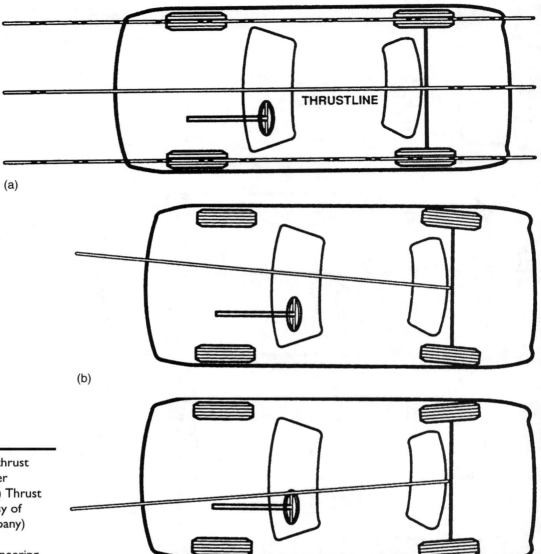

(a)

(b)

(c)

Figure 11–49 (a) Zero thrust angle. (Courtesy of Hunter Engineering Company) (b) Thrust line to the right. (Courtesy of Hunter Engineering Company) (c) Thrust line to the left. (Courtesy of Hunter Engineering Company)

■ TRACKING

The rear wheels should track directly behind the front wheels. If the vehicle has been involved in an accident, it is possible that the frame or rear axle mounting could cause dog tracking.

To check the frame for possible damage, two diagonal measurements of the frame and/or body are required. The diagonal measurements from known points at the front and the rear should be within 1/8″ (3 mm) of each other (see Figure 11–50).

■ FOUR-WHEEL ALIGNMENT

Four-wheel alignment refers to the checking and/or adjustment of all four wheels. Four-wheel alignment is important for proper handling and tire wear, to check the camber and the toe of the rear wheels of front-wheel-drive vehicles. Some rear-wheel-drive vehicles, equipped with independent rear suspension, can be adjusted for camber and toe. Rear-wheel caster cannot be measured or adjusted because to measure caster, the wheels must be turned from straight ahead. Since rear wheels are securely attached, a caster *sweep* (turning the wheels to take a caster reading) is not possible. While rear camber can cause tire wearing problems, by far the greatest tire wear occurs due to toe settings. *Unequal* toe in the rear can cause the vehicle to pull or lead. The rear camber and toe are always adjusted first before adjusting the front caster, camber, and toe. This procedure ensures that the thrust line and centerline of the vehicle are the same.

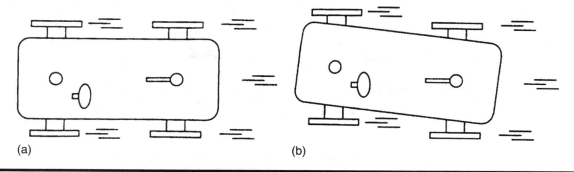

Figure 11–50 (a) Proper tracking. (Courtesy of Hunter Engineering Company) (b) Front wheels steering toward thrust line. (Courtesy of Hunter Engineering Company)

■ SUMMARY

1. The need for a wheel alignment results from wear or damage to suspension and steering components.

2. Low or unequal tire pressures can often cause symptoms such as wander, pull, and excessive tire wear.

3. Camber is both a pulling angle (if not equal side to side) as well as a tire wearing angle (if not set to specifications).

4. Incorrect camber can cause tire wear and pulling if camber is not within 1/2 degree from one side to the other.

5. Toe is the most important alignment angle because toe is usually the first requiring correction. When incorrect, toe causes severe tire wear.

6. Incorrect toe causes excessive tire wear and creates instability if not within specifications.

7. Caster is the basic stability angle, yet it does not cause tire wear (directly) if not correct or equal side to side.

8. SAI and included angle (SAI and camber added together) are important diagnostic tools.

9. If the toe-out on turns (TOOT) reading is not within specifications, a bent steering spindle (steering knuckle) is the most likely cause.

10. A four-wheel alignment includes aligning all four wheels of the vehicle; a thrust line alignment sets the front toe equal to the thrust line (total rear toe) of the rear wheels.

■ REVIEW QUESTIONS

1. Explain the three basic alignment angles of camber, caster, and toe.

2. Describe what happens to tire wear and vehicle handling if toe, camber, and caster are out of specification or *not* equal side to side.

3. Explain how knowing SAI, TOOT, and included angle can help in the correct diagnosis of an alignment problem.

4. Explain what thrust angle means.

■ ASE CERTIFICATION-TYPE QUESTIONS

1. Technician A says that upper control arms are not part of the steering system and, therefore, cannot cause play in the steering. Technician B says that a defective universal joint between the steering column and the steering gear box stub shaft can cause excessive steering wheel play. Which technician is correct?
   a. Technician A only
   b. Technician B only
   c. Both Technician A and B
   d. Neither Technician A nor B

2. Technician A says that a vehicle will pull (or lead) to the side with the most camber (or least negative camber). Technician B says that a vehicle will pull (or lead) to the side with the most front toe. Which technician is correct?
   a. Technician A only
   b. Technician B only
   c. Both Technician A and B
   d. Neither Technician A nor B

3. Technician A says that the vehicle will pull to the side with the most toe in the rear. Technician B says that the rear toe angle determines the thrust angle. Which technician is correct?
   a. Technician A only
   b. Technician B only
   c. Both Technician A and B
   d. Neither Technician A nor B

4. Strut rods adjust
   a. Toe
   b. Camber
   c. Caster
   d. Toe-out on turns

5. If metal shims are used for alignment adjustment in the front, they adjust
   a. Camber
   b. Caster
   c. Toe
   d. a and b only

6. Technician A says that as wear occurs, camber usually becomes negative. Technician B says that as steering linkage wear occurs, toe usually becomes toe-out from an original toe-in specification. Which technician is correct?
   a. Technician A only
   b. Technician B only
   c. Both Technician A and B
   d. Neither Technician A nor B

Use the following information to answer question 7:

Specifications:	Min.	Preferred	Max.
Camber (degree)	0	1.0	1.4
Caster (degree)	0.8	1.5	2.1
Toe (inch)	−0.10	0.06	0.15

Results:	L	R
Camber (degree)	−0.1	0.6
Caster (degree)	1.8	1.6
Toe (inch)	1.12	+0.12

7. The vehicle above will
   a. Pull toward the right and feather edge both tires
   b. Pull toward the left
   c. Wear the outside of the left tire and the inside of the right tire
   d. None of the above

Use the following information to answer questions 8 and 9:

Specifications:	Min.	Preferred	Max.
Camber (degree)	−1/4	+1/2	1
Caster (degree)	0	+2	+4
Toe (inch)	−1/16	1/16	3/16

Results:	L	R
Camber (degree)	−0.3	−0.1
Caster (degree)	3.6	1.8
Toe (inch)	−0.16	+0.32

8. The vehicle above will
   a. Pull toward the left
   b. Pull toward the right
   c. Wander
   d. Lead to the left slightly

9. The vehicle above will
   a. Wander
   b. Wear tires, but will not pull
   c. Will pull, but not wear tires
   d. Pull toward the left and cause feather-edge tire wear

10. Which alignment angle is most likely to need correction and cause the most tire wear?
    a. Toe
    b. Camber
    c. Caster
    d. SAI/KPI

Alignment Diagnosis and Service

Objectives: After studying Chapter 12, the reader should be able to:

1. List the many checks that should be performed before aligning a vehicle.
2. Describe the proper alignment setup procedure.
3. Explain how to correct for memory steer, torque steer, pull, drift (lead), and wander.
4. Describe the use of unit conversion and diagnostic charts.
5. Discuss tolerance alignment and how to check for accident damages.

Proper wheel alignment of all four wheels is important for the safe handling of any vehicle. When all four wheels are traveling the same path and/or being kept nearly vertical, tire life and fuel economy are maximized and vehicle handling is sure and predictable. A complete wheel alignment is a complex process that includes many detailed steps and the skill of a highly trained technician.

■ PRE-ALIGNMENT CORRECTION TECHNIQUES

There are four basic steps in the correction of any problem:

1. *Verify.* What, when, where, and to what extent does the problem occur?
2. *Isolate.* Eliminate known good parts and systems. Always start with the simple things first. For example, checking and correcting tire pressure and rotating tires should be the first things performed when trying to isolate the cause of an alignment-related problem.
3. *Repair the problem.* This step involves replacing any worn or damaged components and making sure that the alignment is within factory specifications (see Figure 12–1).
4. *Recheck.* **Always** test-drive the vehicle after making a repair. Never allow the customer to be the first to drive the vehicle after any service work.

Figure 12–1 The owner of this Honda thought that all it needed was an alignment. Obviously, something more serious than an alignment caused this left rear wheel to angle inward at the top.

Align and Replace at the Same Time

Magnetic bubble-type camber/caster gauges can be mounted directly on the hub or on an adapter attached to the wheel or spindle nut on front-wheel-drive vehicles (see Figure 12–2). Besides being used as an alignment setting tool, a magnetic alignment head is a great tool to use whenever replacing suspension components.

Any time a suspension component is replaced, the wheel alignment should be checked and corrected as necessary. An easy way to avoid having to make many adjustments is to use a magnetic alignment head on the front wheels to check camber with the vehicle hoisted in the air *before* replacing front components, such as new MacPherson struts. Then, before tightening all of the fasteners, check the front camber readings again to make sure they match the original setting. This is best done when the vehicle is still off the ground. For example, a typical front-wheel-drive vehicle with a MacPherson strut suspension may have a camber reading of +1/4° on the ground and +2° while on the hoist with the wheels off the ground. After replacing the struts, simply return the camber reading to +2° and it should return to the same +1/4° when lowered to the ground.

Although checking and adjusting camber before and after suspension service work does not guarantee a proper alignment, it does permit the vehicle to be moved around with the alignment fairly accurate until a final alignment can be performed.

Figure 12–2 Magnetic bubble-type camber/caster gauge. To help it keep its strong magnetism, it is best to keep it stored stuck to a metal plate as shown here.

A thorough inspection of the steering, suspension, and tires should be performed *before* the alignment of the vehicle is begun. (See Chapters 3, 8, and 10 for inspection procedures and test results.)

■ PRE-ALIGNMENT CHECKS

Before checking or adjusting the front-end alignment, the following items should be checked and corrected if necessary:

1. Check all the tires for proper inflation pressures. Tires should be approximately the same size and tread depth, and the recommended size for the vehicle (see Figure 12–3).

> **NOTE:** Some alignment technicians think that the vehicle must have new tires installed before an accurate alignment can be performed. Excessively worn tires, especially if only one tire is worn, can cause the vehicle to lean slightly. It is this unequal ride height that is the important fact to consider. If, for example, all four tires are equally worn, then the vehicle *can* be properly aligned. (Obviously, excessively worn tires should be replaced, and it would be best to align the vehicle with the replacement tires installed to be assured of an accurate alignment.)

2. Check the wheel bearings for proper adjustment.
3. Check for loose ball joints or torn ball joint boots.
4. Check the tie rod ends for damage or looseness.
5. Check the center link or rack bushings for play.
6. Check the pitman arm for any movement.
7. Check for runout of the wheels and the tires (see Chapter 3).
8. Check for vehicle ride height (should be level front to back as well as side to side). Make sure that the factory load leveling system is functioning correctly, if the vehicle is so equipped. Check height according to manufacturers' specifications (see Figure 12–4).

> **NOTE:** Manufacturers often have replacement springs or spring spacers that can be installed between the coil spring and the spring seat to restore proper ride level.

9. Check for steering gear looseness at the frame.
10. Check for improperly operating shock absorbers.
11. Check for worn control arm bushings.
12. Check for loose or missing stabilizer bar attachments.
13. Check the trunk for excess loads (see Figure 12–5).
14. Check for dragging brakes.

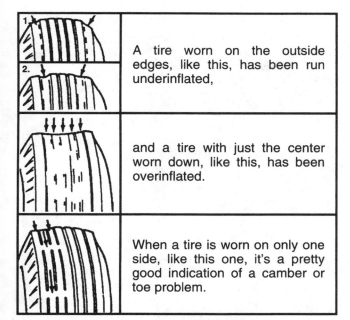

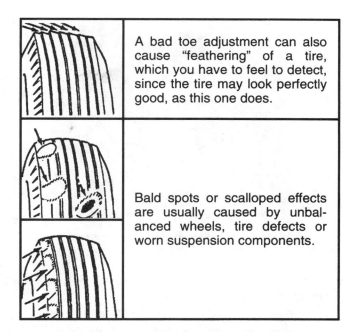

| | A tire worn on the outside edges, like this, has been run underinflated, |
| A bad toe adjustment can also cause "feathering" of a tire, which you have to feel to detect, since the tire may look perfectly good, as this one does. |

and a tire with just the center worn down, like this, has been overinflated.

When a tire is worn on only one side, like this one, it's a pretty good indication of a camber or toe problem.

Bald spots or scalloped effects are usually caused by unbalanced wheels, tire defects or worn suspension components.

Figure 12–3 Typical tire wear chart as found in a service manual. Abnormal tire wear usually indicates a fault in a steering or suspension component that should be corrected or replaced before an alignment is performed. (Courtesy of Ford Motor Company)

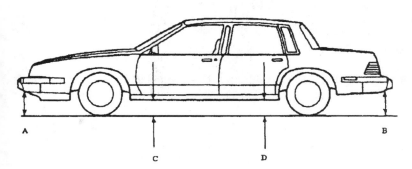

Figure 12–4 Measuring points for ride (trim) height vary by manufacturer. (Courtesy of Hunter Engineering Company)

Figure 12–5 Heavy or unequal loads can affect alignment angles. (Courtesy of Hunter Engineering Company)

NOTE: Checking for dragging brakes is usually performed when installing alignment heads to the wheels prior to taking an alignment reading. A brake dragging can cause the vehicle to pull or lead toward the side with the dragging brake.

■ LEAD/PULL

Diagnosis

Many alignment requests come from customers attempting to have a lead or pull condition corrected. Before aligning the vehicle, verify the customer complaint first, then perform a careful inspection.

Figure 12–6 The bulge in this tire was not noticed until it was removed from the vehicle as part of a routine brake inspection. After replacing this tire, the vehicle stopped pulling and vibrating.

1. Inspect all tires for proper inflation. Both tires on the same axle (front and rear) should be the same size and brand (see Chapter 2). A lead/pull problem could be due to a defect or condition in one or more of the tires, as shown in Figure 12–6. Before attempting to correct the lead/pull condition with changing alignment angles, try rotating the tires front to back or side to side.
2. Road-test the vehicle on a straight, level road away from traffic, if possible. Bring the vehicle to about 40 mph (65 km/h) and shift into neutral and feel for a **pull** in the steering, either to the left or to the right. A **lead** or **drift** is less severe than a pull and may occur only if you momentarily remove your hands from the steering wheel while driving.
3. If the lead/pull problem is sometimes toward the left and other times toward the right, check for a **memory steer** condition. If the lead/pull problem occurs during acceleration and deceleration, check for a **torque steer** condition.

■ MEMORY STEER

Diagnosis

Memory steer is a term used to describe the lead or pull of a vehicle caused by faults in the steering or suspension system. Often a defective upper strut bearing or steering gear can cause a pulling condition in one direction after making a turn in the same direction. It is as if the vehicle had a memory and pulled in the same direction. To test for memory steer, follow these simple steps during a test drive:

1. With the vehicle stopped at an intersection or in a parking area, turn the steering wheel completely to the left stop, and then straighten the wheel without going past the straight-ahead position.

Diagnostic Story

The Five-Wheel Alignment

The steering wheel should always be straight when driving on a straight, level road. If the steering wheel is not straight, the customer will often think that the wheel alignment is not correct. One such customer complained that the vehicle pulled to the right while driving on a straight road. The service manager test-drove the vehicle and everything was perfect, except that the steering wheel was not perfectly straight, even though the toe setting was correct. Whenever driving on a straight road, the customer would "straighten the steering wheel" and, of course, the vehicle went to one side. After adjusting toe with the steering wheel straight, the customer and the service manager were both satisfied. The technician learned that regardless of how accurate the alignment, the steering wheel *must* be straight; it is this "fifth wheel" that the customer notices most.

NOTE: Many vehicle manufacturers now include the maximum allowable steering wheel angle variation from straight. This specification is commonly ±3° (plus or minus 3 degrees) or less.

2. Lightly accelerate the vehicle and note any tendency of the vehicle to lead or pull toward the left.
3. Repeat the procedure, turning the steering wheel to the right.

If the vehicle first pulls to the left, then pulls to the right, the vehicle has a memory steer condition.

Correction

A binding suspension or steering component is the most likely cause of memory steer. Disconnect each wheel from its tie rod end and check for free rotation of movement of each wheel. Each front wheel should rotate easily without binding or roughness. Repair or replace components as necessary to eliminate the binding condition.

NOTE: One of the most common causes of memory steer is the installation of steering or suspension components while the front wheels were turned. Most steering and suspension parts contain rubber, which has a memory. If the memory steer condition is only in one direction, then this is the most likely cause. The rubber component exerts a force on the suspension or steering that causes the vehicle to pull toward the side that the wheels were turned toward when the part was installed.

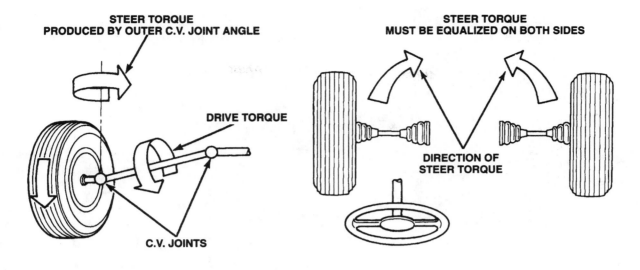

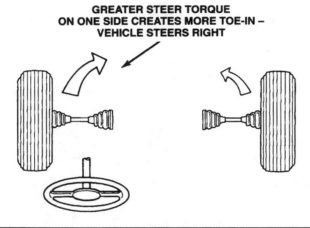

Figure 12–7 Equal outer CV joint angles produce equal steer torque (toe-in). If one side receives more engine torque, that side creates more toe-in and the result is a pull toward one side especially during acceleration. (Courtesy of General Motors)

■ TORQUE STEER

Diagnosis

Torque steer occurs in front-wheel-drive vehicles when engine torque causes a front wheel to change its angle from straight ahead (see Figure 12–7). This resulting pulling effect is most noticeable during rapid acceleration, especially whenever upshifting of the transmission creates a sudden change in torque being applied to the front wheels. When turning and accelerating at the same time, torque steer has a tendency to straighten the vehicle, so more steering effort may be required to make the turn. Then, if the accelerator is released, a reversing force is applied to the front wheels. Now the driver must take corrective steering motions to counteract the change in steering effect of engine torque. To summarize:

Torque to wheel	a toe-in condition
More torque	more toe-in
Unequal torque	unequal toe-in
Unequal drive shaft angles	unequal torque to the wheels

Most manufacturers try to reduce torque steer in the design of their vehicles by keeping drive axle angles low and equal side to side. If the engine and transaxle are level and the drive axle shafts are kept level, then the torque from the engine will be divided equally between the front wheels.

Correction

The service technician cannot change the design of a vehicle, but the technician can, and should, check and correct problems that often cause torque steer. **Check to be sure that the condition is not normal.** It is normal for front-wheel-drive vehicles to exert a tug on the steering wheel and steer toward one side (usually to the right) during acceleration. This is especially noticeable when the

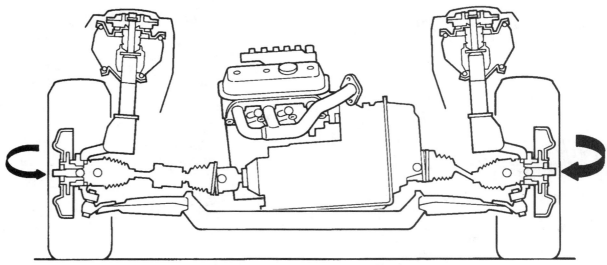

UNEQUAL DRIVE ANGLES

CAUSED BY ENGINE ROLL
PRODUCE STRONGER
LEFT-WHEEL STEER TORQUE

Figure 12–8 Broken or defective engine or transaxle mounts can cause the power train to sag causing unequal drive axle shaft CV joint angles. (Courtesy of Oldsmobile)

transmission shifts from first to second gear under heavy acceleration. To determine how severe the problem is, place a strip of masking tape at the top of the steering wheel. Drive the vehicle and observe the amount of movement required to steer the vehicle straight during heavy acceleration. Repeat the test with a similar make and model vehicle. If the torque steer is excessive, determine and correct the cause by carefully following the prealignment inspection and checking for a level power train.

A defective engine mount can cause the entire drivetrain to sag on one end. If the engine and transaxle of any front-wheel-drive vehicle is not level, the drive axle shaft angles will not be equal, as shown in Figure 12–8.

Hold a straightedge along the engine supporting frame and measure up to points along the transaxle pan rail or the drive axle shaft. Side-to-side distances should be equal. Standard alignment shims can be used to shim the mounts and level the drivetrain.

If torque steer is still excessive, check all alignment angles, including SAI and included angle. Unequal alignment angles can cause a pull or a lead condition. SAI and included angle should be within 1/2 degree (0.5 degree) side to side for best results. A vehicle will tend to pull toward the side with the least SAI.

■ ALIGNMENT SPECIFICATIONS

Before attempting any alignment:

1. Determine the make, model, and year of the vehicle.
2. Determine if the vehicle is equipped with power steering or manual steering. (Some older models

TECH TIP

Set Everything to Zero?

An apprentice service technician observed that the experienced alignment technician seldom looked at the specifications for the vehicle being aligned. When questioned, the technician said that for best tire life, the tires should rotate perpendicular to the road. After studying alignment specifications, the technician noticed that almost every camber and toe specification for both front and rear included zero within the range of the specifications. Caster, of course, varies from one vehicle to another and should be checked and adjusted to specifications. The beginning technician learned that zero camber and zero toe will be acceptable and within specifications on almost all vehicles, and is easy to remember!

use lower caster specifications for manual steering to reduce steering effort.)
3. Determine the correct specifications (if possible, check the specifications from two different sources to ensure correct readings).
4. Compensate for the lack of a full gas tank by placing an equal amount of weight in the luggage compartment. (Gasoline weighs 6 lb. per gallon [0.7 kg per l]—a 20-gallon gas tank, when full, weighs 120 lb. [80 l weigh 54 kg].)
5. Determine the correct specifications for the *exact* vehicle being checked.

ALIGNMENT SPECIFICATIONS AT CURB HEIGHT

FRONT WHEEL ALIGNMENT	ACCEPTABLE ALIGNMENT RANGE AT CURB HEIGHT	PREFERRED SETTING
CAMBER . . All*...............................	-0.6° to +0.6°	+0.0°
*Side To Side Differential	0.7° or less	0.0°
TOTAL TOE All Vehicles (See Note)		
Specified In Degrees	0.4° In to 0.0°	0.2° In
CASTER*...	REFERENCE ANGLE	
All Models...	+2.0° to +4.0°	+3.0°
*Side To Side Caster Differential		
Not to Exceed...............................	1.0° or less	0.0°
REAR WHEEL ALIGNMENT	**ACCEPTABLE ALIGNMENT RANGE AT CURB HEIGHT**	**PREFERRED SETTING**
CAMBER . . All Models.................................	-0.6° to +0.4°	-0.1°
TOTAL TOE* All Vehicles (See Note)		
Specified In Degrees	0.2° Out to 0.4° In	0.1° In
THRUST ANGLE...................................	-0.15° to +0.15°	
*TOE OUT When Backed On Alignment		
Rack Is TOE IN When Driving.		

NOTE: Total toe is the arithmetic sum of the left and right wheel toe settings. Positive is Toe-in, negative is Toe-out. Total Toe must be equally split between each front wheel to ensure a centered steering wheel. Left and Right toe must be equal to within 0.02 degrees.

Figure 12–9 Curb height is ride height or trim height as measured at the curb weight. Curb weight is when the vehicle has a full tank of fuel and all other fluids filled. This alignment chart specifies the acceptable range for each alignment setting as well as the preferred setting. Every effort should be made to set the alignment to the preferred setting. (Courtesy of Chrysler Corporation)

NOTE: Some alignment specifications are published as guidelines for acceptable values for state or local vehicle inspections. Be sure to use the *service* or *set to* specifications.

Reading Alignment Specifications

There are several methods used by vehicle manufacturers and alignment equipment manufacturers to specify alignment angles.

Maximum/Minimum/Preferred Method This method indicates the preferred setting for each alignment angle and the minimum and maximum allowable value for each. The alignment technician should always attempt to align the vehicle to the preferred setting (see Figure 12–9).

Plus or Minus Method This method indicates the preferred setting with the lowest and highest allowable value indicated by a negative (−) and positive (+) sign, as in Figure 12–10. For example, if a camber reading is specified as +1/2 degree with a + and − value of 1/2 degree, it could be written as +1/2° ±1/2°. The minimum value would be 0° (1/2° − 1/2° = 0°), and the maximum value would be +1° (+1/2° + 1/2° = 1°). The range would be from 0 degree to 1 degree.

NOTE: The angle is assumed positive unless labeled with a negative (−) sign in front of the number.

Degrees, Minutes, and Fractions

Specifications are often published in fractional or decimal degrees, or in degrees and minutes. There are 60 minutes (written as 60′) in 1 degree.

Angle-Unit Conversions

Units	Conversions		
Fractional degrees	1/4°	1/2°	3/4°
Decimal degrees	0.25°	0.50°	0.75°
Degrees and minutes	0°15′	0°30	0°45′

WHEEL ALIGNMENT SPECIFICATIONS

	CASTER	CROSS CASTER (LH-RH)	CAMBER	CROSS CAMBER (LH-RH)	TOE (TOTAL IN) DEGREES	STEERING WHEEL ANGLE	THRUST ANGLE
FRONT	+3°±.5°	0°±.75°	+.2° ±.5°	0°±.75°	0°±.3°	0°±3°	- -
REAR	- -	- -	-.3° ±.5°	0°±.75°	+.1° ±.2°	- -	0°±.1°

NT076

Figure 12–10 This alignment chart indicates the preferred setting with a plus or minus tolerance. (Courtesy of Oldsmobile)

TECH TIP

Keep the Doors Closed, but the Window Down

An experienced alignment technician became upset when a beginning technician opened the driver's door to lock the steering wheel in a straight-ahead position on the vehicle being aligned. The weight of the open door caused the vehicle to sag. This disturbed the level position of the vehicle and changed all the alignment angles.

The beginning technician learned an important lesson that day: Keep the window down on the driver's door so that the steering wheel and brakes can be locked without disturbing the vehicle weight balance by opening a door. The brake pedal must be locked with a pedal depressor to prevent the wheels from rolling as the wheels are turned during a caster sweep. The steering must be locked in the straight-ahead position when adjusting toe.

See Appendix 6 for complete unit conversions.

To help visualize the amount of these various units, think of decimal degrees as representing money or cents (100 cents = 1 dollar).

$$0.75 = 75 \text{ cents, or } 3/4 \text{ of a dollar}$$

Minutes can be visualized as minutes in an hour (60′ = 1 hour).

$$45' = 3/4 \text{ of an hour}$$

Now which is larger, 35′ or 0.40 degree? The larger angle is 35′ because this is slightly greater than 1/2 degree, whereas 0.40 degree is less than 1/2 degree.

Finding the Midpoint of Specifications

Many manufacturers specify alignment angles within a range. If you are using equipment that requires a midpoint to be entered, use the following method to determine easily the midpoint of specifications.

Example I Specification: 55′ to 2° 25′

Step I The first step is to determine the specification range or span, which is the total angle value from lowest to highest:

$$\begin{array}{r} 2° \ 25' \\ -\ 55' \\ \hline 1° \ 30' \text{ specification range} \end{array}$$

Step 2 Dividing the specification range by 2 will give the midpoint of the range:

$$\frac{1° \ 30'}{2} = \frac{90'}{2} = 45'$$

Step 3 To find the midpoint of the specifications, add the midpoint of the range to the smaller specification (or subtract from the larger specification):

$$\begin{array}{r} 45' \text{ midpoint of range} \\ +\ 55' \text{ lowest specification} \\ \hline 1° \ 40' \text{ midpoint of specification} \end{array}$$

Example 2 Specification: $-0.5°$ to $+0.80°$

Step I The total range of the specification is determined by adding 0.5 (1/2) to 0.80, totaling 1.30 degree.

> **NOTE:** Since the lower specification is a negative number, we had to *add* the 0.5 degree to bring the lower range of the specification to zero. Then the total is simply the upper range specification added to the number required to bring the lower end of the range to zero.

Step 2 Dividing the specification range by 2 gives the midpoint of the range:

$$\frac{1.30°}{2} = 0.65°$$

Step 3 To find the midpoint of the specification, add the midpoint of the range to the smaller specification:

$$\begin{array}{r} 0.65° \text{ midpoint of range} \\ +(-0.50°) \text{ lowest specification} \\ \hline +0.15° \text{ midpoint of specification} \end{array}$$

Figure 12–11 Drive onto the alignment rack as straight as possible with the turn plates positioned so that the center of the wheel is directly over the center of the turn plates.

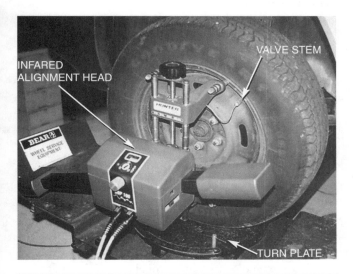

Figure 12–12 This wheel sensor has a safety wire that screws to the valve stem to keep the sensor from falling onto the ground if the clamps slip on the wheel lip.

■ ALIGNMENT SETUP PROCEDURES

After confirming that the tires and all steering and suspension components are serviceable, the vehicle is ready for an alignment. Setup procedures for the equipment being used must always be followed. Typical alignment procedures include:

Step 1 Drive onto the alignment rack straight and adjust the ramps and/or turn plates so that they are centered under the tires of the vehicle (see Figure 12–11).

Step 2 Use chocks for the wheels to keep the vehicle from rolling off the alignment rack.

Step 3 Attach and calibrate the wheel sensors to each wheel as specified by the alignment equipment

manufacturer (see Figure 12–12). The calibration procedure is required whenever the head of the machine is attached to the wheel of the vehicle. All alignment angles and measurements are taken from the readings of the wheel sensors. Calibration of these wheel sensors is needed for two reasons:

a. *The wheel may be bent.* If the wheel (rim) is bent, even slightly, this small amount of tilt would be read as an angle of the suspension.

> **NOTE:** Compensating or calibrating the wheel sensor for a bent wheel does *not correct* (or repair) the bent wheel! The bent wheel is still present and can result in a shimmy-type vibration.

b. *The sensor may not be identically installed on the wheel.* Most sensors use three or four wheel mounting locations. It is not possible to install all wheel sensors perfectly at the same depth on the wheel in all locations. If a sensor must be removed during the alignment process and reinstalled, it should be recalibrated. Always follow the manufacturer's recommended procedure for compensating the sensors.

Step 4 Unlock all rack or turn plates.

Step 5 Lower the vehicle and jounce the vehicle by pushing down on the front, then rear, bumper. This motion allows the suspension to become centered.

Step 6 Following the procedures for the alignment equipment, determine all alignment angles.

■ MEASURING CAMBER, CASTER, SAI, TOE, AND TOOT
Camber

Camber is measured with the wheels in the straight-ahead position on a level platform. Since camber is a

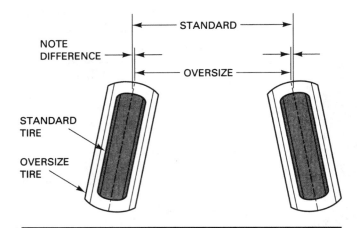

Figure 12–13 If toe for an oversize tire is set by distance, the toe angle will be too small. Toe angle is the same regardless of tire size.

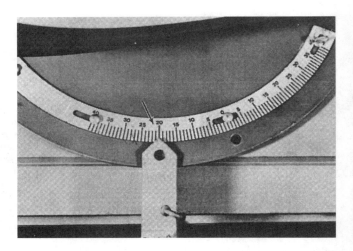

Figure 12–14 The protractor scale on the front turn plates allows the technician to test the turning radius by turning one wheel to an angle specified by the manufacturer and observing the angle of the other front wheel.

vertical reference angle, alignment equipment reads camber directly.

Caster

Caster is measured by moving the front wheels through an arc inward, then outward, from straight ahead. This necessary movement of the front wheels to measure caster is called *caster sweep*. What the alignment measuring equipment is actually doing is measuring the camber at one wheel sweep and measuring the camber again at the other extreme of the caster sweep. *The caster angle itself is the difference between the two camber readings.*

SAI

Steering axis inclination (SAI) is also measured by performing a caster sweep of the front wheels. While this angle can be read at the same time as caster on many alignment machines, most experts recommend that SAI be measured separately from the caster reading. When measuring SAI separately, the usual procedure involves raising the front wheels off the ground and leveling and locking the wheel sensors before performing a caster sweep. The reason for raising the front wheels is to allow the front suspension to extend to its full droop position. When the suspension is extended, the SAI is more accurately determined because the angle itself is expanded.

Toe

Toe is determined by measuring the angle of both front and/or both rear wheels from the straight-ahead (0°) position. Most alignment equipment reads the toe angle for each wheel *and* the combined toe angle of both wheels on the same axle. This combined toe is called **total toe.** Toe angle is more accurate than the center-to-center distance, especially if oversize tires are installed on the vehicle (see Figure 12–13).

Toot

Toe-out on turns (TOOT) is a diagnostic angle and is normally not measured as part of a regular alignment, but it is recommended to be performed as part of a total alignment check. TOOT is measured by recording the angle of the front wheels as indicated on the front turn plates (see Figure 12–14).

If, for example, the inside wheel is turned 20 degrees, then the outside wheel should indicate about 18 degrees on the turn plate. The exact angles are usually specified by the vehicle manufacturer. The turning angle should be checked only after the toe is correctly set. *The turning angle for the wheel on the outside of the turn should not vary more than 1.5 degrees from specifications.* For example, if the specification calls for the right wheel to be steered into the turn 20°, the outside wheel should measure 18 degrees. This should be within 1.5 degrees (16.5° to 19.5°). If the TOOT is not correct, a bent steering arm is the usual cause. If TOOT is not correct, tire squealing noise is usually noticed while cornering and excessive tire wear may occur.

NOTE: Some front-wheel-drive vehicles use a non-symmetrical (unequal) turning angle design. The design is found on various makes and models of vehicles to assist in controlling torque steer. The test procedure is the same except that the turning angle specifications include left wheel and right wheel angles when turned inward and outward.

SPECIFICATIONS VERSUS ALIGNMENT READINGS

Secure both the alignment specifications from the manufacturer and the alignment readings and compare the two. If the specifications and the alignment machine are using two different units, use the unit conversion charts in Appendix 6 to convert to the same units. Before start-ing an alignment, the smart technician checks the SAI, included angle, setback, and toe-out on turns to make sure that there is no hidden damage such as a bent spindle or strut that was not found during the pre-alignment inspection. *Setback is also a diagnostic angle and should be less than 0.5" (13 cm or 1/2°).* If setback is greater than 0.5" (13 cm or 1/2°), check the body, frame, and cradle for accident damage or improper alignment.

NOTE: If the SAI or included angle is unequal—suggesting a possible problem such as a bent strut—check the front and rear toe readings. Some alignment equipment cannot show accurate SAI readings if the front or rear toe readings are not within specifications. If the front and rear toe readings are okay and the alignment readings indicate a bent strut, go ahead with the diagnosis and correction as explained later in this chapter.

CHECKING FOR BENT STRUTS, SPINDLES, OR CONTROL ARMS

Even a minor bump against a curb can bend a spindle or a strut housing (see Figure 12–15).

Before attempting to correct an alignment, check all the angles and use the appropriate diagnostic chart to check for hidden damage that a visual inspection may miss.

The following charts (Figures 12–16 through 12–20 on pages 314–315) can be used to determine what is wrong if the alignment angles are known. Simply use the chart that correctly identifies the type of suspension on the problem vehicle.

At spec   The alignment angle is within specifications.
Over spec   The alignment angle is greater or higher than specified by the manufacturer.
Under spec The alignment angle is less than or lower than specified by the manufacturer.

CHECKING FRAME ALIGNMENT OF FRONT-WHEEL-DRIVE VEHICLES

Many front-wheel-drive vehicles mount the drivetrain (engine and transaxle) and lower suspension arms to a subframe or cradle. If the frame is shifted either left or right, this can cause differences in SAI, included angle and camber (see Figures 12–21 and 12–22 on page 316).

Adjust the frame if SAI and camber angles are different left and right side, yet the included angles are equal.

Figure 12–15 Notice the difference in angle of the strut and spindle between the original on the right and the new replacement strut housing on the left.

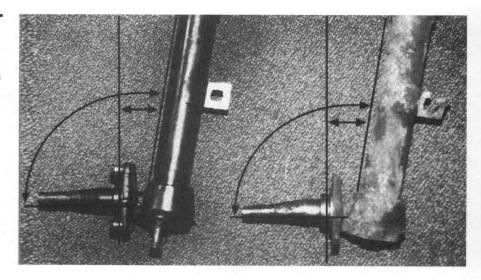

Strut-type suspension diagnostic chart

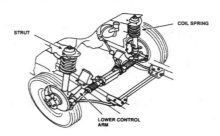

Figure 12–16 (Courtesy of Hunter Engineering Company)

SAI	Camber	Included Angle	Possible Cause
At spec	Under spec	Under spec	Bent spindle or strut
At spec	Over spec	Over spec	Bent spindle or strut
Under spec	Over spec	At spec	Bent transverse link or tower out at top
Under spec	Over spec	Over spec	Bent transverse link or strut tower out at top as well as bent spindle or strut
Under spec	Over spec	Under spec	Bent transverse link or strut tower out at top as well as bent spindle or strut
Over spec	Under spec	At spec	Strut tower in at top
Over spec	Over spec	Over spec	Strut tower in at top and bent spindle or strut

SLA-type suspension diagnostic chart

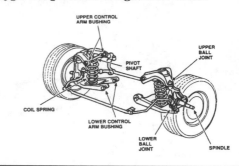

Figure 12–17 (Courtesy of Hunter Engineering Company)

SAI	Camber	Included Angle	Possible Cause
At spec	Under spec	Under spec	Bent spindle
Under spec	Over spec	At spec	Bent lower transverse link
Under spec	Over spec	Over spec	Bent lower transverse link and spindle
Over spec	Under spec	At spec	Bent upper transverse link

Multi-link-type suspension diagnostic chart

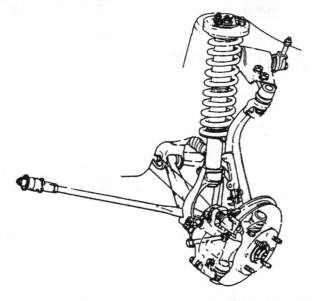

Figure 12–18 (Courtesy of Hunter Engineering Company)

SAI	Camber	Included Angle	Possible Cause
At spec	Under spec	Under spec	Steering knuckle
At spec	Over spec	Over spec	Steering knuckle
Under spec	Over spec	At spec	Bent transverse link, upper link out, or third link out at bottom
Under spec	Under spec	Under spec	Bent steering knuckle and either bent transverse link or upper link out
Under spec	At spec	Under spec	Bent steering knuckle and either bent transverse link or upper link out
Over spec	Under spec	At spec	Upper link in
Over spec	Over spec	Over spec	Upper link in and bent steering knuckle

Straight (mono) axle diagnostic chart

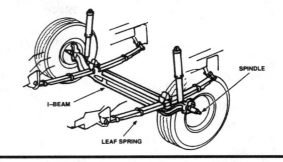

Figure 12–19 (Courtesy of Hunter Engineering Company)

SAI	Camber	Included Angle	Possible Cause
At spec	Over spec	Over spec	Bent spindle/ assembly
Over spec	Under spec	At spec	Bent axle housing
Under spec	Over spec	At spec	Bent axle housing
Under spec	Over spec	Over spec	Bent spindle and axle housing

Twin I-beam diagnostic chart

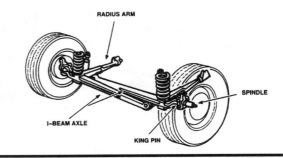

Figure 12–20 (Courtesy of Hunter Engineering Company)

SAI	Camber	Included Angle	Possible Cause
At spec	Over spec	Over spec	Bent spindle assembly
Over spec	Under spec	At spec	Bent I-beam
Under spec	Over spec	At spec	Bent I-beam
Under spec	Over spec	Over spec	Bent I-beam and spindle assembly

Figure 12–21 In this example, both SAI and camber are far from being equal side to side. However, both sides have the same included angle indicating that the frame may be out of alignment. An attempt to align this vehicle by adjusting the camber on both sides with either factory or aftermarket kits would result in a totally incorrect alignment. (Courtesy of Oldsmobile)

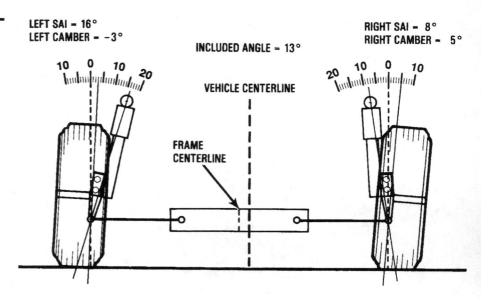

Figure 12–22 This is the same vehicle as shown in Figure 12–21, except now the frame (cradle) has been shifted over and correctly positioned. Notice how both the SAI and camber become equal without any other adjustments necessary. (Courtesy of Oldsmobile)

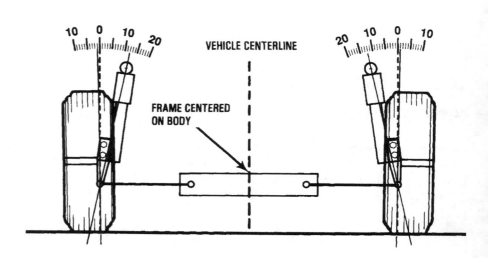

■ TYPES OF ALIGNMENTS

There are three types of alignment: geometric centerline, thrust line, and total four-wheel alignment.

Geometric Centerline

Until the 1980s, most wheel alignment concerned only the front wheels. Vehicles, such as sports cars, that had independent rear suspensions were often aligned by backing the vehicle onto the alignment rack and adjusting the rear camber and/or toe. This type of alignment was simply an alignment that uses the geometric centerline of the vehicle as the basis for all measurements of toe (front or rear). (See Figure 12–23.)

This method is now considered to be obsolete.

Thrust Line

A thrust line alignment uses the thrust angle of the rear wheels and sets the front wheels parallel to the thrust line (see Figure 12–24).

The thrust line is the bisector of rear total toe or the actual direction in which the rear wheels are pointed. The rear wheels of any vehicle *should* be pointing parallel to the geometric centerline of the vehicle. However, if the rear toe angles of the rear wheels do not total exactly zero (perfectly in a line with the centerline of the vehicle), a thrust condition exists. The front wheels will automatically steer to become parallel to that condition. A crooked steering wheel may also result from an improper thrust condition.

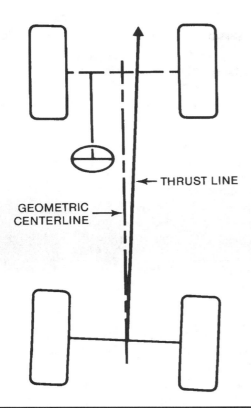

Figure 12–23 Geometric centerline–type alignment sets the front toe readings based on the geometric centerline of the vehicle and does not consider the thrust line of the rear wheel toe angles. (Courtesy of Hunter Engineering Company)

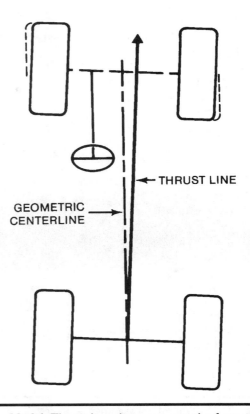

Figure 12–24 Thrust line alignment sets the front toe parallel with the rear wheel toe. (Courtesy of Hunter Engineering Company)

> **HINT:** It has often been said that while the front wheels steer the vehicle, the rear wheels determine the direction in which the vehicle will travel. Think of the rear wheels as a rudder on a boat. If the rudder is turned, the direction of the boat changes due to the angle change at the rear of the boat.

Thrust line alignment is *required* for any vehicle with a nonadjustable rear suspension. If a vehicle has an adjustable rear suspension, then a total four-wheel alignment is necessary to ensure proper tracking.

Total Four-Wheel Alignment

A total four-wheel alignment is the most accurate method and is necessary to ensure maximum tire wear and vehicle handling. The biggest difference between a thrust line alignment and a total four-wheel alignment is that the rear toe is adjusted to bring the thrust line to zero. In other words, the rear toe on both rear wheels is adjusted equally so that the actual direction in which the rear wheels are pointed is the same as the geometric centerline of the vehicle (see Figure 12–25).

The procedure for a total four-wheel alignment includes these steps:

1. Adjust the rear camber (if applicable).
2. Adjust the rear toe (this should reduce the thrust angle to near zero).
3. Adjust the front camber and caster.
4. Adjust the front toe, being sure that the steering wheel is in the straight-ahead position.

■ SAMPLE ALIGNMENT SPECIFICATIONS AND READINGS

The service technician must know not only all of the alignment angles but also the interrelationship that exists among the angles. As an aid toward understanding these relationships, two examples are presented: Example 1 gives the front-wheel angles that are acceptable when compared with the specifications; Example 2 gives four-wheel alignment and readings of a vehicle that are not within specifications.

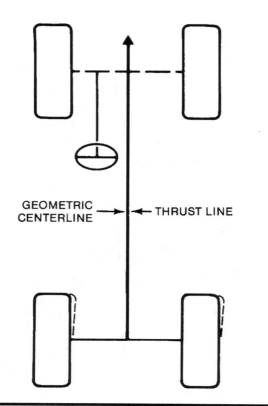

GEOMETRIC CENTERLINE → ← THRUST LINE

Figure 12–25 Four-wheel alignment corrects for any rear-wheel toe to make the thrust line and the geometric centerline of the vehicle both the same. (Courtesy of Hunter Engineering Company)

Example 1: Wheel Alignment

Alignment Specifications

	Left	Right
Camber =	+1/2° ±1/2°	+1/2° ±1/2°
Caster =	+ 1° ±1/2°	+ 1° ±1/2°
Toe (total) =	1/8″ ±1/16″	

Actual Reading

	Left	Right
Camber =	+1/4°	0°
Caster =	+ 1°	+1 1/4°
Toe (total) =	1/8″	

Answer

Alignment is perfect.

   a. No tire wear
   b. No pulling

Explanation

Camber is within specifications.

Camber is within 1/2 degree difference side-to-side.

Camber is not equal, but there is more camber on the left (1/4 degree) than on the right, thereby helping to compensate for the road crown.

Caster is within specifications.

Caster is within 1/2 degree difference side-to-side.

Caster is not equal, but there is more caster on the right (1/4 degree) than on the left, thereby helping (very slightly) to compensate for road crown.

Toe is within specifications.

Example 2: Four-Wheel Alignment

Specifications

	Left	Right
Front camber =	+1/4° ±1/2°	0° ±1/2°
Front caster =	+3° ±1/2°	+1/2° ±1/2°
Front toe (total) =	3/16″ ±1/16″	
Rear camber =	0° ±1/4°	0° ±1/4°
Rear toe =	0″ ±1/16″	

Actual Reading

	Left	Right
Front camber =	−1/2°	−1°
Front caster =	+2 3/4°	+2°
Front toe =	−1/8″ (toe-out)	
Rear camber =	0°	−1/4°
Rear toe =	+1/2° (toe-in)	

Answer

Alignment is incorrect.

   a. Left front tire will wear slightly on the inside edge.
   b. Right front tire will wear on the inside edge.
   c. Vehicle could tend to pull slightly to the left due to the camber difference (may not pull at all due to the pulling effect of the road crown).
   d. Vehicle could tend to pull slightly to the right due to the caster difference (3/4 degree more caster on the left).
   e. Overall pull could be slight toward the left because it requires four times the caster difference to have the same pulling forces as camber.
   f. Tires could wear slightly on both the inside edges due to toe-out.
   g. The negative camber of the present alignment puts a heavy load on the outer wheel bearing, if RWD, because the load is being carried by the smaller outer wheel bearing instead of the larger inner wheel bearing.

Conclusion

The vehicle would wear the inside edges of both front tires.

The vehicle may or may not pull slightly to the left.

The vehicle would not act as stable while driving—possible wander.

Figure 12–26 The rear camber is adjustable on this vehicle by rotating the eccentric cam and watching the alignment machine display.

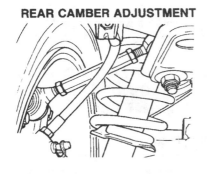

Figure 12–27 Some vehicles use a threaded fastener similar to a tie rod to adjust camber on the rear suspension. (Courtesy of FMC)

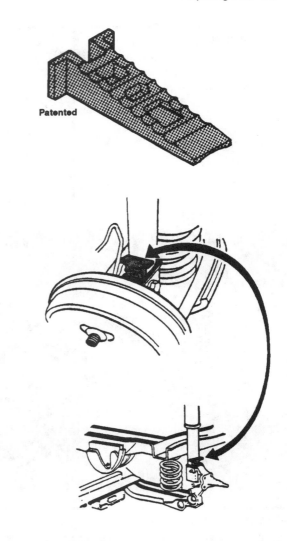

Figure 12–28 Aftermarket alignment parts or kits are available to change the rear camber. (Courtesy of Shimco International, Inc.)

■ ADJUSTING REAR CAMBER

Adjusting rear camber is the first step in the total four-wheel alignment process. Rear camber is rarely made adjustable, but can be corrected by using aftermarket alignment kits or shims. If rear camber is not correct, vehicle handling and tire life are affected. Before attempting to adjust or correct rear camber, carefully check the body and/or frame of the vehicle for accident damage, including:

1. Weak springs, torsion bars, or overloading (check ride height).
2. Bowed rear axle, trailing arm, or rear control arm.
3. Suspension mount or body dimension not in proper location.
4. Incorrectly adjusted camber from a previous repair.

The cause of the incorrect rear camber could be accident related, and the body or frame may have to be pulled into correct position. See Figures 12–26 through 12–28 for examples of various methods used to adjust rear camber.

Simple to Use

1. Take alignment readings and determine change needed.
2. Select proper shims using easy-to-read chart.
3. Using template provided, mark and remove tabs to create proper mounting bolt pattern.
4. Install shim.

■ ADJUSTING REAR TOE

Many vehicle manufacturers provide adjustment for rear toe on many vehicles that use an independent rear suspension. Rear toe adjusts the thrust angle. A thrust angle

Figure 12–29 Full contact plastic or metal shims can be placed between the axle housing and the brake backing plate to change rear camber or toe or both. (Courtesy of Northstar Manufacturing Company, Inc.)

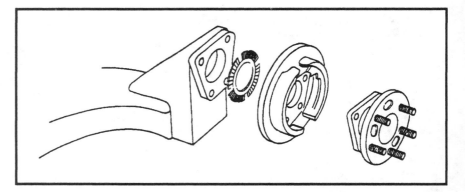

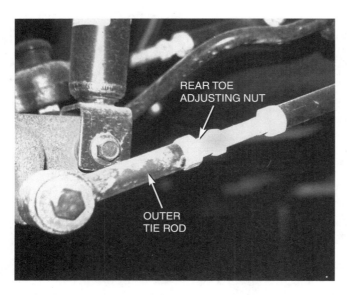

REAR TOE ADJUSTING NUT

OUTER TIE ROD

Figure 12–30 The rear toe was easily set on this vehicle. The adjusting nuts were easy to get to and turn. Rear toe is not this easy on every vehicle.

TECH TIP

The Gritty Solution

Many times it is difficult to loosen a torx bolt, especially those used to hold the backing plate onto the rear axle on many GM vehicles (see Figure 12–29).

A technique that always seems to work is to place some valve-grinding compound on the fastener. The gritty compound keeps the torx socket from slipping up and out of the fastener, and more force can be exerted to break loose a tight bolt. Valve-grinding compound can also be used on Phillips-head screws as well as on other types of bolts, nuts, and sockets.

method for the vehicle being aligned. Most solid rear axles do not have a method to adjust rear toe except for aftermarket shims or kits (see Figures 12–30 through 12–32 on pages 320–322).

NOTE: On vehicles equipped with four-wheel steering, refer to the service manual for the exact procedure to follow to lock or hold the rear wheels in position for a proper alignment check.

■ GUIDELINES FOR ADJUSTING FRONT CAMBER/SAI AND INCLUDED ANGLE

If the camber is adjusted at the base of the MacPherson strut, camber and included angle are changed and SAI remains the same (see Figure 12–33 on page 323).

If camber is adjusted by moving the upper strut mounting location, included angle remains the same, but SAI and camber change (see Figure 12–34 on page 323).

This is the reason to use the factory alignment methods before using an aftermarket alignment adjustment kit. SAI and included angle should be within 1/2 degree (0.5°) side to side. If these angles differ, check the frame mount location before attempting to correct differences in camber. As the frame is changed, camber and SAI change, but the included angle remains the same. Cross camber/caster means the difference between the camber or caster on one side of the vehicle and the camber or caster on the other side of the vehicle.

Alignment Angle	Recommended Maximum Variation
SAI	Within 1/2° (0.5°) side to side
Included angle	Within 1/2° (0.5°) side to side
Cross camber	Within 1/2° (0.5°) side to side
Cross caster	Within 1/2° (0.5°) side to side

CAUTION: Do not attempt to correct a pull condition by increasing cross camber or cross caster beyond the amount specified by the vehicle manufacturer.

that exceeds 1/2 degree (0.5°) on a vehicle with a solid axle is an indication that a component may be damaged or out of place in the rear of the vehicle. Rear toe is often adjusted using an adjustable tie rod end or an eccentric cam on the lower control arm. Check a service manual for the exact

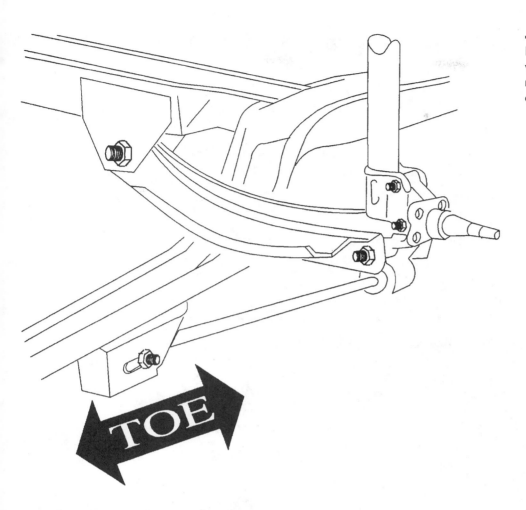

Figure 12–31 By moving various rear suspension members, the rear toe can be changed. (Courtesy of FMC)

■ FRONT CAMBER/CASTER ADJUSTMENT METHODS

Many vehicles are constructed with only limited camber/caster factory adjustment. See Figure 12–35 on page 324 for a summary of which adjustments are *generally* possible for various types of vehicles and suspension systems.

> *CAUTION:* Most vehicle manufacturers warn technicians not to adjust camber by bending the strut assembly. Even though several equipment manufacturers make tools that are designed to bend the strut, most experts agree that it can cause harm to the strut itself.

■ ADJUSTING FRONT CAMBER/CASTER

Most SLA-type suspensions can be adjusted for caster and camber. Most manufacturers recommend adjusting caster, then camber, before adjusting the toe. As the caster is changed, such as when the strut rod is adjusted as shown in Figure 12–36 on page 325, the camber and toe also change.

If the camber is then adjusted, the caster is unaffected. Many technicians adjust caster and camber at the same time using shims (Figures 12–37 and 12–38 on pages 325–326), slots (Figure 12–39 on page 326), or eccentric cams (Figure 12–40 on page 326).

Always follow the manufacturer's recommended alignment procedure. For example, many manufacturers include a *shim chart* in their service manual that gives the thickness and location of the shim changes based on the alignment reading. Shim charts are used to set camber and caster at the same time. Shim charts are designed for each model of vehicle (see Figure 12–41 on page 327).

Regardless of the methods or procedures used, toe is always adjusted after all the angles are set because caster and camber both affect the toe.

■ SETTING TOE

Front toe is the last angle that should be adjusted and is the most likely to need correction. This has led to many sayings in the alignment field:

"Set the toe and let it go."

"Do a toe and go."

"Set the toe and collect the dough."

As wear occurs at each steering joint, the forces exerted on the linkage by the tire tend to cause a toe-out condition.

HOW TO SELECT THE CORRECT ALIGNMENT SHIM

1. The fraction or whole number embossed on each shim is the approximate degree of correction.

2. The shim drawings illustrate the position in which the shims could be installed on the rear axle.

3. **TO REDUCE CAMBER,** the thick part of the shim should be at the bottom (six o'clock) of the axle.

4. **TO INCREASE CAMBER,** the thick part of the shim should be at the top (twelve o'clock) of the axle.

5. **TO REDUCE TOE-OUT,** the thick part of the shim should point towards the rear of the vehicle.

6. **TO INCREASE TOE-OUT,** the thick part of the shim should point towards the front of the vehicle.

7. Fine adjustment can be obtained by rotating the shim right or left with the slots provided.

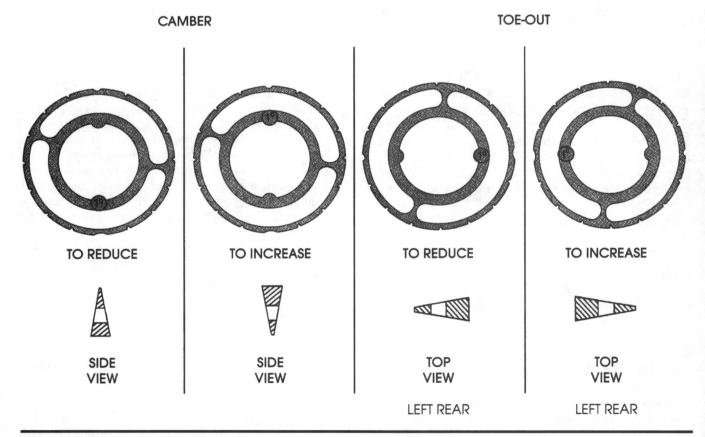

CAMBER

TO REDUCE — SIDE VIEW

TO INCREASE — SIDE VIEW

TOE-OUT

TO REDUCE — TOP VIEW — LEFT REAR

TO INCREASE — TOP VIEW — LEFT REAR

Figure 12–32 The use of these plastic or metal shims requires that the rear wheel as well as the hub assembly and/or backing plate be removed. Proper torque during reassembly is critical to avoid damage to the shims. (Courtesy of Shimco International, Inc.)

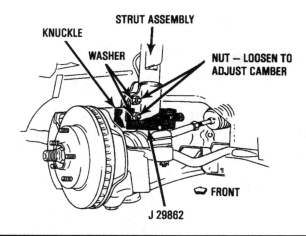

Figure 12–33 Many struts allow camber adjustment at the strut-to-knuckle fasteners. Here a special tool is being used to hold and move the strut into alignment with the fasteners loosened. Once the desired camber angle is achieved, the strut nuts are tightened and the tool removed. (Courtesy of Oldsmobile)

Front-wheel-drive (FWD) vehicles transmit engine power through the front wheels. Many manufacturers of FWD vehicles specify a toe-out setting. This toe-out setting helps compensate for the slight toe-in effect of the engine torque being transferred through the front wheels.

Most newer alignment equipment displays in degrees of toe instead of inches of toe. (See the toe unit conversion chart.) Just remember that positive (+) toe means toe-in and negative (−) toe means toe-out.

Toe Unit Conversions

Units	Conversions			
Fractional inches	1/16″	1/8″	3/16″	1/4″
Decimal inches	0.062″	0.125″	0.188″	0.250″
Millimeters	1.60 mm	3.18 mm	4.76 mm	6.35 mm
Decimal degrees	0.125°	0.25°	0.375°	0.5°
Degrees & minutes	0°8′	0°15′	0°23′	0°30′
Fractional degrees	1/8°	1/4°	3/8°	1/2°

HELPFUL HINT: To convert from degrees to decimal inches, simply divide by 2. For example, if the total toe is 0.25 degree, then one-half (divided by 2) is equal to 0.125″ (1/8″). Toe is usually specified in degrees because it more accurately reflects the toe angle regardless of the size of the wheels/tires.

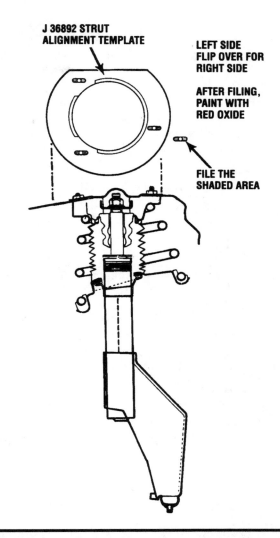

Figure 12–34 Some struts require modification of the upper mount for camber adjustment. (Courtesy of Oldsmobile)

To make sure the steering wheel is straight after setting toe, the steering wheel *must* be locked in the straight-ahead position while the toe is being adjusted. Another term used to describe steering wheel position is **spoke angle.** To lock the steering wheel, always use a steering wheel lock that presses against the seat and the outer rim of the steering wheel. *Do not* use the locking feature of the steering column to hold the steering wheel straight. Always unlock the steering column, straighten the steering wheel, and install the steering wheel lock (see Figure 12–42 on page 328).

NOTE: If the vehicle is equipped with power steering, the engine must be started and the steering wheel straightened with the engine running to be assured a straight steering wheel. Lock the steering wheel with the steering lock tool before stopping the engine.

METHODS OF ADJUSTMENT

Tools and adjustment devices may be available from aftermarket suppliers to perform adjustments in cases where manufacturers do not make such provisions.

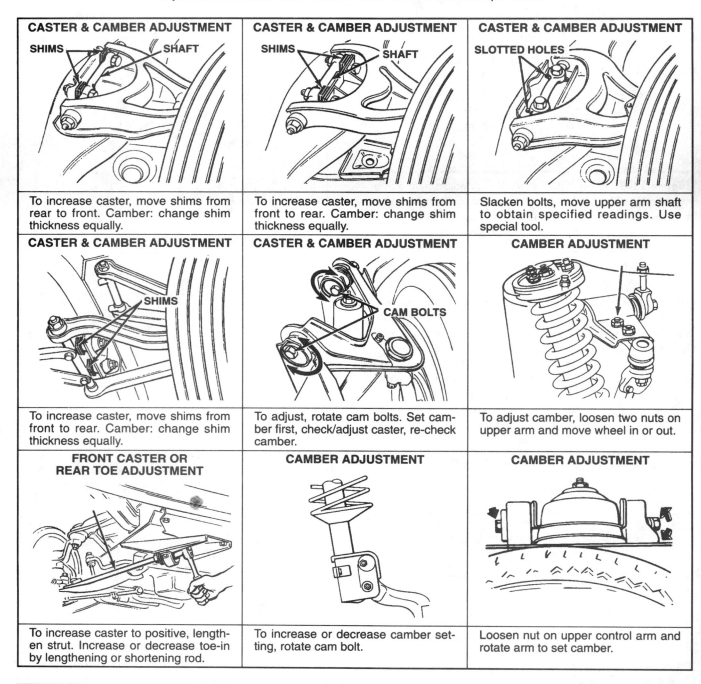

CASTER & CAMBER ADJUSTMENT	CASTER & CAMBER ADJUSTMENT	CASTER & CAMBER ADJUSTMENT
To increase caster, move shims from rear to front. Camber: change shim thickness equally.	To increase caster, move shims from front to rear. Camber: change shim thickness equally.	Slacken bolts, move upper arm shaft to obtain specified readings. Use special tool.
CASTER & CAMBER ADJUSTMENT	**CASTER & CAMBER ADJUSTMENT**	**CAMBER ADJUSTMENT**
To increase caster, move shims from front to rear. Camber: change shim thickness equally.	To adjust, rotate cam bolts. Set camber first, check/adjust caster, re-check camber.	To adjust camber, loosen two nuts on upper arm and move wheel in or out.
FRONT CASTER OR REAR TOE ADJUSTMENT	**CAMBER ADJUSTMENT**	**CAMBER ADJUSTMENT**
To increase caster to positive, lengthen strut. Increase or decrease toe-in by lengthening or shortening rod.	To increase or decrease camber setting, rotate cam bolt.	Loosen nut on upper control arm and rotate arm to set camber.

Figure 12–35 An example of the many methods that are commonly used to adjust front caster and camber. (Courtesy of FMC)

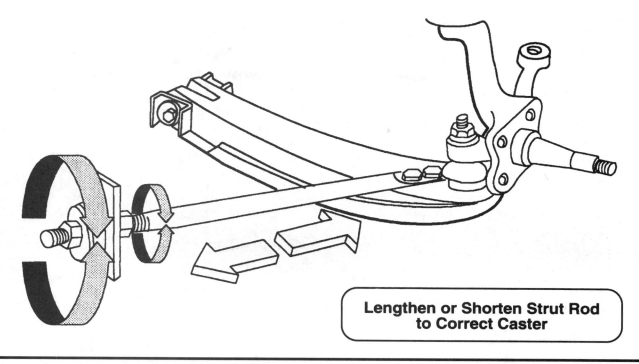

**Lengthen or Shorten Strut Rod
to Correct Caster**

Figure 12–36 If there is a nut on both sides of the strut rod bushing, then the length of
the rod can be adjusted to change caster. (Courtesy of FMC)

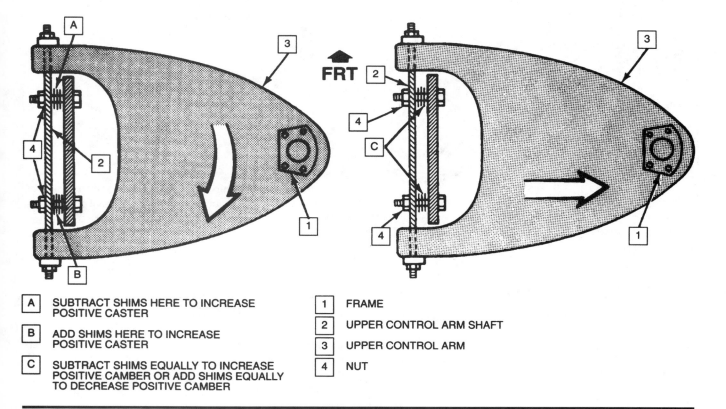

A	SUBTRACT SHIMS HERE TO INCREASE POSITIVE CASTER
B	ADD SHIMS HERE TO INCREASE POSITIVE CASTER
C	SUBTRACT SHIMS EQUALLY TO INCREASE POSITIVE CAMBER OR ADD SHIMS EQUALLY TO DECREASE POSITIVE CAMBER

1	FRAME
2	UPPER CONTROL ARM SHAFT
3	UPPER CONTROL ARM
4	NUT

Figure 12–37 Placing shims between the frame and the upper control arm pivot shaft
is a popular method of alignment for many SLA suspensions. Both camber and caster can
be easily changed by adding or removing shims. (Courtesy of Oldsmobile)

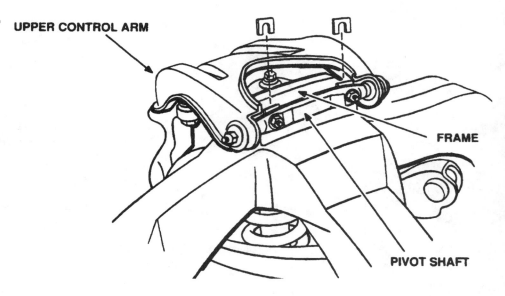

Figure 12–38 The general rule of thumb is that an 1/8″ shim added or removed from *both* shim locations changes the camber angle about 1/2 degree. Adding or removing an 1/8″ shim from *one* shim location changes the caster by about 1/4 degree. (Courtesy of Hunter Engineering Company)

UPPER CONTROL ARM

FRAME

PIVOT SHAFT

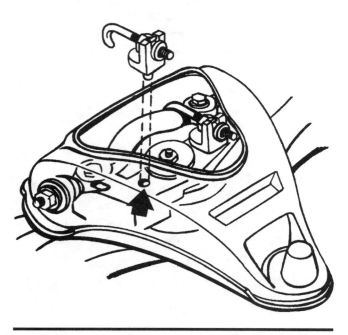

Figure 12–39 Some SLA-type suspensions use slotted holes for alignment angle adjustments. When the pivot shaft bolts are loosened, the pivot shaft is free to move unless held by special clamps as shown. By turning the threaded portion of the clamps, the camber and caster can be set and checked before tightening the pivot shaft bolts. (Courtesy of FMC)

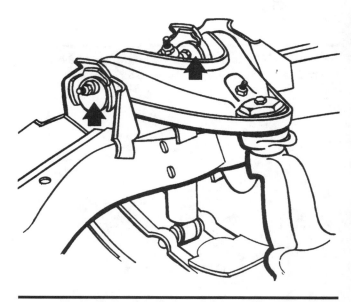

Figure 12–40 When the nut is loosened and the bolt on the eccentric cam is rotated, the upper control arm moves in and out. By adjusting both eccentric cams, both camber and caster can be adjusted. (Courtesy of FMC)

■ CENTERING THE STEERING WHEEL

Centerline steering *should* be accomplished by adjusting the tie rod length on both sides of the vehicle while the toe is set. Some vehicles, however, do not provide two adjustment points for toe. Therefore, centering the steering wheel is not possible on vehicles such as certain models and years of Ford trucks and vans as well as Volkswagen Rabbit, Scirocco, Golf, etc. When servicing these vehicles, the toe should be adjusted and the

After straightening the steering wheel, turn the tie rod adjustment until the toe for both wheels is within specifications (see Figures 12–43 through 12–46 on pages 329–330).

Test-drive the vehicle for proper handling and centerline steering. *Centerline steering* is a centered steering wheel with the vehicle traveling a straight course.

DEGREES CASTER

DEGREES CAMBER	BOLT	+4.9°	+4.7°	+4.5°	+4.3°	+4.1°	+3.9°	+3.7°	+3.5°	+3.3°	+3.1°	+2.9°	+2.7°	+2.5°	+2.3°	+2.1°
+2.2°	FRONT	+300	+211	+211	+210	+210	+201	+201	+201	+200	+200	+111	+111	+110	+110	+110
	REAR	+101	+101	+110	+110	+111	+200	+200	+201	+210	+210	+211	+211	+300	+301	+301
+2.0°	FRONT	+210	+210	+210	+201	+201	+200	+200	+111	+110	+110	+110	+110	+101	+101	+100
	REAR	+011	+100	+101	+101	+110	+110	+111	+200	+201	+201	+210	+210	+211	+300	+300
+1.8°	FRONT	+201	+201	+200	+200	+111	+111	+110	+110	+110	+101	+101	+100	+100	+100	+011
	REAR	+010	+011	+011	+100	+101	+101	+110	+110	+111	+200	+200	+201	+210	+210	+211
+1.6°	FRONT	+200	+200	+111	+111	+110	+110	+101	+101	+100	+100	+100	+011	+011	+010	+010
	REAR	+001	+001	+010	+011	+011	+100	+100	+101	+110	+110	+111	+200	+200	+201	+210
+1.4°	FRONT	+111	+110	+110	+101	+101	+100	+100	+011	+011	+011	+010	+010	+010	+001	+001
	REAR	-000	+000	+001	+001	+010	+011	+011	+100	+100	+101	+110	+110	+111	+200	+200
+1.2°	FRONT	+110	+101	+101	+100	+100	+011	+011	+010	+010	+010	+001	+001	+000	+000	-000
	REAR	-010	-001	+000	+001	+001	+010	+010	+011	+011	+100	+101	+101	+110	+111	+111
+1.0°	FRONT	+100	+100	+011	+011	+010	+010	+010	+001	+001	+000	+000	-000	-001	-001	-010
	REAR	-011	-010	-010	-001	+000	+000	+010	+010	+011	+011	+011	+100	+101	+101	+110
+0.8°	FRONT	-011	+011	-010	+001	+001	+000	+000	-000	-000	-001	-001	-010	-010	-010	-011
	REAR	-100	-100	-011	-010	-010	-001	-001	+000	+001	+001	+010	+011	+011	+100	+101
+0.6°	FRONT	+010	+000	+001	+001	+000	-000	-001	-001	-010	-010	-010	-011	-011	-100	-100
	REAR	-101	-101	-100	-100	-011	-010	-010	-001	+000	+000	+001	+001	+010	+011	+011
+0.4°	FRONT	+001	+000	-001	-100	-100	-010	-010	-010	-011	-011	-100	-100	-100	-101	-101
	REAR	-111	-110	-111	-101	-101	-100	-011	-010	-010	-001	-000	+000	+001	+010	+010
+0.2°	FRONT	-001	-001	-001	-010	-010	-011	-011	-100	-010	-010	-101	-101	-110	-110	-110
	REAR	-200	-111	-111	-110	-101	-101	-100	-100	-100	-010	-010	-001	-000	+000	+001
0.0°	FRONT	-010	-010	-011	-011	-100	-100	-100	-101	-110	-110	-110	-111	-111	-111	-200
	REAR	-201	-201	-200	-111	-110	-110	-101	-101	-100	-100	-011	-010	-010	-001	-000
-0.2°	FRONT	-011	-011	-100	-100	-101	-101	-110	-110	-111	-111	-111	-200	-200	-201	-201
	REAR	-210	-210	-201	-201	-200	-111	-111	-101	-101	-101	-100	-100	-110	-010	-010
-0.4°	FRONT	-100	-101	-101	-110	-110	-110	-111	-111	-200	-200	-201	-201	-201	-210	-210
	REAR	-300	-211	-210	-210	-201	-201	-200	-111	-111	-110	-101	-101	-100	-011	-011

INSTRUCTIONS FOR USING ALIGNMENT CHART

1. Determine vehicle's current caster and camber measurements.
2. Using the current caster reading, read down the appropriate column to the lines corresponding to the current camber reading.
3. Correction values will be given for the front and rear bolts.
   EXAMPLE: Current reading +1.6° caster +0.4° camber. By reading down the chart from +1.6° caster to +0.4° camber you will find that the front bolt requires an adjustment of −101 and the rear bolt requires an adjustment of +010.

+ = Shim Addition
− = Shim Removal

Correction Value Example:

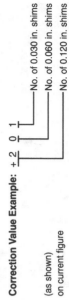

+ 2   0   1

— No. of 0.030 in. shims
— No. of 0.060 in. shims
— No. of 0.120 in. shims

(as shown)
on current figure

Figure 12–41 Typical shim alignment chart. Shims 1/8″ (0.125) can be substituted for the 0.120″ shims; 1/16″ (0.0625) shims can be substituted for the 0.060″ shims; and 1/32″ (0.03125) shims can be substituted for the 0.030″ shims. (Courtesy of Oldsmobile)

HIGH-PERFORMANCE TIP

Race Vehicle Alignment

Vehicles used in autocrossing (individual timed runs through cones in a parking lot) or road racing usually perform best if the following alignment steps are followed:

- *Increase caster (+)*. Not only will the caster provide a good solid feel for the driver during high speed on a straight section of the course, but it will also provide some lean into the corners due to the camber change during cornering. A setting of 5° to 9° positive caster is typical depending on the type of vehicle and the type of course.
- *Adjust for 1° to 2° of negative camber*. As a race vehicle corners, the body and chassis lean. As the chassis leans, the top of the tire also leans outward. By setting the camber to 1° to 2° negative, the tires will be neutral while cornering, thereby having as much rubber contacting the road as possible.

> **NOTE:** While setting negative camber on a street-driven vehicle will decrease tire life, the negative setting on a race vehicle is used to increase cornering speeds, and tire life is not a primary consideration.

- *Set toe to a slight toed-out position*. When the front toe is set negative (toed out), the vehicle is more responsive to steering commands from the driver. With a slight toed-out setting, one wheel is already pointed in the direction of a corner or curve. Set the toe-out to −3/8° to −1/2° depending on the type of vehicle and the type of race course.

vehicle driven on a straight, level road. Note the location of the steering wheel spokes. To correctly center the steering wheel, the steering wheel must be removed and relocated onto the steering column splined shaft.

> **CAUTION:** Do not attempt to straighten the steering wheel by relocating the wheel on the steering column on a vehicle with two tie rod end adjusters. The steering wheel is positioned at the factory in the center of the steering gear, regardless of type. If the steering wheel is not in the center, then the variable ratio section of the gear will not be in the center as it is designed. Another possible problem with moving the steering wheel from its designed straight-ahead position is that the turning radius may be different for right- and left-hand turns.

■ STEERING WHEEL REMOVAL

If the steering wheel *must* be removed, first disconnect the airbag wire connector at the base of the steering col-

TECH TIP

Locking Pliers to the Rescue

Many vehicles use a jam nut on the tie rod end. This jam nut must be loosened to adjust the toe. Because the end of the tie rod is attached to a tie rod end that is movable, loosening the nut is often difficult. Every time force is applied to the nut, the tie rod end socket moves and prevents the full force of the wrench from being applied to the nut. To prevent this movement, simply attach locking pliers (Vise Grips®) to hold the tie rod. Wedge the pliers against the control arm to prevent any movement of the tie rod. By preventing the tie rod from moving, full force can be put on a wrench to loosen the jam nut without doing any harm to the tie rod end.

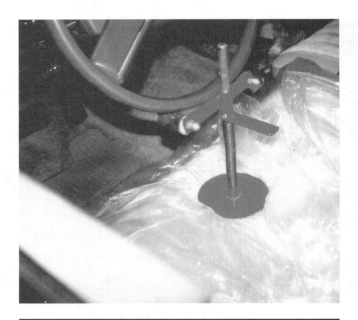

Figure 12–42 Many procedures for setting toe specify that the steering wheel be held in the straight-ahead position using a steering wheel lock, as shown. One method recommended by Hunter Engineering sets toe without using a steering wheel lock.

umn. This reduces the chance of personal injury and prevents accidental airbag deployment.

> **CAUTION:** Always follow the manufacturer's recommended procedures whenever working on or around the steering column.

Remove the center section of the steering column by removing the retaining screws, including the inflator module on vehicles equipped with an airbag.

After removal of the inflator module, remove the steering wheel retaining nut. Note the locating marks on

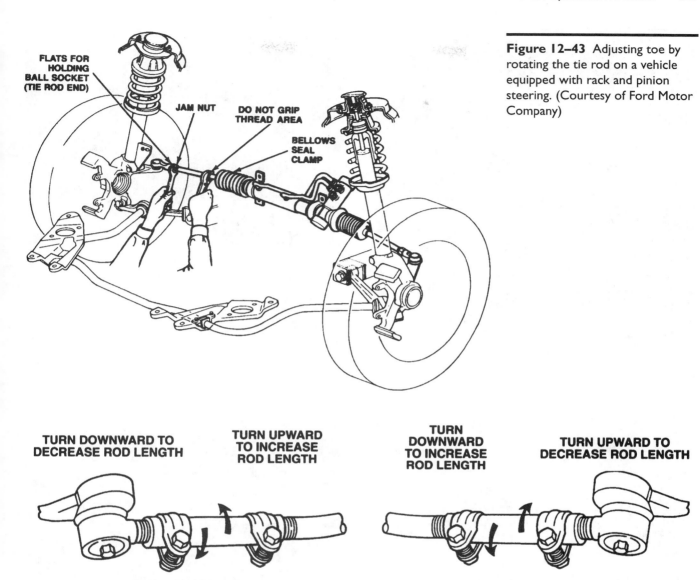

FLATS FOR
HOLDING
BALL SOCKET
(TIE ROD END)

JAM NUT

DO NOT GRIP
THREAD AREA

BELLOWS
SEAL
CLAMP

Figure 12–43 Adjusting toe by rotating the tie rod on a vehicle equipped with rack and pinion steering. (Courtesy of Ford Motor Company)

TURN DOWNWARD TO DECREASE ROD LENGTH

TURN UPWARD TO INCREASE ROD LENGTH

TURN DOWNWARD TO INCREASE ROD LENGTH

TURN UPWARD TO DECREASE ROD LENGTH

LEFT-HAND SLEEVE

RIGHT-HAND SLEEVE

Figure 12–44 Toe is adjusted on a parallelogram-type steering linkage by turning adjustable tie rod sleeves. Special tie rod sleeve adjusting tools should be used that grip the slot in the sleeve and will not crush the sleeve while it is being rotated. (Courtesy of Ford Motor Company)

the steering wheel and steering shaft (see Figure 12–47). These marks indicate the proper position of the steering wheel for centerline steering. This means that the steering wheel spoke angle is straight and in line with the centerline position of the steering gear or rack and pinion steering unit.

Most steering wheels are attached to the steering shaft with a spline and a taper. After removing the steering wheel nut, use a steering wheel puller to remove the steering wheel from the steering shaft (see Figure 12–48).

To reinstall the steering wheel, align the steering wheel in the desired straight-ahead position and slip it down over the splines. Install and tighten the retaining nut to specifications.

HINT: Because of the taper, it is easier to remove a steering wheel if the steering wheel puller is struck with a dead-blow hammer. The shock often releases the taper and allows the easy removal of the steering wheel. Some technicians simply use their hands and pound the steering wheel from the taper without using a puller.

■ TOLERANCE ADJUSTMENT PROCEDURE

Many vehicles are designed and built without a method to change caster or camber, or both. (All vehicles have

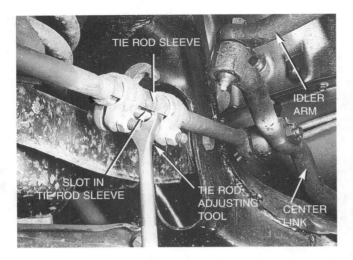

Figure 12–45 Special tie rod adjusting tools should be used to rotate the tie rod adjusting sleeves. The tool grips the slot in the sleeve and allows the service technician to rotate the sleeve without squeezing or damaging the sleeve.

Figure 12–47 Most vehicles have alignment marks made at the factory on the steering shaft and steering wheel to help the service technician keep the steering wheel in the center position.

WHEN TOE IS CORRECT TURN BOTH CONNECTING ROD SLEEVES DOWNWARD TO ADJUST SPOKE POSITION.

WHEN TOE IS CORRECT TURN BOTH CONNECTING ROD SLEEVES UPWARD TO ADJUST SPOKE POSITION.

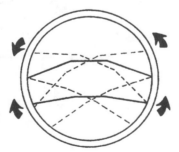

WHEN TOE IS NOT CORRECT LENGTHEN LEFT ROD TO DECREASE TOE-IN SHORTEN RIGHT ROD TO INCREASE TOE-IN.

WHEN TOE IS NOT CORRECT SHORTEN LEFT ROD TO INCREASE TOE-IN LENGTHEN RIGHT ROD TO DECREASE TOE-IN.

ADJUST BOTH RODS EQUALLY TO MAINTAIN NORMAL SPOKE POSITION.

Figure 12–46 Illustration shows which way to turn which tie rod to achieve a straight steering wheel. Most alignment machines will indicate straight-ahead steering, and this information will not be needed to successfully perform a good alignment. (Courtesy of Ford Motor Company)

Diagnostic Story

Left Thrust Line, but a Pull to the Right!

A new four-door sport sedan had been aligned several times at the dealership in an attempt to solve a pull to the right. The car had front-wheel drive and four-wheel in-dependent suspension. The dealer rotated the tires without its making any difference. The alignment angles of all four wheels were in the center of specifications. The dealer even switched all four tires from another car in an attempt to solve the problem.

In frustration, the owner took the car to an alignment shop. Almost immediately the alignment technician discovered that the right rear wheel was slightly toed in. This caused a pull to the right (see Figure 12–49).

The alignment technician adjusted the toe on the right rear wheel and reset the front toe. The car drove beautifully.

The owner was puzzled why the new car dealer was unable to correct the problem. It was later discovered that the alignment machine at the dealership was out of calibration by the exact amount that the right rear wheel was out of specification. The car pulled to the right because the independent suspension created a rear steering force toward the left that caused the front to pull to the right. Alignment equipment manufacturers recommend that alignment equipment be calibrated regularly.

Figure 12–48 A puller being used to remove a steering wheel after the steering wheel retaining nut has been removed.

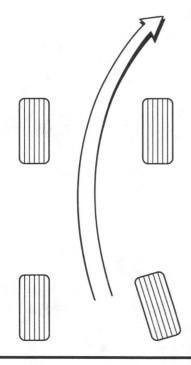

Figure 12–49 The toe-in on the right wheel creates a turning force toward the right.

Figure 12–50 As older vehicles age, crossmembers and springs sag resulting in excessive negative (−) camber. This aftermarket upper control arm pivot shaft is designed with a greater offset to give more positive camber than the stock shaft.

an adjustment for toe.) Before trying an aftermarket alignment correction kit, many technicians first attempt to correct the problem by moving the suspension attachment points within the build tolerance. All vehicles are constructed with a slight amount of leeway or tolerance; slight corrections can be made because bolt holes are almost always slightly larger than the bolt diameter, allowing for slight movement. When several fasteners are involved, such as where the power train cradle (subframe) attaches to the body of the front-wheel-drive vehicle, a measurable amount of alignment change (often

over 1/2°) can be accomplished without special tools or alignment kits. The steps include:

Step 1 Determine which way the suspension members have to be moved to accomplish the desired alignment—for example, the right front may require more positive camber to correct a pulling or tire wear problem.

Step 2 Locate and loosen the cradle (subframe) bolts about four turns each. DO NOT REMOVE ANY OF THE BOLTS.

Step 3 Using pry bars, move the cradle in the direction that will result in an improvement of the alignment angles. Have an assistant tighten the bolts as pressure is maintained on the cradle.

Step 4 Measure the alignment angles and repeat the above procedure if necessary.

■ AFTERMARKET ALIGNMENT METHODS

Accurate alignments are still possible on vehicles without factory methods of adjustment by using alignment kits or parts. Aftermarket alignment kits are available for most vehicles. Even when there are factory alignment methods, sometimes the range of adjustment is not enough to compensate for sagging frame members or other normal or accident-related faults (see Figures 12–50 through 12–52 on pages 332–333). See Appendix 7 for names and addresses of companies manufacturing alignment kits.

■ HIDDEN STRUCTURAL DAMAGE DIAGNOSIS

Many accidents can cause hidden structural damage that can cause alignment angles to be out of specification. If alignment angles are out of specification tolerances, then

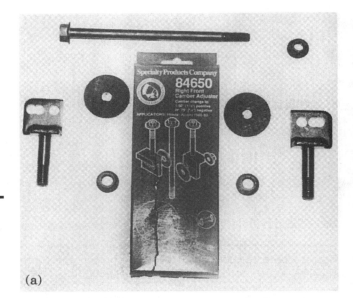

Figure 12–51 (a) Aftermarket camber kit designed to provide some camber adjustments for a vehicle that does not provide any adjustment. (b) Installation of this kit requires that the upper control arm shaft be removed. Note that the upper control arm was simply rotated out over the wheel pivoting on the upper ball joint.

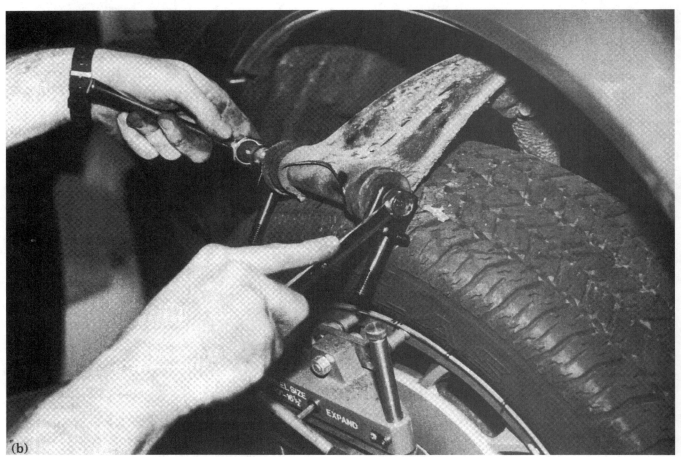

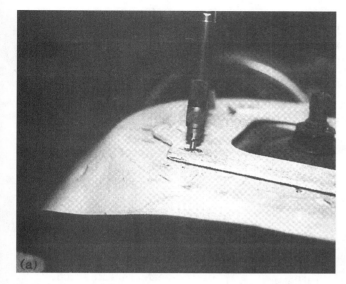

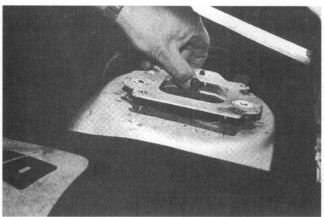

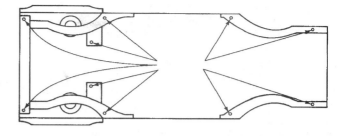

Figure 12–53 Jig holes used at the assembly plant to locate suspension and drivetrain components.

T E C H   T I P

TSBs Can Save Time

Technical service bulletins (TSBs) are issued by vehicle and aftermarket manufacturers to inform technicians of a situation or technical problem and give the corrective steps and a list of parts needed to solve the problem.

TSBs are often released by new-vehicle manufacturers to the dealership service department. They usually concern the current year vehicle of a particular model. While many of these TSBs concern minor problems covering few vehicles, many contain very helpful solutions to hard-to-find problems.

Most TSBs can be purchased directly from the manufacturer, but the cost is usually very high. TSBs can also be purchased through aftermarket companies that are licensed to include TSBs on CD-ROM computer discs. Factory TSBs can often save the technician many hours of troubleshooting.

Figure 12–52 (a) The installation of some aftermarket alignment kits requires the use of special tools such as this cutter being used to drill out spot welds on the original alignment plate on a strut tower. (b) Original plate being removed. (c) Note the amount of movement the upper strut bearing mount has around the square openings in the strut tower. An aftermarket plate can now be installed to allow both camber and caster adjustment.

accident damage should be suspected. Look for evidence of newly replaced suspension parts, body work, or repainted areas of the body. While a body and/or frame of a vehicle can be straightened, it must be done by a knowledgeable person using body-measuring equipment.

The first thing that must be done is to determine a *datum plane.* Datum means a basis on which other measurements can be based. The datum plane is the horizontal plane.

However, most alignment technicians do not have access to body/frame alignment equipment. The service technician can use a common steel rule to measure several points of the vehicle to determine if the vehicle is damaged or needs to be sent to a frame shop for repair.

Frame/Body Diagonals

If the frame or body is perfectly square, then the diagonal measurements should be within 1/8″ (3 mm) of each other (see Figure 12–53).

While there are specified measurement points indicated by the manufacturer, the diagonal measurements can be made from almost any point that is repeated exactly on the other side, such as the center of a bolt in the suspension mounting bracket.

■ ALIGNMENT TROUBLESHOOTING

The following table lists common alignment problems and their probable causes.

Alignment Troubleshooting Guide

Problems	Probable Causes
Pull left/right	Uneven tire pressure, tire conicity, mismatched tires, unequal camber, unequal caster, brake drag, setback, suspension/frame sag, unbalanced power assist, bent spindle, bent strut, worn suspension components (front or rear), rear suspension misalignment.
Incorrect steering wheel position	Incorrect individual or total toe, rear-wheel misalignment, excessive suspension or steering component play, worn rack and pinion attachment bushings, individual toe adjusters not provided.
Hard steering	Improper tire pressure, binding steering gear or steering linkage, low P/S fluid, excessive positive caster, lack of lubrication, upper strut mount(s), worn power steering pump, worn P/S belt.
Loose steering	Loose wheel bearings, worn steering or suspension components, loose steering gear mount, excessive steering gear play, loose or worn steering coupler.

Problems	Probable Causes
Excessive road shock	Excessive positive caster, excessive negative camber, improper tire inflation, too wide wheel/tire combination for the vehicle, worn or loose shocks, worn springs.
Poor returnability	Incorrect camber or caster, bent spindle or strut, binding suspension or steering components, improper tire inflation.
Wander/instability	Incorrect alignment, defective or improperly inflated tires, worn steering or suspension parts, bent spindle or strut, worn or loose steering gear, loose wheel bearings.
Squeal/scuff on turns	Defective or improperly inflated tires, incorrect turning angle (TOOT), bent steering arms, excessive wheel setback, poor driving habits (too fast for conditions), worn suspension or steering parts.
Excessive body sway	Loose or broken stabilizer bar links or bushings, worn shocks or mountings, broken or sagging springs, uneven vehicle load, uneven or improper tire pressure.
Memory steer	Binding steering linkage, binding steering gear, binding upper strut mount, ball joint or king pin.
Bump steer	Misalignment of steering linkage, bent steering arm, frame, defective or sagged springs, uneven load, bent spindle or strut.
Torque steer	Bent spindle or strut, bent steering arm, misaligned frame, worn torque strut, defective engine or transaxle mounts, drive axle misalignment, mismatched or unequally inflated tires.

PHOTO SEQUENCE   Total Four-Wheel Alignment Procedures

P12–1 Begin the alignment procedure by first driving the vehicle onto the alignment rack as straight as possible.

P12–2 Position the front tires in the center of the turn plates. These turn plates can be moved inward and outward to match any width vehicle.

P12–3 Check and adjust tire pressures and perform the other prealignment checks necessary to be assured of a proper alignment.

P12–4 Raise the vehicle and perform a dry park test to determine whether steering and/or suspension parts may need replacement before continuing with the alignment.

P12–5 Position both the front and rear rack jacking systems under the suspension system.

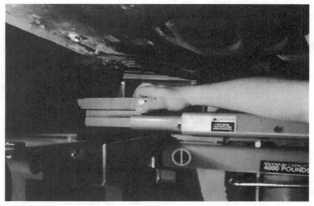

P12–6 Move the pads of the lifting unit under the suspension so that the vehicle can be raised off the drive-on surface of the alignment rack.

P12–7 Lower the alignment rack floor supports before lowering the alignment rack.

P12–8 When the alignment rack is lowered, the support arms should contact the floor or the bottom of the hoist in the case of this scissor-type alignment rack.

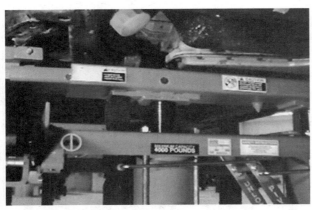

P12–9 With the alignment rack firmly supported by the support legs, raise the vehicle off the alignment rack using the air/hydraulic jacks previously placed under the front and rear suspension.

P12–10 With the wheels off the rack, install the alignment heads. Position the alignment heads with the valve core located in the one o'clock position so that the safety cable can be installed to the valve core.

P12–11 Remove the tire valve cap and either put it in your pocket or place it in a location where it will not be lost. Screw the safety cable for the alignment head to the tire valve.

P12–12 Connect all of the cables and lines necessary. In this situation, the alignment heads are battery powered and communicate to the alignment machine via radio frequency signals. To power up this type of alignment head, simply turn it on.

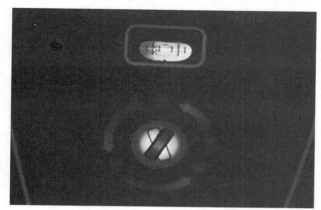

P12–13 The bubble level and three lights in the circle are used during the compensation process.

P12–14 Compensate all four alignment heads according to the alignment machine specified procedures. This compensation process allows for correct alignment of the wheels even if the alignment heads are not all installed to exactly the same depth or if there is a bent wheel.

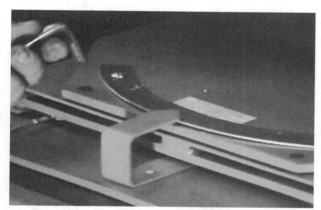

P12–15 Remove the pins from the turn plates (both front and rear).

P12–16 Lower the vehicle onto the turn plates.

P12–17 Now is a good time to check for toe-out on turns (TOOT). The front wheels are turned until the turn plate under the outside wheel reads 20°.

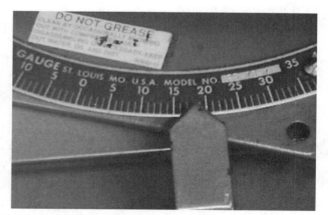

P12–18 A check on the other side of the vehicle indicates that the inside turn plate had rotated 18°, which was within specifications for this vehicle. A bent or damaged steering arm (knuckle) is the most likely cause if the TOOT is not within specifications, and a tire squeal while turning a corner at low speeds is the usual symptom.

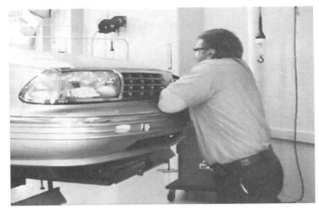

P12–19 After lowering the vehicle and making sure the wheels are turned in the straight-ahead position, jounce (bounce) the vehicle at both front and rear to center the suspension.

P12–20 Some alignment machines such as this Hunter P211 have a built-in ride height gauge. The readout shows the actual measurement and the specifications so the technician will know if replacement springs are necessary before the alignment is begun.

P12–21 To prevent the wheels from rotating during the checking procedures, most alignment equipment manufacturers specify that the brakes should be kept applied. A brake pedal depressor tool connects between the seat and the brake pedal.

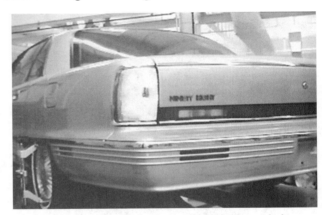

P12–22 Whenever the brake pedal is depressed, the brake lights will also be on with most vehicles. To prevent the brake lights from draining the battery, the brake light fuse could be removed or the connector to the rear lights disconnected.

P12–23 To measure caster, the front wheels must be steered first one direction, then the other direction, as directed by on-screen instructions. Most experts recommend that the front wheels be turned using the steering wheel because this creates the same forces on the steering system as is normally exerted during vehicle operation.

P12–24 After the caster angle has been determined, raise the front wheels off the rack and allow the wheels to droop. This allows a more accurate measurement for SAI than if the vehicle was kept on the rack.

P12–25 The front wheels are again rotated to the right and then to the left following the directions on the alignment machine display to measure the steering axis inclination (SAI).

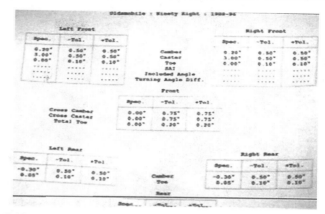

P12–26 After all of the angles have been measured, the alignment results can be printed out and compared with specifications.

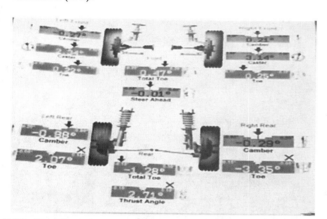

P12–27 This printout shows that most angles are out of specification—not unusual for this training vehicle that has been used by students practicing replacing suspension and steering components.

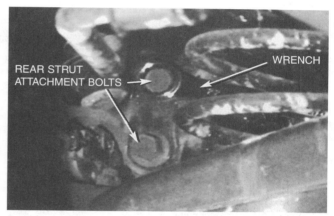

P12–28 Start correcting the alignment by adjusting rear camber. The rear camber is adjusted on this vehicle by loosening the strut attachment bolts and moving the bottom portion of the rear strut.

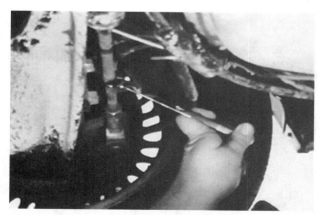

P12–29 After the rear camber has been adjusted, the rear toe is then brought back into factory specifications by rotating the rear tie rod after loosening the jam nut. Many vehicles require aftermarket shims to adjust the rear toe and/or camber.

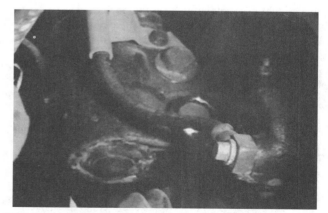

P12–30 After the rear camber and toe have been adjusted, the front camber is now being adjusted by loosening and moving the lower strut mount. The caster on this vehicle was okay and did not require adjustment. If it had required adjustment, the caster should be adjusted before adjusting the camber.

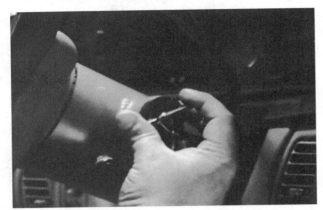

P12–31 Before setting the front toe, start the engine if the vehicle is equipped with power steering.

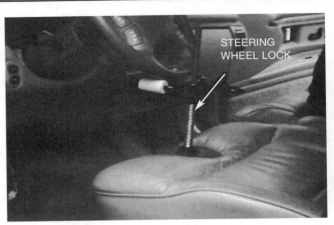

P12–32 Straighten the steering wheel and use a steering wheel lock to hold the steering wheel in the straight-ahead position as shown. Some methods of adjusting front toe do not require that the steering wheel be locked. Follow the recommended procedures as specified by the alignment equipment manufacturer.

STEERING WHEEL LOCK

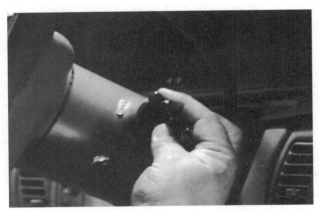

P12–33 After centering the steering wheel and locking it in the straight-ahead position, turn the engine and ignition off.

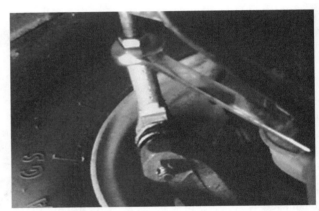

P12–34 Loosen the jam nut to allow the tie rod to be lengthened or shortened to adjust the front toe.

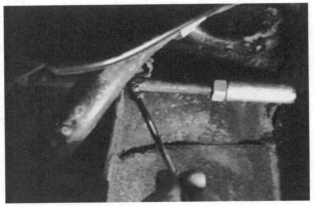

P12–35 Use a wrench or the flats of the tie rod (if equipped) to rotate the tie rod to bring the toe into factory specifications.

P12–36 After the toe has been adjusted, hold the tie rod with a wrench while tightening the lock (jam) nut to prevent any change in the toe setting.

Total Four-Wheel Alignment Procedures—continued

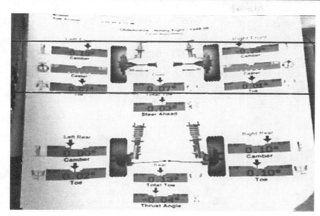

P12–37 After the alignment is completed, print out the results so the customer can see that all angles are within factory specifications.

P12–38 After completing the alignment, carefully disconnect the alignment heads from the wheels. Reinstall the valve caps and wheel covers if necessary.

P12–39 Lower the vehicle.

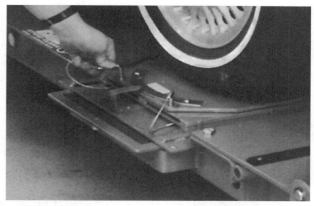

P12–40 Install the pins in the turn plates before driving the vehicle off the alignment rack.

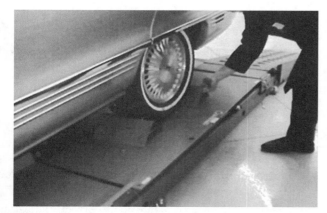

P12–41 Remove any chocks (blocks) used to keep the vehicle from moving on the rack.

P12–42 Carefully drive the vehicle off the alignment rack. The vehicle should be driven to check for proper vehicle handling and double-check that the steering wheel is straight before returning the vehicle to the customer.

■ SUMMARY

1. Before attempting to align any vehicle, it must be checked for proper ride height (trim height), tire conditions, and tire pressures. A thorough inspection of all steering and suspension components must also be made.

2. Memory steer is a condition that causes the vehicle to lead or pull to the same direction it was last steered. Binding steering or suspension components are the most frequent causes of memory steer.

3. Torque steer is the pull or lead caused by engine torque being applied to the front wheels unevenly on a front-wheel-drive vehicle. Out-of-level drivetrain, suspension components, or tires are the most common causes of excessive torque steer.

4. Lead/pull diagnosis involves a thorough road test and careful inspection of all tires.

5. There are three types of alignment: geometric centerline, thrust line, and total four-wheel alignment. Only total four-wheel alignment should be used on a vehicle with an adjustable rear suspension.

6. The proper sequence for a complete four-wheel alignment is rear camber, rear toe, front camber and caster, and front toe.

■ REVIEW QUESTIONS

1. List ten pre-alignment checks that should be performed before the wheel alignment is checked and/or adjusted.

2. Describe the difference between a lead (drift) and a pull.

3. Explain the causes and possible corrections for torque steer.

4. Explain the causes and possible corrections for memory steer.

5. List the steps necessary to follow for a four-wheel alignment.

■ ASE CERTIFICATION-TYPE QUESTIONS

1. When performing an alignment, which angle is the most important for tire wear?
   a. Toe
   b. Camber
   c. Caster
   d. SAI (KPI)

2. Replacement rubber control arm bushings should be
   a. Tightened while the control arm is held in a vise
   b. Torqued with the vehicle on the ground in normal driving position
   c. Tightened with the control arm resting on the frame
   d. Lubricated with engine oil before tightening

3. Which alignment angle is adjustable on all vehicles?
   a. Camber
   b. Caster
   c. Toe
   d. SAI (KPI)

4. Positive (+) toe is
   a. Toe-in
   b. Toe-out

5. If the top of the steering axis is tilted 2 degrees toward the rear of the vehicle, this is
   a. Positive camber
   b. Negative camber
   c. Negative caster
   d. Positive caster

6. Which angle is largest?
   a. 0.55°
   b. 1/4°
   c. 45′
   d. 1/2°

7. If the turning radius (toe-out on turns, or TOOT) is out of specification, what part or component is defective?
   a. The strut is bent
   b. The steering arm is bent
   c. The spindle is bent
   d. The control arm is bent

8. Which angle determines the thrust angle?
   a. Front toe
   b. Rear toe
   c. Rear camber
   d. Front caster, SAI, and included angle

9. The proper order in which to perform a four-wheel alignment is
   a. Front camber, caster toe, rear camber, then rear toe
   b. Rear camber, front camber, caster, front toe, then rear toe
   c. Rear camber and toe, front camber, caster, then front toe
   d. Front toe, camber, front caster, rear camber, then rear toe

10. Centerline steering is achieved by correctly adjusting
    a. Rear and front toe
    b. SAI and included angle
    c. Caster and SAI
    d. Rear camber

Vibration and Noise Diagnosis and Correction

<table>
<tr><td>

Objectives: After studying Chapter 13, the reader should be able to:

1. Discuss how to perform a road test for vibration and noise diagnosis.
2. List the possible vehicle components that can cause a vibration or noise.
3. Describe the use of a reed tachometer or electronic vibration analyzer in determining the frequency of the vibration.
4. Discuss the procedures used in measuring and correcting driveshaft angles.
5. List the items that should be checked or adjusted to prevent or repair power steering and/or noise under the vehicle.

</td></tr>
</table>

Vibration and noise are two of the most frequent complaints from vehicle owners and drivers. If something is vibrating, it can move air; changes in air pressure (air movement) are what we call *noise*. While anything that moves vibrates, wheels and tires account for the majority of vehicle vibration problems.

■ CAUSES OF VIBRATION AND NOISE

Vehicles are designed and built to prevent vibrations and to dampen out any vibrations that cannot be elimi-

nated. For example, engines are designed and balanced to provide smooth power at all engine speeds. Some engines, such as a large four-cylinder or 90° V6s, require special engine mounts to absorb or dampen any remaining oscillations or vibrations. Dampening weights are also fastened to engines or transmissions in an effort to minimize **noise, vibration,** and **harshness** (called **NVH**).

If a new vehicle has a vibration or noise problem, then the most likely cause is an assembly or parts problem. This is difficult to diagnose because the problem could be almost anything, and a careful analysis procedure should be followed as outlined later in this chapter.

If an older vehicle has a vibration or noise problem, the first step is to question the vehicle owner as to when the problem first appeared. Some problems and possible causes include:

Problem	Possible Causes
Vibration at idle	Engine mount could be defective or not reinstalled correctly after an engine or transmission repair.
Noise/vibration	Exhaust system replacement or repair (see Figure 13–1).

> **NOTE:** A typical exhaust system can "grow" or lengthen about 2″ [1 cm] when warm, compared with room temperature. Always inspect an exhaust system when warm, if possible, being careful to avoid being burned by the hot exhaust components.

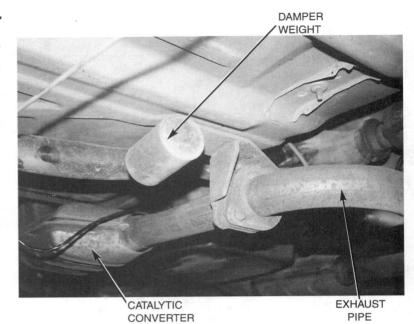

Figure 13–1 Many vehicles, especially those equipped with four-cylinder engines, use a damper weight attached to the exhaust system to dampen out certain frequency vibrations.

DAMPER WEIGHT

CATALYTIC CONVERTER

EXHAUST PIPE

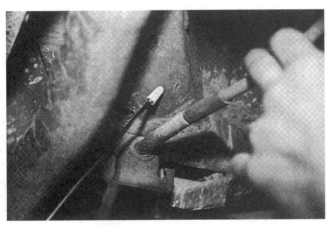

Figure 13–2 This parking brake cable was hitting on the underneath of the body. It was found by looking for evidence of "witness marks."

Vibration at higher vehicle speeds	Incorrect driveshaft angles could be the result of a change in the U-joints, springs, transmission mounts, or anything else that can cause a change in driveshaft angles.
Noise over rough roads	Exhaust system or parking brake cables are often causes of noise while driving over rough road surfaces (see Figure 13–2). Defective shock absorbers or shock absorber mountings are also a common cause of noise.

■ TEST DRIVE

The first thing a technician should do when given a vibration or noise problem to solve is to duplicate the condition. This means to drive the vehicle and observe when and where the vibration is felt or heard the chart in Figure 13–3).

Although there are many possible sources of a vibration, some simple observations that may help to locate the problem quickly.

1. If the vibration is felt or seen in the steering wheel, dash, or hood of the vehicle, the problem is most likely to be caused by defective or out-of-balance *front* wheels or tires (see Figure 13–4).

2. If the vibration is felt in the seat of the pants or seems to be all over the vehicle, the problem is most likely to be caused by defective or out-of-balance *rear* wheels or tires. In a rear-wheel-drive vehicle, the driveshaft (propeller shaft) and related components might also be the cause.

While on the test drive, try and gather as much information about the vibration or noise complaint as possible.

Step 1 Determine the vehicle speed (mph or km/h) or engine speed (rpm) where the vibration occurs. Drive on a smooth, level road and accelerate up to highway speed, noting the vehicle speed or speeds at which the vibration or noise occurs.

Step 2 To help pin down the exact cause of the vibration, accelerate to a speed slightly above the point of maximum vibration. Shift the vehicle into neutral and allow it to coast down through the speed of maximum vibration. If the vibration still exists, then the cause of the problem could be wheels, tires, or other rotating components, *except* the engine.

If the vibration is eliminated when shifted out of gear, the problem is engine or transmission related.

HINT: If the engine or transmission has been removed from the vehicle, such as during a clutch replacement, carefully observe the location and condition of the mounts. If an engine or transmission mount is defective or out of location, engine and driveline vibrations are often induced and transmitted throughout the vehicle.

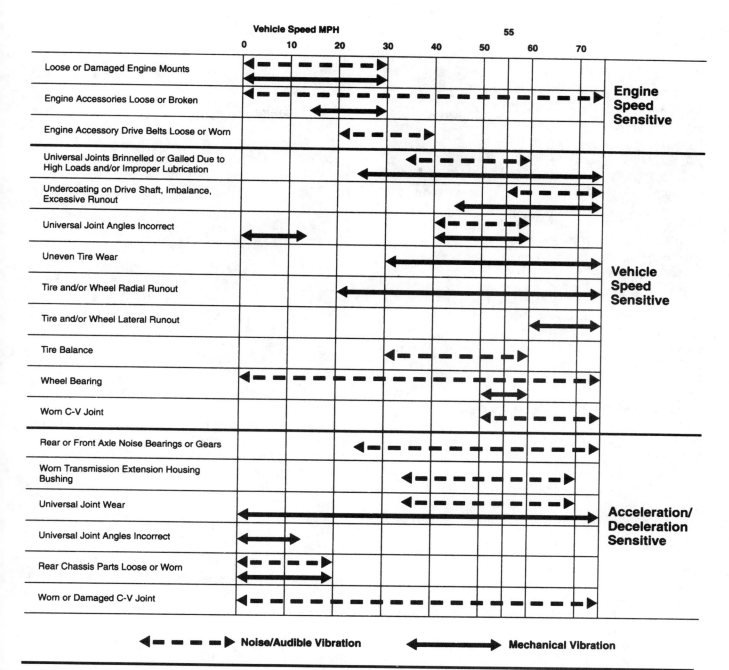

Figure 13–3 A chart showing the typical vehicle and engine speeds at which various components will create a noise or vibration and under what conditions. (Courtesy of Dana Corporation)

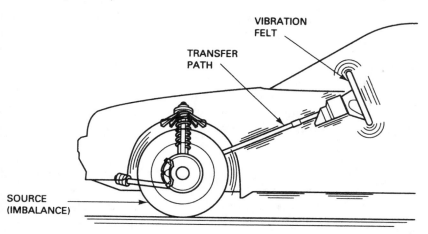

Figure 13–4 Vibration created at one point is easily transferred to the passenger compartment. MacPherson strut suspensions are more sensitive to tire imbalance than SLA-type suspensions.

The Duct Tape Trick

A clicking noise heard at low speeds from the wheels is a common complaint about noise. This noise is usually most noticeable while driving with the windows lowered. This type of noise is caused by loose disc brake pads or noisy wheel covers. Wire wheel covers are especially noisy. To confirm exactly what is causing the noise, simply remove the wheel covers and drive the vehicle. If the clicking noise is still present, check the brakes and wheels for faults. If the noise is gone with the wheel covers removed, use duct tape over the inner edge of the wheel covers before installing them onto the wheels. The duct tape will cushion and dampen the wheel cover and help reduce the noise. The sharp prongs of the wheel cover used to grip the wheel will pierce the duct tape and still help retain the wheel covers.

■ NEUTRAL RUN-UP TEST

The neutral run-up test is to determine if the source of the vibration is engine related. With the transmission in neutral or park, slowly increase the engine rpm and with a tachometer observe the rpm at which the vibration occurs. DO NOT EXCEED THE MANUFACTURER'S RECOMMENDED MAXIMUM ENGINE RPM.

■ VIBRATION DURING BRAKING

A vibration during braking usually indicates out-of-round brake drums, warped disc brake rotors, or other braking system problems. The **front** rotors are the cause of the vibration if the steering wheel is also vibrating (moving) during braking. The **rear** drums or rotors are the cause of the vibration if the vibration is felt throughout the vehicle and brake pedal, but *not* the steering wheel. Another way to check if the vibration is due to rear brakes is to use the parking brake to stop the vehicle. If a vibration occurs while using the parking brake, the rear brakes are the cause.

NOTE: Wheels should *never* be installed using an air impact wrench. Even installation torque is almost impossible to control, and overtightening almost always occurs. The use of impact wrenches causes the wheel, hub, and rotor to distort, resulting in vibrations and brake pedal pulsations. Always tighten wheel lugs in the proper sequence and with proper torque value using a torque wrench or torque-limiting adapter bars.

S-10 Pickup Truck Frame Noise

The owner of a Chevrolet S-10 pickup truck complained of a loud squeaking noise, especially when turning left. Several technicians attempted to solve the problem and replaced shock absorbers, ball joints, and control arm bushings without solving the problem. The problem was finally discovered to be the starter motor hitting the frame. A measurement of new vehicles indicated that the clearance between the starter motor and the frame was about 1/8″ (0.125″) (0.3 cm)! The sagging of the engine mount and the weight transfer of the engine during cornering caused the starter motor to rub up against the frame. The noise was transmitted through the frame throughout the vehicle and made the source of the noise difficult to find.

■ VIBRATION SPEED RANGES

Vibration describes an oscillating motion around a reference position. The number of times a complete motion cycle takes place during a period of one second is called **frequency** and is measured in **Hertz (Hz)** (named for Heinrich R. Hertz, a 19th-century German physicist) (See Figures 13–5 and 13–6).

The unit of measure for frequency was originally **cycles per second (CPS)**. This was changed to Hz in the 1960s.

To help understand frequency, think of the buzzing sound made by some fluorescent light fixtures. That 60-Hz hum is the same frequency as the alternating current. A 400-Hz sound is high pitched. In fact, most people can only hear sounds between 20 and 15,000 Hz. Generally, low-frequency oscillations between 1 and 80 Hz are the most disturbing to vehicle occupants.

Tire and Wheel Vibrations

Typical vehicle components that can cause vibration, in specific frequency ranges at 50 mph (80 km/h), include:

Low frequency (5–20 Hz). This frequency range of vibration is very disturbing to many drivers because this type of vibration can be seen and felt in the steering wheel, seats, mirrors, and other components. Terms used to describe this type of vibration include *nibble, shake, oscillation, shimmy,* and *shudder.*

Tires and wheels are the most common source of vibration in the low-frequency range. To determine the *exact* frequency for the vehicle being checked, the following formula and procedure can be used.

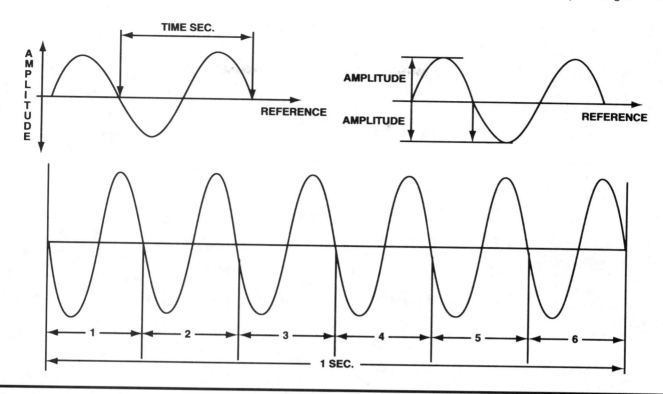

Figure 13–5 Hertz means *cycles* per *second*. If six cycles occur in one second, then the frequency is 6 Hz. The amplitude refers to the total movement of the vibrating component. (Courtesy of Hunter Engineering Company)

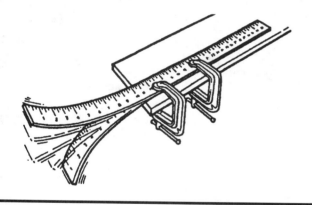

Figure 13–6 Every time the end of a clamped yardstick moves up and down, it is one cycle. The number of cycles divided by the time equals the frequency. If the yardstick moves up and down ten times (ten cycles) in two seconds, the frequency is 5 Hertz (10 ÷ 2 = 5). (Courtesy of Hunter Engineering Company)

Tire rolling frequency is

$$\text{Hz} = \frac{\text{mph} \times 1.47}{\text{tire circumference in ft}}$$

This formula works for all vehicles regardless of tire size. The circumference (distance around the tread) can be measured by using a tape measure around the tire. The **rolling circumference** of the tire is usually shorter due to the contact patch. To determine the rolling circumference follow these easy steps:

Step 1 Inflate the tire(s) to the recommended pressure. Park the vehicle with the valve stem pointing straight down and mark the location on the floor directly below the valve stem.

Step 2 Slowly roll the vehicle forward (or rearward) until the valve stem is again straight down. Mark the floor below the valve stem.

Step 3 Measure the distance between the marks in feet (see Figure 13–7). To change inches to feet, divide by 12 (for example, 77″ divided by 12″ = 6.4 ft). To determine the rolling frequency of this tire at 60 mph, use 6.4 feet as the tire circumference in the formula.

$$\text{Frequency} = \frac{60 \text{ mph} \times 1.47}{6.4 \text{ ft}} = 13.8 \text{ Hz}$$

NOTE: Tire circumference is critical on four-wheel-drive vehicles. The transfer case can be damaged and severe vibration can occur if the rolling circumference is different by more than 0.6″ (15 mm) on the same axle, or more than 1.2″ (30 mm) front to rear.

Figure 13–7 Determining the rolling circumference of a tire.

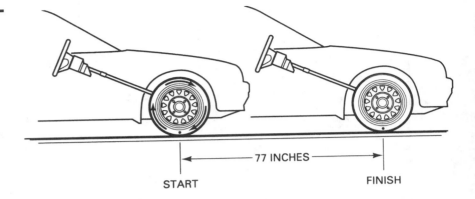

START — 77 INCHES — FINISH

The Vibrating Van

After the engine was replaced in a rear-wheel-drive van, a vibration that felt like an engine miss was noticed by the driver. Because the vibration was not noticed before the engine was replaced, the problem was thought to be engine related. Many tests failed to find anything wrong with the engine. Even the ignition distributor was replaced, along with the electronic ignition module, on the suspicion that an ignition misfire was the cause.

After hours of troubleshooting, a collapsed transmission mount was discovered. After replacing the transmission mount, the "engine miss" and the vibration were eliminated. The collapsed mount caused the driveshaft U-joint angles to be unequal, which caused the vibration.

Figure 13–8 A reed tachometer being used to measure the frequency of a vibration in the vehicle. The reed that is moving up and down (arrow) represents the frequency of the vibration (32 Hz).

Driveline Vibrations

Medium frequency (20–50 Hz). This frequency range of vibrations may also be described as a shake, oscillation or shimmy. These higher frequencies may also be called *roughness* or *buzz*. **Components become blurred and impossible to focus on above a vibration of 30 Hz.**

Engine-Related Vibrations

High frequency (50–100 Hz). Vibrations in this range may also be *heard* as a moan or hum. The vibration is high enough that it may be felt as a numbing sensation that can put the driver's hands or feet to sleep. Engine-related vibrations vary with engine speed, regardless of road speed. Frequency of the vibration from an engine is determined from the engine speed in revolutions per minute (rpm).

$$\text{Frequency in Hz} = \frac{\text{engine rpm}}{60}$$

For example, if a vibration occurs at 3000 engine rpm, then the frequency is 50 Hz.

$$\text{Hz} = \frac{3000 \text{ rpm}}{60} = 50 \text{ Hz}$$

■ FREQUENCY

Measuring Frequency

Knowing the frequency of the vibration greatly improves the speed of tracking down the source of the vibration. Vibration can be measured using an **electronic vibration analyzer (EVA)** or a **reed tachometer** (see Figure 13–8).

A reed tachometer is placed on the dash, console, or other suitable location inside the vehicle. The vehicle is then driven on a smooth, level road at the speed where

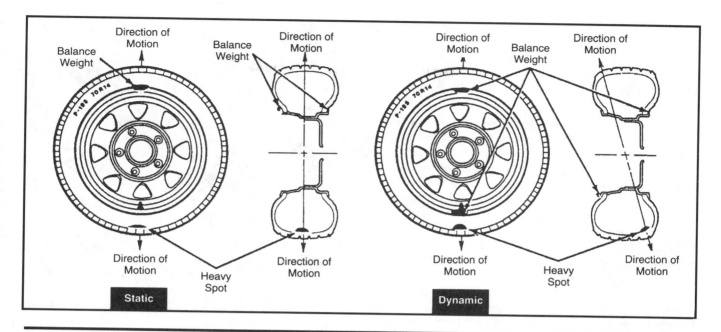

Figure 13–9 Properly balancing all wheels and tires solves most low-frequency vibrations.

Figure 13–10 An out-of-balance tire showing scallops or bald spots around the tire. Even if correctly balanced, this cupped tire would create a vibration.

the vibration is felt the most. The reeds of the reed tachometer vibrate at the frequency of the vibration.

NOTE: Other types of vibration diagnostic equipment may be available from vehicle or aftermarket manufacturers.

Correcting Low-Frequency Vibrations

A low-frequency vibration (5–20 Hz) is usually due to tire/wheel problems, including:

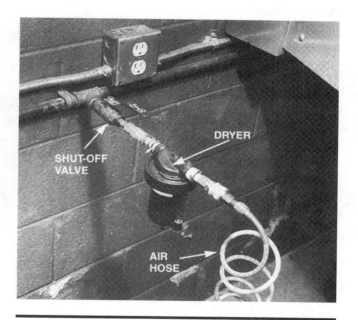

Figure 13–11 All air hoses should be equipped with a water separator to prevent water from being installed into a tire with the compressed air. Any liquid in a tire can cause a vibration that "comes and goes" depending on the location the water is pushed by centrifugal force during rotation.

1. Tire and/or wheel imbalance (see Figures 13–9, 13–10, and 13–11; see Chapter 3 for details)
2. Tire and/or wheel radial or lateral runout (see Chapter 3)
3. Radial force variation within the tire itself (see Chapter 2)
4. If front-wheel drive, a bent or damaged drive axle joint or shaft

Tires that are out of round or defective will be most noticeable at low speeds, get better as speed increases, then vibrate again at highway speeds.

Tires that are out of balance will tend not to be noticeable at low speeds and most noticeable at highway speeds.

> **NOTE:** High-performance tires are manufactured with different carcass and belt package angles as well as a stiffer, harder tread rubber compound than standard tires. While these construction features produce a tire that allows sporty handling, the tire itself causes a stiff ride, often with increased tire noise. The tire also generates and transmits different frequencies than regular tires. Using replacement high-performance tires on a vehicle not designed for this type of tire may create noise and vibration concerns for the driver/owner that the technician cannot correct.

Correcting Medium-Frequency Vibrations

Medium-frequency vibrations (20–50 Hz) can be caused by imbalances of the driveline as well as such things as:

1. Defective U-joints. Sometimes an old or a newly installed U-joint can be binding. This binding of the U-joint can cause a vibration. Often a blow to the joint with a brass hammer can free a binding U-joint. This is often called *relieving* the joint. Hitting the joint using a brass punch and a hammer also works well.
2. Driveshaft imbalance (such as undercoating on the driveshaft) or excessive runout
3. Incorrect or unequal driveshaft angles

Driveline vibrations are usually the result of an imbalance in the rotating driveshaft (propeller shaft) assembly.

The driveshaft of a typical rear-wheel-drive (RWD) vehicle rotates at about three times the speed of the drive wheels. The differential gears in the rear end change the direction of the power flow from the engine and driveshaft, as well as provide for a gear reduction. Front-wheel-drive (FWD) vehicles do not have medium-frequency vibration caused by the driveshaft because the differential is inside the transaxle and the drive axle shafts rotate at the same speed as the wheels and tires.

All driveshafts are balanced at the factory, and weights are attached to the driveshaft, if necessary, to achieve proper balance. A driveshaft should be considered one of the items checked if the medium-frequency vibration is felt throughout the vehicle or in the seat of the pants.

TECH TIP

Squeaks and Rattles

Many squeaks and rattles commonly heard on any vehicle can be corrected by tightening all bolts and nuts you can see. Raise the hood and tighten all fender bolts. Tighten all radiator support and bumper brackets. Open the doors and tighten all hinge and body bolts.

An even more thorough job can be done by hoisting the vehicle and tightening all under-vehicle fasteners, including inner fender bolts, exhaust hangers, shock mounts, and heat shields. It is amazing how much this quiets the vehicle, especially on older models. It also makes the vehicle feel more solid with far less flex in the body, especially when traveling over railroad crossings or rough roads.

> **HINT:** If a vibration is felt during heavy acceleration at low speeds, a common cause is incorrect universal joint angles. This often happens when the rear of the vehicle is heavily loaded or sagging due to weak springs. If the angles of the U-joints are not correct (excessive or unequal from one end of the driveshaft to the other), a vibration will also be present at higher speeds and is usually torque-sensitive.

Correcting High-Frequency Vibrations

High-frequency vibrations (50–100 Hz) are commonly caused by a fault of the clutch, torque converter, or transmission main shaft that rotates at engine speed; in the engine itself it can be caused by items such as:

1. A defective spark plug wire
2. A burned valve
3. Any other mechanical fault that will prevent any one or more cylinders from firing correctly
4. A defective harmonic balancer

If the engine is the cause, run it in neutral at the same engine speed. If the vibration is present, perform a complete engine condition diagnosis. Some engines only misfire under load and will not vibrate while in neutral without a load being placed on the engine, even though the engine is being operated at the same speed (rpm).

Exhaust System Pulses Occur at

4-cylinder engine	2 × engine rpm
6-cylinder engine	3 × engine rpm
8-cylinder engine	4 × engine rpm

If the exhaust system is touching the body, it will transfer these pulses as a vibration. Exhaust system

MAGNETIC MOUNT

DRIVE SHAFT

DIAL INDICATOR

Figure 13–12 A magnetic base dial indicator (gauge) being used to check driveshaft runout.

vibrations vary with engine speed and usually increase as the load on the engine increases.

■ CORRECTING DRIVE-LINE ANGLES

Incorrect drive line angles are usually caused by one or more of the following:

1. Worn, damaged, or improperly installed U-joints
2. Worn, collapsed, or defective engine or transmission mount(s)
3. Incorrect vehicle ride height (As weight is added to the rear of a rear-wheel-drive vehicle, the front of the differential rises and changes the working angle of the rear U-joint.)

■ CHECKING DRIVE-SHAFT RUNOUT

Check to see that the driveshaft is not bent by performing a runout test using a dial indicator (see Figure 13–12). Runout should be measured at three places along the length of the driveshaft.

The maximum allowable runout is 0.030″ (0.76 mm). If runout exceeds 0.030″, remove the driveshaft from the rear end and reindex the driveshaft onto the companion flange 180 degrees from its original location. Remeasure the driveshaft runout. If the runout is still greater than 0.030″, the driveshaft is bent and needs replacement *or* the companion flange needs replacement.

■ MEASURING DRIVE SHAFT U-JOINT PHASING

Driveshaft U-joint phasing is checking to see if the front and rear U-joints are directly in line or parallel with each other. With the vehicle on a **drive-on** lift, or if using a frame contact hoist, support the weight of the vehicle on stands placed under the rear axle. Place an inclinometer on the front U-joint bearing cup and rotate the driveshaft until horizontal; note the inclinometer reading. Move the inclinometer to the rear U-joint. The angles should match. If the angles are not equal, the driveshaft is **out of phase** and should be replaced. Incorrect phasing is usually due to a twisted driveshaft or an incorrectly welded end yoke.

> **NOTE:** Some high-performance General Motors vehicles were built with a slight difference in driveshaft phasing. This was done to counteract the twisting of the driveshaft during rapid acceleration.

■ COMPANION FLANGE RUNOUT

The companion flange is splined to the rear axle pinion shaft and provides the mounting for the rear U-joint of the driveshaft. Two items should be checked on the companion flange while diagnosing a vibration:

1. The companion flange should have a maximum runout of 0.006″ (0.15 mm) while being rotated. If the flange had been pounded off with a hammer during a previous repair, the deformed flange could cause a vibration.
2. Check the companion flange for a missing balance weight. Many flanges have a balance weight that is spot-welded onto the flange. If the weight is missing, a driveline vibration can result.

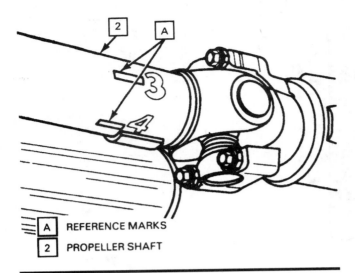

A REFERENCE MARKS

2 PROPELLER SHAFT

Figure 13-13 When checking the balance of a driveshaft, make reference marks around the shaft so that the location of the unbalance may be viewed when using a strobe light. (Courtesy of Oldsmobile)

■ BALANCING THE DRIVESHAFT

If the driveshaft (propeller shaft) is within runout specification and a vibration still exists, the balance of the shaft should be checked and corrected as necessary.

Checking for driveshaft balance is usually done with a strobe balancer. The strobe balancer was commonly used to balance tires on the vehicle and was very popular before the universal use of computer tire balancers. A strobe balancer uses a magnetic sensor that is attached to the pinion housing of the differential. The sensor causes a bright light to flash (strobe) whenever a shock force is exerted on the sensor. The procedure for testing driveshaft balance using a strobe balancer includes:

Step 1 Raise the vehicle and mark the driveshaft with four equally spaced marks around its circumference. Label each mark with a 1, 2, 3, and 4 (see Figure 13–13).

Step 2 Attach the strobe balancer sensor to the bottom of the differential housing as close to the companion flange as possible (see Figure 13–14).

> **NOTE:** The sensor does not rotate with the driveshaft but picks up the vibration of the driveshaft through the differential housing.

Step 3 With the vehicle securely hoisted and the drive wheels off the ground, start the engine and put the

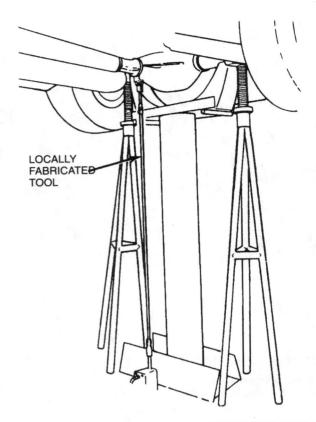

LOCALLY FABRICATED TOOL

NOTE: LOCALLY FABRICATED EXTENSION FOR BALANCER PICK-UP CONSISTS OF 3/8" TUBE AND COMPRESSION FITTINGS

Figure 13-14 Using a strobe balancer to check for driveline vibration requires that an extension be used on the magnetic sensor. Tall safety stands are used to support the rear axle to keep the driveshaft angles the same as when the vehicle is on the road. (Courtesy of Oldsmobile)

transmission into gear to allow the drive wheels to rotate.

Step 4 Hold the strobe light close to the marks on the driveshaft.

   a. If the light does *not* flash, the driveshaft is balanced and no corrective action is necessary.

   b. If the light *does* flash, observe what number mark is shown by the flashing light.

Step 5 Apply hose clamps so the screw portion of the clamp(s) is *opposite* the number seen with the strobe light. The screw portion of the hose clamp is the corrective weight. Remember, the strobe light sensor was mounted to the *bottom* of the differential housing. The strobe light flashes when the heavy part of the driveshaft is facing *downward*. If the heavy part of the driveshaft is down, then corrective weight must be added to the opposite side of the driveshaft (see Figures 13–15 and 13–16).

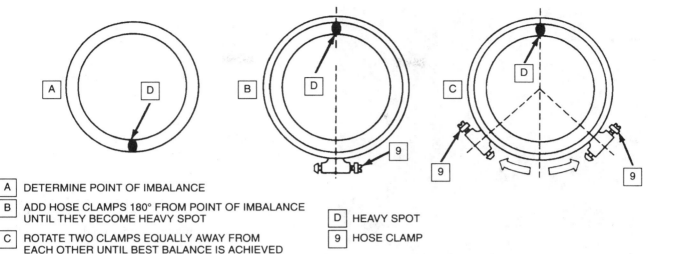

A	DETERMINE POINT OF IMBALANCE
B	ADD HOSE CLAMPS 180° FROM POINT OF IMBALANCE UNTIL THEY BECOME HEAVY SPOT
C	ROTATE TWO CLAMPS EQUALLY AWAY FROM EACH OTHER UNTIL BEST BALANCE IS ACHIEVED

| D | HEAVY SPOT |
| 9 | HOSE CLAMP |

Figure 13–15 Typical procedure to balance a driveshaft using hose clamps. (Courtesy of Oldsmobile)

FACTORY BALANCE WEIGHT

Figure 13–16 Two clamps were required to balance this front driveshaft of a four-wheel-drive vehicle. Be careful when using hose clamps that the ends of the clamps do not interfere with the body or other parts of the vehicle.

■ NOISE DIAGNOSIS

Noise diagnosis is difficult because a noise is easily transmitted from its source to other places in the vehicle. For example, if a rear shock absorber mount is loose, the noise may be heard as coming from the middle or even the front of the vehicle. As the axle moves up and down, the noise is created where metal touches metal between the shock absorber bolt and the axle shock mount. The noise is then transmitted throughout the frame of the vehicle and, therefore,

Diagnostic Story

Everything Is Okay until I Hit a Bump

The owner of an eight-year-old car asked that the vibration in the steering wheel be repaired. It seemed that the car drove fine until the front wheels hit a bump in the road; then the steering wheel shimmied for a few seconds.

This problem is typical of a vehicle with excessive steering linkage free play. When driving straight, centrifugal (rolling) force on the tires tends to force the front wheels outward (toe-out). When one or both wheels hit a bump, the play in the linkage becomes apparent, causing the steering wheel to shimmy until the rolling force again equalizes the steering.

The service technician performed a test drive and a careful steering system inspection and discovered free play in both inner tie rod end sockets of the rack and pinion unit. The steering unit also had some power steering leakage at the tie rod bellows. A replacement-remanufactured power rack and pinion steering unit was recommended to the customer. The customer approved the replacement rack and authorized the required re-alignment. A careful test drive confirmed that the problem was corrected.

causes the sound to appear to come from "everywhere." To help pin down the exact location of the sound, perform a thorough test drive, including driving beside parked vehicles or walls with the vehicle windows open. (See the chart for driveline and bearing-type noise diagnosis.)

Figure 13–17 Tire wear caused by improper alignment or driving habits, such as high-speed cornering, can create tire noise. Notice the feather-edged outer tread blocks.

Noise	Diagnostic Procedure
Tire noise	Change tire pressure; if no change, then the problem is not tires but bearings, etc. Drive on various road surfaces— smooth asphalt reduces tire noise. Rotate the tires front to rear if possible. Various tread designs can cause added noise (see Figure 13–17).
Engine/exhaust noise	Operate the engine at various speeds and loads. Drive faster than speed where the noise occurs and place the transmission in neutral and "coast" down through the speed of maximum noise. Determine if the engine speed or vehicle speed is the cause of the noise.
Wheel bearing noise (see Figure 13–18)	Drive the vehicle slowly on a smooth road. Make left and right turns with the vehicle. Wheel bearing noise changes as weight is transferred side to side. If noise occurs when turning to the right, then the left bearing is the cause. If the noise occurs when turning to the left, then the right bearing is the cause. Hoist the vehicle and rotate the wheel by hand to verify the roughness.

Noise	Diagnostic Procedure
Differential side bearing noise	Drive the vehicle slowly on a smooth road. Differential bearing noise is a low-pitch noise that does not change when turning. The noise varies with vehicle speed.
Differential pinion bearing noise	A whine noise increases with the vehicle speed. Drive on a smooth road and accelerate, coast, and hold a steady speed (float). A defective front pinion bearing may be louder on acceleration. A defective rear pinion bearing may be louder on deceleration. Pinion bearing noise usually peaks in a narrow speed range.
U-joint noise	Drive slowly on a smooth road surface. Drive in reverse and forward. U-joints usually make a "chirp, chirp, chirp" noise in reverse because of lack of lubrication and brinelling from driving forward. Driving in reverse changes the force on the needle bearings in the U-joint, and noise is created.
Clutch noise	**Transmission input bearing:** Start the engine with the transmission in neutral and the parking brake set. The clutch should be engaged (foot off the clutch pedal). If the bearing noise is heard, the transmission input bearing is the source (see Figure 13–19). **Release (throw-out) bearing:** Start the engine with the transmission in neutral and the parking brake set. Lightly depress the clutch pedal just enough to take up free play (usually 1″ or less). If the noise is now heard, the source is the release (throw-out) bearing as the clutch fingers make contact with the bearing. **Pilot bearing:** Start the engine with the transmission in neutral and the parking brake set. Push the clutch pedal fully to the floor (disengage the clutch). If the bearing noise is heard with the clutch disengaged, it is caused by the pilot bearing.

Some noises may be normal; a similar vehicle should be driven and compared before replacing parts that may not be defective. Noises usually become louder and easier to find as time and mileage increase. An occasional noise usually becomes a constant noise.

Figure 13–18 This is an outer bearing race (cup) from a vehicle that sat over the winter. This corroded bearing produced a lot of noise and had to be replaced.

Figure 13–19 An inner race from an input shaft bearing. This bearing caused the five-speed manual transmission to be noisy in all gears except fourth gear. In fourth gear, the torque is transferred straight through the transmission, whereas in all other gears the torque is applied to the counter shaft that exerts a side load to the input bearing.

TECH TIP ✔

Rap It

Many technicians who service transmissions and differentials frequently replace *all* bearings in the differential when there is a noise complaint. While this at first may seem overkill, these technicians have learned that one defective bearing may put particles in the lubricant, often causing the destruction of all the other bearings. This practice has been called *RAP* (replace all parts), but in the case of differentials, RAP may not be such a bad idea.

■ NOISE CORRECTION

The proper way to repair a noise is to repair the cause. Other methods that have been used by technicians include:

1. Insulating the passenger compartment to keep the noise from the passengers
2. Turning up the radio!

Although these methods are usually inexpensive, the noise is still being generated, and if a noisy bearing or other vehicle component is not corrected, more expensive damage is likely to occur. Always remember: *Almost all vehicle faults cause noise first—do not ignore the noise because it is the early warning signal of more serious and possibly dangerous problems.*

Some of the things that can be done to correct certain vibrations and noise include:

1. Check all power steering high-pressure lines, being certain that they do not touch any part of the body, frame, or engine except where they are mounted.

Diagnostic Story 📖

Engine Noise

An experienced technician was assigned to diagnose a loud engine noise. The noise sounded like a defective connecting rod bearing or other major engine problem. The alternator belt was found to be loose. Knowing that a loose belt can "whip" and cause noise, the belt was inspected and the alternator moved on its adjustment slide to tighten the belt. After tightening the belt, the engine was started and the noise was still heard. After stopping the engine, the technician found that the alternator belt was still loose. The problem was discovered to be a missing bolt that attached the alternator mounting bracket to the engine. The forces on the alternator caused the bracket to hit the engine. This noise was transmitted throughout the entire engine. Replacing the missing bracket bolt solved the loud engine noise and pleased a very relieved owner. ✎

2. Carefully check, tighten, and lubricate the flexible couplings in the exhaust system. Use a drive-on lift to ensure normal suspension positioning to check the exhaust system clearances. Loosen, then tighten, all exhaust clamps and hangers to relieve any built-up stress.
3. Lubricate all rubber bushings with rubber lube and replace any engine or transmission mounts that are collapsed.
4. Replace and/or tighten all engine drive belts and check that all accessory mounting brackets are tight.

■ SUMMARY

1. Vibration and noise are two of the most frequently heard complaints from vehicle owners. Noise *is* a vibration (vibrations cause the air to move, creating noise).

2. A vibration felt in the steering wheel, dash, or hood is usually due to out-of-balance or defective front tires. A vibration felt in the seat of the pants or throughout the entire vehicle is usually due to out-of-balance or defective rear tires.

3. Defective engine or transmission mounts, warped rotors, or out-of-round brake drums can all cause a vibration.

4. Vibration is measured by an electronic vibration analyzer (EVA) or a reed tachometer and measured in units called Hertz.

5. Low-frequency vibrations (5–20 Hz) are usually due to tires or wheels.

6. Medium-frequency vibrations (20–50 Hz) are usually caused by driveline problems on rear-wheel-drive vehicles.

7. High-frequency vibrations (50–100 Hz) are usually caused by an engine problem.

8. Driveshafts should be inspected for proper U-joint working angles and balance.

■ REVIEW QUESTIONS

1. Describe how you can tell if the source of a vibration is at the front or the rear during a test drive.

2. Explain the terms *cycle* and *Hertz*.

3. List two types of frequency-measuring instruments.

4. Discuss why the balance of a rear-wheel driveshaft is more important than the balance of a front-wheel-drive axle shaft.

5. Explain how to check and balance a driveshaft on a rear-wheel-drive vehicle.

■ ASE CERTIFICATION-TYPE QUESTIONS

1. A vibration that is felt in the steering wheel at highway speeds is usually due to
   a. Defective or out-of-balance rear tires
   b. Defective or out-of-balance front tires
   c. Out-of-balance or bent driveshaft on a RWD vehicle
   d. Out-of-balance drive axle shaft or defective outer CV joints on a FWD vehicle

2. A vibration during braking is usually caused by
   a. Out-of-balance tires
   b. Warped front brake rotors
   c. A bent wheel
   d. An out-of-balance or bent driveshaft

3. The rolling circumference of both tires on the same axle of a four-wheel-drive vehicle should be within
   a. 0.1″ (2.5 mm)
   b. 0.3″ (7.6 mm)
   c. 0.6″ (15 mm)
   d. 1.2″ (30 mm)

4. The maximum allowable driveshaft runout is
   a. 0.030″ (0.8 mm)
   b. 0.10″ (2.5 mm)
   c. 0.50″ (13 mm)
   d. 0.015″ (0.4 mm)

5. A driveshaft can be checked for proper balance by marking the circumference of the shaft in four places and running the vehicle drive wheels to spot the point of imbalance using a
   a. Reed tachometer
   b. Strobe balancer

6. A defective clutch release (throw-out) bearing is usually heard when the clutch is
   a. Engaged in neutral
   b. Disengaged in a gear
   c. Depressed to take up any free play

7. Wheel and tire imbalance is the most common source of vibrations that occur in what frequency range?
   a. 5–20 Hz.
   b. 20–50 Hz
   c. 50–100 Hz

8. Driveline vibrations due to a bent or out-of-balance driveshaft on a rear-wheel-drive vehicle usually produce a vibration that is
   a. Felt in the steering wheel
   b. Seen as a vibrating dash or hood
   c. Felt in the seat or all over the vehicle

9. Rubber is used for exhaust system hangers because the exhaust system gets longer as it gets hot and rubber helps isolate noise and vibration from the passenger compartment.
   a. True
   b. False

10. A vibration is felt in the steering wheel during braking only. A common cause of the vibration is
    a. Worn idler arm
    b. Out-of-balance front tires
    c. Loose or defective wheel bearing(s)
    d. Warped or nonparallel front disc brake rotors

ASE-Style Sample Test

1. A customer complains that the steering lacks power assist all the time. Technician A says that the power steering pump drive belt could be slipping or loose. Technician B says that worn outer tie rod ends could be the cause. Which technician is correct?
   a. Technician A only
   b. Technician B only
   c. Both Technician A and B
   d. Neither Technician A nor B

2. A front-wheel-drive vehicle pulls toward the right during acceleration. The most likely cause is:
   a. Worn or defective tires
   b. Leaking or defective shock absorbers
   c. Normal torque steer
   d. A defective power steering rack and pinion steering assembly

3. When replacing a rubber bonded socket (RBS) tie rod end, the technician should be sure to:
   a. Remove the original using a special tool
   b. Install and tighten the replacement with the front wheels in the straight-ahead position
   c. Grease the joint before installing on the vehicle
   d. Install the replacement using a special clamp vise

4. Whenever installing a tire on a rim, do not exceed:
   a. 25 psi
   b. 30 psi
   c. 35 psi
   d. 40 psi

5. Two technicians are discussing mounting a tire on a wheel. Technician A says that for best balance, the tire should be match mounted. Technician B says that silicone spray should be used to lubricate the tire bend. Which technician is correct?
   a. Technician A only
   b. Technician B only
   c. Both Technician A and B
   d. Neither Technician A nor B

6. Technician A says that radial tires should *only* be rotated front to rear, never side to side. Technician B says that radial tires should be rotated using the modified X method. Which technician is correct?
   a. Technician A only
   b. Technician B only
   c. Both Technician A and B
   d. Neither Technician A nor B

7. For a tire that has excessive radial runout, Technician A says that it should be broken down on a tire-changing machine and the tire rotated 180 degrees on the wheel and retested. Technician B says that the tire should be replaced. Which technician is correct?
   a. Technician A only
   b. Technician B only
   c. Both Technician A and B
   d. Neither Technician A nor B

8. Technician A says that overloading a vehicle can cause damage to the wheel bearings. Technician B says that tapered roller bearings used on a non-drive wheel should be adjusted hand tight only after seating. Which technician is correct?
   a. Technician A only
   b. Technician B only
   c. Both Technician A and B
   d. Neither Technician A nor B

9. Defective wheel bearings usually sound like
   a. A growl
   b. A rumble
   c. Snow tires
   d. All of the above

10. Defective outer CV joints usually make a clicking noise
    a. Only when backing
    b. While turning and moving
    c. While turning only
    d. During braking

11. The proper lubricant usually specified for use in a differential is
    a. SAE 15-40 engine oil
    b. SAE 80W-90 GL-5
    c. STF
    d. SAE 80W-140 GL-1

12. A vehicle owner complained that a severe vibration was felt throughout the entire vehicle only during rapid acceleration from a stop and up to about 20 mph (32 km/h). The most likely cause is
    a. Unequal driveshaft working angles
    b. A bent driveshaft
    c. Defective universal joints
    d. A bent rim or a defective tire

13. To remove a C-clip axle, what step does *not* need to be done?
    a. Remove the differential cover
    b. Remove the axle flange bolts/nuts
    c. Remove the pinion shaft
    d. Remove the pinion shaft lock bolt

14. Driveshaft working angles can be changed by
    a. Replacing the U-joints
    b. Using shims or wedges under the transmission or rear axle
    c. Rotating the position of the driveshaft on the yoke
    d. Tightening the differential pinion nut

15. A driver complains that the vehicle darts or moves first toward one side and then to the other side of the road. Technician A says that bump steer caused by an unlevel steering linkage could be the cause. Technician B says that a worn housing in the spool valve area of the power rack and pinion is the most likely cause. Which technician is correct?
    a. Technician A only
    b. Technician B only
    c. Both Technician A and B
    d. Neither Technician A nor B

16. A vehicle equipped with power rack and pinion steering is hard to steer when cold only. After a couple of miles of driving, the steering power assist returns to normal. The most likely cause of this temporary loss of power assist when cold is
    a. A worn power steering pump
    b. Worn grooves in the spool valve area of the rack and pinion steering unit
    c. A loose or defective power steering pump drive belt
    d. A defective power steering computer sensor

17. A dry park test is performed
    a. On a frame-type lift with the wheels hanging free
    b. By pulling and pushing on the wheels with the vehicle supported by a frame-type lift

    c. On the ground or on a drive-on lift and moving the steering wheel while observing for looseness
    d. Driving in a figure 8 in a parking lot

18. On a parallelogram-type steering linkage, the part that usually needs replacement first is the
    a. Pitman arm
    b. Outer tie rod end(s)
    c. Center link
    d. Idler arm

19. What parts need to be added to a "short" rack to make a "long" rack and pinion steering unit?
    a. Bellows and ball socket assemblies
    b. Bellows and outer tie rod ends
    c. Ball socket assemblies and outer tie rod ends
    d. Outer tie rod ends

20. The adjustment procedure for a typical integral power steering gear is
    a. Overcenter adjustment, then worm thrust bearing preload
    b. Worm thrust bearing preload, then the overcenter adjustment

21. A vehicle is sagging at the rear. Technician A says that standard replacement shock absorbers should restore proper ride (trim) height. Technician B says that replacement springs are needed to properly restore ride height. Which technician is correct?
    a. Technician A only
    b. Technician B only
    c. Both Technician A and B
    d. Neither Technician A nor B

22. Technician A says that indicator ball joints should be loaded with the weight of the vehicle on the ground to observe the wear indicator. Technician B says that the non-indicator ball joints should be inspected *unloaded*. Which technician is correct?
    a. Technician A only
    b. Technician B only
    c. Both Technician A and B
    d. Neither Technician A nor B

23. The maximum allowable axial play in a ball joint is usually:
    a. 0.001″ (0.025 mm)
    b. 0.003″ (0.076 mm)
    c. 0.030″ (0.76 mm)
    d. 0.050″ (1.27 mm)

24. The ball joint used on MacPherson strut suspension is load carrying.
    a. True
    b. False

25. Technician A says that tapered parts, such as tie rod ends, should be tightened to specifications, then loosened 1/4 turn before installing the cotter key. Technician B says that the nut used to retain tapered parts should never be loosened after torquing, but

rather tightened further, if necessary, to line up the cotter key hole. Which technician is correct?
   a. Technician A only
   b. Technician B only
   c. Both Technician A and B
   d. Neither Technician A nor B

26. When should the strut rod (retainer) nut be removed?
   a. After compressing the coil spring
   b. Before removing the MacPherson strut from the vehicle
   c. After removing the cartridge gland nut
   d. Before removing the brake hose from the strut housing clip

27. "Dog tracking" is often caused by broken or damaged
   a. Stabilizer bar links
   b. Strut rod bushings
   c. Rear leaf springs
   d. Track (panhard) rod

28. A pull toward one side during braking is one symptom of (a) defective or worn
   a. Stabilizer bar links
   b. Strut rod bushings
   c. Rear leaf springs
   d. Track (panhard) rod

29. Oil is added to the MacPherson strut housing before installing a replacement cartridge to
   a. Lubricate the cartridge
   b. Transfer heat form the cartridge to the outside strut housing
   c. Act as a shock damper
   d. Prevent unwanted vibrations

30. A vehicle will pull toward the side with the
   a. Most camber
   b. Least camber

31. Excessive toe-out will wear the edges of both front tires.
   a. Inside
   b. Outside

32. A vehicle will pull toward the side with the
   a. Most caster
   b. Least caster

33. If the turning radius (TOOT) is out of specification, what should be replaced?
   a. The outer tie rod ends
   b. The inner tie rod ends
   c. The idler arm
   d. The steering knuckle

34. SAI and camber together form the
   a. Included angle
   b. Turning radius angle
   c. Scrub radius angle
   d. Setback angle

35. The thrust angle is being corrected. The alignment technician should adjust which angle to reduce thrust angle?
   a. Rear camber
   b. Front SAI or included angle and camber
   c. Rear toe
   d. Rear caster

36. Strut rods adjust if there is a nut on both sides of the frame bushings.
   a. Camber
   b. Caster
   c. SAI or included angle, depending on the exact vehicle
   d. Toe

37–40. Questions 37 through 40 will use the following specifications:
front camber $0.5° + -0.3°$
front caster $3.5°$ to $4.5°$
toe $0° + -0.1°$
rear camber $0° + -0.5°$
rear toe $-0.1°$ to $0.1°$
alignment angles
   front camber left $0.5°$
   front camber right $-0.1°$
   front caster left $3.8°$
   front caster right $4.5°$
   front toe left $-0.2°$
   front toe right $+0.2°$
   total toe $0.0°$
   rear camber left $0.15°$
   rear camber right $-0.11°$
   rear toe left $0.04°$
   rear toe right $0.14°$

37. The first angle corrected should be
   a. Right front camber
   b. Right rear camber
   c. Right rear toe
   d. Left front camber

38. The present alignment will cause excessive tire wear to the inside of both front tires.
   a. True
   b. False

39. The present alignment will cause excessive tire wear to the rear tires.
   a. True
   b. False

40. With the present alignment, the vehicle will
   a. Pull toward the right
   b. Go straight
   c. Pull toward the left

Lug Nut Tightening Torque Chart

To be used as a guide only. Consult the factory service manual or literature for the exact specifications and exceptions for the vehicle being serviced.

Name	Model	Years	Lb. Ft. Torque
Acura	All	86–99	80
American Motors	All	70–87	75
Audi	All	78–99	81
BMW	All except the following	78–99	65–79
	320I	77–83	59–65
	528L	79–81	59–65
Buick	All except the following	76–99	100
	Century	76–81	80
	Regal	78–86	80
	LeSabre	76–85	80
Cadillac	All except	76–99	100
	1976 Seville	1976	80
Chevrolet	Geo Prizm	92–99	100
	Geo Prizm	89–91	76
	Geo Storm	90–99	86.5
	Sprint	85–88	50
	Spectrum	85–88	65
	Chevette	82–87	80
	Chevette	76–81	70
	Nova	85–89	76
	Vega and Monza	76–80	80
	Cavalier	82–99	100
	Celebrity	82–90	100

Name	Model	Years	Lb. Ft. Torque
	Citation	80–86	100
	Camaro	89–99	100
	Camaro	78–88	80
	Malibu and Monte Carlo	76–88	80
	Malibu Wagon	76–86	80
	Impala and Caprice Sedan	77–90	80
	Caprice	91–93	100
	Corvette	84–99	100
	Corvette	76–83	90
	Corsica & Beretta	87–98	100
Chevrolet/ GMC Light Trucks and Vans	Geo Tracker	92–93	60
	Geo Tracker	89–91	37–58
	Lumina APV	90–99	100
	Astro/Safari Van	85–99	100
	S/10 & S/15 Pickup	80–88	80
	T/10 & T/15 Pickup	88–99	100
	C/K Pickup all except:	88–99	120
	C/K Pickup Dual Rear Wheels	88–99	140

Name	Model	Years	Lb. Ft. Torque
	V10 (4WD Full Size) Suburban and Blazer (Aluminum Wheels)	88–89	100
	V10 (4WD Full Size) Suburban and Blazer (Steel Wheels)	88–89	90
	V10 (4WD Full Size) Suburban (All)	90–99	100
	R10 (2WD Full Size) Suburban and Blazer	1989	100
	R/V20 (2WD, 4WD Full Size) Suburban	1989	120
	G10, 20 (Full Size) Van	88–99	100
	G30 (Full Size) Van except:	88–99	120
	G30 (Full Size) Van (Dual Rear Wheels)	88–99	140
	El Camino/Caballero, Sprint	67–87	90
	Luv Pickup	76–82	90
	C/K10 Blazer and Jimmy	71–87	90
	Chevy and GMC Pickups 10/15, 20/25, 30/35 (Single Rear Wheel with 7/16 & 1/2 Studs)	71–87	90
	Chevy and GMC Pickups 10/15, 20/25, 30/35 (Single Rear Wheel with 9/16 Studs)	71–87	120
Chrysler	Concorde	93–99	95
	Chrysler T/C by Maserati	89–91	95
	Conquest	87–89	65–80
	LeBaron (FWD)	84–93	95
	LeBaron (FWD)	82–83	80
	New Yorker (FWD)	83–99	95
	Town and Country (FWD)	84–88	95
	Town and Country (FWD)	82–83	80
	Fifth Avenue (RWD)	83–90	85
	New Yorker (RWD)	76–82	85
	New Yorker (FWD)	93–99	95
	Laser	84–86	95
	Limousine	85–86	95
	Executive Sedan	84–85	95
	E-Class	83–84	80
	Imperial	90–93	95
	Imperial (RWD)	81–83	85
	Cordoba	76–83	85
	LeBaron (RWD)	78–81	85
	Town and Country (RWD)	78–81	85
	Newport	76–81	85
Daihatsu	Charade	88–91	65–87
Daihatsu Light Trucks and Vans	Rocky (All)	90–91	65–87
Dodge	Intrepid	93–99	95
	Stealth	91–99	87–101
	Spirit	89–93	95
	Shadow	87–93	95
	Colt	76–93	65–80
	Lancer	85–89	95
	Aries	84–89	95
	Aries	81–83	80
	Charger	84–87	95
	Charger	82–83	80
	Daytona	84–93	95
	Omni	84–90	95
	Omni	78–83	80
	Vista	84–93	50–57
	600	84–88	95

Name	Model	Years	Lb. Ft. Torque	Name	Model	Years	Lb. Ft. Torque
	600	1983	80		Festiva	89–93	65–87
	Diplomat	78–89	80		Escort	81–83	80–105
	Dynasty	88–93	95		EXP	82–83	80–105
	Monaco	90–91	54–72		Fiesta	78–80	63–85
	Conquest	1986	65–80		Mustang	79–83	80–105
	Conquest	84–85	50–57		Pinto	76–80	80–105
	400	82–83	80		Fairmont	78–83	80–105
	Challenger	78–83	51–58		Granada	76–82	80–105
	Mirada	80–83	85		LTD	79–83	80–105
	St. Regis	79–81	85		LTD	76–78	70–115
	Aspen	76–80	85		Torino	76–79	80–105
Dodge Light Trucks and Vans	Caravan, Ram Van (FWD)	84–99	95		LTD Crown Victoria	1983	80–105
	Rampage (FWD)	82–84	90		Country Sedan and Squire	79–83	80–105
	Ramcharge AD, AW100	79–93	105		Thunderbird	80–83	80–105
	Wagons B100/150	72–99	105		Thunderbird	76–79	70–115
	Wagons B200/250	72–99	105	**Ford Light Trucks and Vans**	E150/F150 and Bronco	88–99	100
	Wagons B300/350 1/2″ Studs	69–99	105		E250/E350, F250, F350	88–99	140
	Wagons B300/350 5/8″ Studs	79–99	200		Aerostar	86–99	100
	D50 Pickup	78–86	55		Bronco	88–93	135
	D50 Pickup	87–99	95		Bronco	72–87	100
	D100/150 Pickup	72–99	105		Explorer	91–99	100
	D200/250 Pickup	81–99	105		Club Wagon E100/150	75–87	100
	D300/350 Pickup 1/2″ Studs	79–99	105		Club Wagon E200/250	76–87	100
	D300/350 Pickup 5/8″ Studs	79–99	200		Club Wagon E300/350 (Single Rear Wheels)	76–87	145
	W100/150 Pickup	79–99	105		Club Wagon E300/350 (Dual Rear Wheels)	76–87	220
	W200/250 Pickup	79–99	105		Econoline Van E100/150	75–81	100
	W300/350 Pickup 1/2″ Studs	79–99	105		Econoline Van E200/250	76–87	100
	W300/350 Pickup 5/8″ Studs	79–99	200		Econoline Van E300/350 (Single Rear Wheels)	76–87	145
	Dakota	87–99	85		Econoline Van E300/350 (Dual Rear Wheels)	76–87	220
Ford	All except the following	84–99	85–105		Ranger Pickup	84–87	100
	Probe	89–91	65–87				

Name	Model	Years	Lb. Ft. Torque
	Courier Pickup	77–83	65
	F100/150 Pickup	75–87	100
	F200/250 Pickup	76–87	100
	F300/350 Pickup (Single Rear Wheels)	76–87	145
	F300/350 Pickup (Dual Rear Wheels)	76–87	220
Honda	All except the following	84–99	80
	Civic All	73–83	58
	Accord All	82–99	80
	Accord All	76–81	58
	Prelude All	79–99	80
Hyundai	All except the following	90–99	65–80
	Excel	86–89	50–57
Infiniti	All	90–99	72–87
Isuzu	Stylus	1991	87
	Impulse	83–91	87
	I Mark	87–89	65
	I Mark	1986	90
	I Mark	1985	50
	I Mark	82–84	50
Isuzu Light Trucks and Vans	Pickup	91–99	72
	Pickup	1990	58–87
	Amigo	1991	72
	Amigo	89–90	58–87
	Rodeo	1991	72
	Trooper	84–91	58–87
Jaguar	All	89–91	65–75
	XJ6 and XJS	1988	75
	All	81–87	80
Eagle	Vision	1993	95
	Premier	89–91	54–72
	Talon	90–99	87–101
	Summit	89–91	65–80
	Medallion	1988	67
Jeep Light Trucks and Vans	Grand Cherokee & Grand Wagoneer	92–99	88
	Wrangler, YJ	90–99	80
	Cherokee, Comanche	84–91	75

Name	Model	Years	Lb. Ft. Torque
	Wagoneer and Grand Wagoneer	84–91	75
	Trucks (under 8400 GVW)	84–89	75
	Trucks (over 8400 GVW)	84–89	130
	CJ Series	84–86	80
	CJ Series	81–83	65–80
	Cherokee, Wagoneer	81–83	65–90
	Trucks (under 8400 GVW)	81–83	65–90
	Trucks (over 8400 GVW)	81–83	110–150
Lexus	All	90–99	76
Lincoln	All	84–99	85–105
	Mark IV	80–83	80–105
	Continental	76–83	80–105
	Town Car	81–83	80–105
	Versailles	77–80	80–105
Mazda	All except the following	88–99	65–87
	Navajo	91–99	100
	323	86–87	65–87
	GLC	81–85	65–80
	GLC	77–80	65–80
	GLC Wagon	84–85	65–87
	626	84–87	65–87
	626	1983	65–80
	626	79–82	65–80
	Cosmo	76–78	65–72
	808	76–77	65–72
	RX7	84–87	65–87
	RX7	79–83	65–80
	RX7	76–78	65–72
	RX3	76–78	65–72
Mazda Light Trucks and Vans	B2600	87–99	65–87
	B2200	86–99	65–87
	B2000/ B2200	80–85	72–80
Mercedes	All	76–99	81
Mercury	All	84–99	85–105
	All	76–83	80–105
Mitsubishi	Sigma V6	89–90	65–80
	Mirage	85–91	65–80
	Precis	87–89	51–58
	Cordia/ Tredia	83–88	50–57

Name	Model	Years	Lb. Ft. Torque
	Eclipse	90–99	87–101
	Galant	85–86	50–57
	Galant	1987	65–80
	Galant	88–99	65–80
	Starion	1983	50–57
	Starion	84–89	50–57
Mitsubishi Light Trucks and Vans	Van/Wagons	89–90	87–101
	Montero	89–91	75–87
	Pickups	89–91	72–87
	Pickups	83–87	65
	Montero	83–87	65
Nissan/ Datsun	All	91–99	72–87
	Maxima	89–90	72–87
	Maxima	87–88	72–89
	Maxima	85–86	58–72
	Pulsar SE, SE	87–90	72–87
	Pulsar	83–86	58–72
	Sentra	83–86	58–72
	Stanza All	87–90	72–87
	Stanza	82–86	58–72
	210	79–82	58–72
	310	79–82	58–72
	510	78–81	58–72
	810	1981	58–72
	810	77–80	58–65
	200 SX All	87–88	87–108
	200 SX	80–86	58–72
	200 SX	77–79	58–65
	280 ZX	79–83	58–72
	300 ZX and ZX Turbo	1990	72–87
	300 ZX and ZX Turbo	87–89	87–108
	Axxess	1990	72–87
	240 SX	1989	72–87
Nissan/ Datsun Light Trucks and Vans	All Pickups and Pathfinder	89–99	87–108
	Van	87–88	72–87
	Van	1990	72–87
Oldsmobile	All except the following	76–99	100
	Starfire	76–80	80
	Cutlass (RWD)	76–88	80
	Cutlass Supreme (FWD)	88–93	100
	Delta 88	77–85	80

Name	Model	Years	Lb. Ft. Torque
Oldsmobile Light Trucks and Vans	Silhouette	90–99	100
	Bravado	92–99	95
Plymouth	Colt	83–93	65–80
	Sundance	87–93	95
	Acclaim	89–93	95
	Laser	90–91	87–101
	Caravelle	85–88	95
	Horizon	84–90	95
	Horizon	78–83	80
	Turismo	84–87	95
	Turismo	82–83	80
	Vista	84–91	50–57
	Reliant	84–89	80
	Reliant	81–83	80
	Gran Fury	80–89	85
	Conquest	1986	50–57
	Conquest	84–85	50–57
	Sapporo	78–83	51–58
	Champ	79–82	51–58
Plymouth Light Trucks and Vans	Voyager	84–99	95
Pontiac	All except the following	80–88	100
	T-1000	81–87	80
	Sunbird	76–80	80
	Firebird	76–88	80
	Grand Prix	76–87	80
	Lemans	89–93	65
	Catalina	76–86	80
	Bonneville	76–86	80
	Parisienne	83–86	80
Pontiac Light Trucks and Vans	Trans Sport	90–99	100
Porsche	All	79–99	94
Range Rover	All	91–99	90–95
Saab	All	88–99	80–90
	All	76–87	65–80
Saturn	All	91–99	100
Sterling	All	87–91	53
Subaru	All	76–99	58–72
Toyota	All except the following	84–99	76
	Celica	70–85	66–86
	Supra	79–85	66–86

Name	Model	Years	Lb. Ft. Torque
	Corolla	80–83	66–86
	Corona	75–82	66–86
	Cressida	78–85	66–86
	Corona MK II	1976	65–94
	Tercel	80–85	66–86
	Starlet	81–85	66–86
Toyota Light Trucks and Vans	All except the following	88–99	76
	Land Cruiser	88–91	116
	Pickups	75–87	75
	Land Cruiser	75–84	75
	Van Wagon	84–86	75
Volkswagen	All except		
	Van	88–99	81
	Golf	85–87	81
	Rabbit All	76–84	73–87
	Jetta	81–87	81

Name	Model	Years	Lb. Ft. Torque
	Scirocco	76–84	73–87
	Quantum	82–86	81
	Dasher	76–81	65
	Scirocco	85–87	81
Volkswagen Light Trucks and Vans	Vanagon	80–99	123
	Pickups	79–84	81
	Transporter	77–79	95
Volvo	All	89–99	63
	740 Series	85–88	63
	760 Series	83–88	63
	GLE	76–82	72–94
	260 Series	75–84	72–95
	240 Series	81–88	63
	240 Series	75–80	72–95
Yugo	All with steel wheels	86–90	63
	All with alloy wheels	88–90	81

DOT Tire Codes

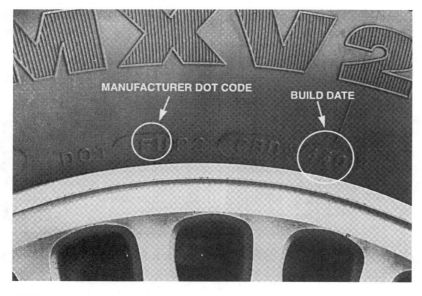

The DOT manufacturer's code on this tire is FU (always the first two letters or numbers after the DOT lettering).

Code No.	Tire Manufacturer
AC	General Tire Company Charlotte, NC, USA
AD	General Tire Company Mayfield, KY, USA
AE	General Fabrica Espanola Del Caucho S.A. (General Tire International) Torrelavego, Santander, Spain
AF	Manufactura Nacional De Borracha S.A.R.L. (General Tire International) Porto, Portugal
AH	General Popo S.A. (General Tire International) Mexico City, Mexico

Code No.	Tire Manufacturer
AJ	Uniroyal Tire Company Detroit, MI, USA
AK	Uniroyal Tire Company Chicopee Falls, MA, USA
AL	Uniroyal Goodrich Tire Co. Eau Claire, WI, USA
AM	Uniroyal Tire Company Los Angeles, CA, USA
AN	Uniroyal Goodrich Tire Co. Opelika, AL, USA
AO	Wearwell Tire & Tube Co. Bhopal (MP) India
AP	Uniroyal Goodrich Tire Co. Ardmore, OK, USA

Code No.	Tire Manufacturer
AT	Avon Rubber Co., Ltd. Melksham, Wiltshire, England
AU	Uniroyal Goodrich Tire Co. Kitchener, Ontario, Canada
AV	Seiberling Tire & Rubber Co. Barberton, OH, USA
AW	Samson Tire & Rubber Co., Ltd. Tel Aviv, Israel
AX	Phoenix Gummiwerke, A.G. Hamburg, Germany
AY	Phoenix Gummiwerke, A.G. Reinsdorf, Germany
A1	Pneumatics Michelin Potiers, France
A2	Lee Tire & Rubber Company Sao Paulo, Brazil
A3	General Tire & Rubber Company Mount Vernon, IL, USA
A4	Hung-A Industrial Co., Ltd. Pusan, Korea
A5	Debickie Zaklady Opon Samochodowych Debica, Poland
A6	Apollo Tyres, Ltd. Cochin, India
A7	Thai Bridgestone Tire Co. Ltd. Changwad Patoom-Thani, Thailand
A8	P.T. Bridgestone Tire Indonesia Factory Jawa Barat, Indonesia
A9	General Tire Company Bryan, OH, USA
BA	Uniroyal Goodrich Tire Company Akron, OH, USA
BB	B.F. Goodrich Tire Company Miami, OK, USA
BC	B.F. Goodrich Tire Company Oaks, PA, USA
BD	B.F. Goodrich Tire Company Miami, OK, USA
BE	Uniroyal Goodrich Tire Co. Tuscaloosa, AL, USA
BF	Uniroyal Goodrich Tire Co. Woodburn, IN, USA
BH	Uniroyal Goodrich Tire Co. Kitchener, Ontario, Canada
BK	B.F. Goodrich do Brasil Campinas, Brazil

Code No.	Tire Manufacturer
BL	Industria Colombiana de Liantas S.A. (B.F. Goodrich International) Bogata, Colombia
BM	B.F. Goodrich Australia Pty., Ltd. Campbellfield, Victoria, Australia
BN	B.F. Goodrich Philippines, Inc. Makti, Rizal. Philippines
BO	General Tire Company Casablanca, Morocco
BT	Semperit, A.G. Traiskirehen, Austria
BU	Semperit Ireland Ltd. Dublin, Ireland
BV	International Rubber Industries Jeffersontown, KY, USA
BW	The Gates Rubber Company Denver, CO, USA
BX	The Gates Rubber Company Nashville, TN, USA
BY	The Gates Rubber Company Littleton, CO, USA
B1	Pneumatics Michelin La Rochesur-Yon, France
B2	Dunlop Malaysian Industries Berhad Selongor, Malaysia
B3	Michelin Tire Mfg. Company of Canada, Ltd. Bridgewater, Nova Scotia
B4	Taurus Hungarian Rubber Works Budapest, Hungary
B5	Olsztynskle Zaklady Opon Samochodowych Olsztyn, Poland
B6	Michelin Tire Corporation Spartanburg, SC, USA
B7	Michelin Tire Corporation Dothan, AL, USA
B8	Cia Brasiliera de Pneumaticos Michelin Industria Rio De Janeiro, Brazil
B9	Michelin Tire Corporation Lexington, SC, USA
CA	The Mohawk Rubber Company Akron, OH, USA
CB	The Mohawk Rubber Company Helena, AR, USA
CC	The Mohawk Rubber Company Salem, VA, USA
CD	Alliance Tire & Rubber Co., Ltd. Hadera, Israel

Code No.	Tire Manufacturer
CF	The Armstrong Rubber Company Des Moines, IA, USA
CH	The Armstrong Rubber Company Hanford, CA, USA
CJ	Inque Rubber Company, Ltd. Atsuta-ku, Nagoyo, Japan
CK	The Armstrong Rubber Company Madison, TN, USA
CL	Continental A.C. Hannover, Germany
CM	Continental Gummi-Werke, A.G. Hannover, Germany
CN	Usine Francaise des Pneumatics Sarreguemines, France
CO	Goodyear Lastikleri Tas Adopozare, Turkey
CP	Continental Gummi-Werke, A.G. Korbach, Germany
CT	Continental Gummi-Werke, A.C. Dannenberg, Germany
CU	Continental Gummi-Werke, A.C. Hannover-Stocken, Germany
CV	The Armstrong Rubber Company Natchez, MS, USA
CW	The Toyo Rubber Industry Co., Ltd. Itami, Hyogo, Japan
CX	The Toyo Rubber Industry Co., Ltd. Natori-Gun, Miyagi, Japan
CY	McCreary Tire & Rubber Company Indiana, PA, USA
C1	Michelin (Nigeria) Ltd. Port Harcourt, Nigeria
C2	Kelly-Springfield Tire Company Americana, Sao Paulo, Brazil
C3	McCreary Tire & Rubber Company Baltimore, MD, USA
C4	Dico Tire, Inc. Clinton, TN, USA
C5	Poznanskie Zaklady Opon Samochodowych Poznan, Poland
C6	MITAS, n.p. Praha Praque, Czechoslovakia
C7	Ironsides Tire & Rubber Company Louisville, KY, USA
C8	Hsin Chu Plant—Bridgestone Hsin Chu, Taiwan
C9	Seven Star Rubber Co., Ltd. Chang-Hua, Taiwan

Code No.	Tire Manufacturer
DA	The Dunlop Tire & Rubber Co. Buffalo, NY, USA
DB	The Dunlop Tire & Rubber Corp. Huntsville, AL, USA
DC	Dunlop Tire Canada, Ltd. Whitby, Ontario, Canada
DD	The Dunlop Company, Ltd. Birmingham, England
DE	The Dunlop Company, Ltd. Washington, Durham, England
DK	Dunlop S.A. Montlucon, France
DL	Dunlop S.A. Amiens, France
DM	Dunlop A.G. Hanau Am Main, Germany
DN	Dunlop A.G. Wittlich, Germany
DO	Kelly-Springfield Lastikleri Tas Adapazari, Turkey
DV	N.V. Nederlandsch-Amerikaansche Enschnede, The Netherlands
DW	Rubberfabriek Vredestein Doetinchem, The Netherlands
DX	N.V. Bataafsche Rubber Ind. Radium Maastricht, The Netherlands
DY	Denman Tire Corporation Leavittsburg, OH, USA
D1	Viking Askim Askim, Norway
D2	Bridgestone/Firestone, Inc. LaVergne, TN, USA
D3	United Tire & Rubber Mfg. Company Cobourg, Ontario, Canada
D4	Dunlop India, Ltd. West Bengal, India
D5	Dunlop India, Ltd. Madras, India
D6	Borovo Borovo, Yugoslavia
D7	Dunlop South Africa Ltd. Natal, Republic of South Africa
D8	Dunlop South Africa Ltd. Natal, Republic of South Africa
D9	United Tire & Rubber Co., Ltd. Rexdale, Ontario, Canada
EA	Metzler, A.G. Munchen, Germany

Code No.	Tire Manufacturer
EB	Metzler, A.G. Zweigwerk Neustadt Neustadt Odenwald, Germany
EC	Metzler, A.G. Hochst, Germany
ED	Michelin-Okamoto Tire Corporation Tokyo, Japan
EE	Nitto Tire Company, Ltd. Kanagawa-ken, Japan
EF	Hung Ah Tire Company, Ltd. Seoul, Korea
EH	Bridgestone Tire Company, Ltd. Fukuoka-ken, Japan
EJ	Bridgestone Tire Company, Ltd. Saga-ken, Japan
EK	Bridgestone Tire Company, Ltd. Fukuoka-ken, Japan
EL	Bridgestone Tire Company, Ltd. Shiga-ken, Japan
EM	Bridgestone Tire Company, Ltd. Tokyo, Japan
EN	Bridgestone Tire Company, Ltd. Tochigi-ken, Japan
EO	Lee Lastikleri Tas Adapazari, Turkey
EP	Bridgestone Tire Company, Ltd. Tochigi-ken, Japan
ET	Sumitomo Rubber Industries Fuklai, Kobe, Japan
EU	Sumitomo Rubber Industries Alchi Prefecture, Japan
EW	Pneumatiques Kleber, S.A. Toul (Meurthe-et-Moselle), France
EX	Pneumatiques Kleber, S.A. La Chapella St. Luc, France
EY	Pneumatiques Kleber, S.A. St. Ingbert (Saar), Germany
E1	Chung Hsing Industrial Co., Ltd. Taichung Hsien, Taiwan
E2	Industria de Pneumaticos Firestone Sao Paulo, Brazil
E3	Seiberling Tire & Rubber Company Lavergne, TN, USA
E4	The Firestone Tire and Rubber Co. of New Zealand, Ltd. Papanui, Christ Church, New Zealand
E5	Firestone South Africa (Pty) Ltd. Port Elizabeth, South Africa
E6	Firestone-Tunisie S.A. Menzel-Bourguiba, Tunisia

Code No.	Tire Manufacturer
E7	Firestone East Africa Ltd. Nairobi, Kenya
E8	Firestone Ghana Ltd. Accra, Ghana
E9	Firestone South Africa (Pty) Ltd. Brits, South Africa
FA	The Yokohama Rubber Company, Ltd. Hiratsuka, Kanagawa-Pref, Japan
FB	The Yokohama Rubber Company, Ltd. Watari-gun, Miye-Pref, Japan
FC	The Yokohama Rubber Company, Ltd. Mishima, Shizuoka-Pref, Japan
FD	The Yokohama Rubber Company, Ltd. Shinsharo, Aichi-Pref, Japan
FE	The Yokohama Rubber Company, Ltd. Ageo, Saitama-Pref, Japan
FF	Manufacture Francaise des Pneumatiques Michelin Clermont-Ferrand, France
FH	Manufacture Francaise des Pneumatics Michelin Clermont-Ferrand, France
FJ	Manufacture Francaise des Pneumatics Michelin Bourges, France
FK	Manufacture Francaise des Pneumatics Michelin Cholet, France
FL	Manufacture Francaise des Pneumatics Michelin Montceau-les-Mines, France
FM	Manufacture Francaise des Pneumatics Michelin Mesmin, Orleans, France
FN	Manufacture Francaise des Pneumatics Michelin Tours, France
FP	Ste. d'Applications Techniques Ind. Algers, Algeria
FT	Michelin Reifenwerke, A.G. Bad Kreuznach, Germany
FU	Michelin Reifenwerke, A.G. Bamberg, Germany
FV	Michelin Reifenwerke, A.G. Homburg, Germany
FW	Michelin Reifenwerke, A.G. Karlsruhe, Germany
FX	S.A. Belge du Pneumatique Michelin Zuen, Belgium

Code No.	Tire Manufacturer
FY	N.V. Nederlandsche Banden, Industries Michelin S'Hertogenbosch, Bois-le-duc, Holland
F1	Michelin Tyre Company, Ltd. Dundee, Scotland
F2	C.A. Firestone Venezolana Valencia, Venezuela
F3	Manufacture Francaise Pneumatiques Michelin Roanne, France
F4	CNB-Companhia Nacional de Borrachas Oporto, Portugal
F5	FATE S.A.I.C.I. Buenos Aires, Argentina
F6	Torrelavega Torrelavega, Spain
F7	Puente San Miguel Firestone Torrelavega, Spain
F8	Vikrant Tyres Ltd. Mysore, Karnataka, India
F9	Dunlop New Zealand, Ltd. Upper Hutt, New Zealand
F0	Fidelity Tire Mfg. Company Natchez, MS, USA
HA	S.A.F.E.N.M. (Michelin) Aranda, Spain
HB	S.A.F.E.N.M. Lasarte (Michelin) Lasarte, Spain
HC	S.A.F.E.N.M. (Michelin) Vitoria, Spain
HD	Societa per Azoni, Michelin Cuneo, Italy
HE	Societa per Azoni, S.P.A., Michelin Marengo, Italy
HF	Societa per Azoni, Michelin Turin (Dora), Italy
HH	Societa per Azoni, Michelin Turin (Stura), Italy
HJ	Michelin Tyre Company, Ltd. Ballymena, North Ireland
HK	Michelin (Belfast) Ltd. Belfast, North Ireland
HL	Michelin Tyre Company, Ltd. Burnley, England
HM	Michelin Tyre Company, Ltd. Stoke-on-Trent, England
HN	Michelin Tire Mfg. Co. of Canada New Glasgow, Nova Scotia
HP	Manufacture Saigonnaise des Pneumatiques Michelin Saigon, Vietnam

Code No.	Tire Manufacturer
HT	Ceat, S.P.A. Pneumatici via Leoncavallo Torino, Italy
HU	Ceat 10036 Settimo Torinese, Italy
HV	Gentyre S.P.A. Frosinone, Italy
HW	Barum Tire Co. Otokovice, Czechoslovakia
HX	The Dayton Tire & Rubber Company Dayton, OH, USA
HY	Bridgestone-Firestone Inc. Oklahoma City, OK, USA
H1	DeLa S.A.F.E. Neumaticos Michelin Valladolid, Spain
H2	Kumho & Co., Inc. Tire Division Kwangsan-gun, Chonnam, Korea
H3	Sava Industrija Gumijevih Skofjeloska, Kranj, Yugoslavia
H4	Bridgestone-Houf Yamaguchi-Ken, Japan
H5	Hutchinson-MAPA Chalette Sur Loing, France
H6	Shin Hung Rubber Co. Ltd. Kyung Nam, Korea
H7	Li Hsin Rubber Industrial Co. Chi-Hu, Chang-Hwa, Taiwan
H9	Reifen-Berg Clevischer Ring, Germany
H0	The General Tyre & Rubber Co. Karachi, Pakistan
JA	The Lee Tire & Rubber Co. (Goodyear Co., Plant 1) Akron, OH, USA
JB	The Lee Tire & Rubber Co. (Goodyear Co., Plant 2) Akron, OH, USA
JC	The Lee Tire & Rubber Company Conshohocken, PA, USA
JD	The Lee Tire & Rubber Company (Kelly-Springfield Tire Co.) Cumberland, MD, USA
JE	The Lee Tire & Rubber Company (Goodyear Tire & Rubber Co.) Danville, VA, USA
JF	The Lee Tire & Rubber Company (Goodyear Tire & Rubber Co.) Fayetteville, NC, USA
JH	The Lee Tire & Rubber Company (Goodyear Tire & Rubber Co.) Freeport, IL, USA

Code No.	Tire Manufacturer
JJ	The Lee Tire & Rubber Company (Goodyear Tire & Rubber Co.) Gadsden, AL, USA
JK	The Lee Tire & Rubber Company (Goodyear Tire & Rubber Co.) Jackson, MI, USA
JL	The Lee Tire & Rubber Company (Goodyear Tire & Rubber Co.) Los Angeles, CA, USA
JM	The Lee Tire & Rubber Company (Goodyear Tire & Rubber Co.) New Bedford, MA, USA
JN	The Lee Tire & Rubber Company (Goodyear Tire & Rubber Co.) Topeka, KS, USA
JP	The Lee Tire & Rubber Company (Goodyear Tire & Rubber Co.) Tyler, TX, USA
JT	The Lee Tire & Rubber Company (Goodyear Tire & Rubber Co.) Union City, TN, USA
JU	The Lee Tire & Rubber Company (Goodyear Tire & Rubber Co.) Medicine Hat, AB, Canada
JV	The Lee Tire & Rubber Company (Goodyear Tire & Rubber Co.) Toronto 14, Ontario, Canada
JW	The Lee Tire & Rubber Company (Seiberling Rubber Co. of Canada) Toronto 9, Ontario, Canada
JX	The Lee Tire & Rubber Company (Goodyear Tire & Rubber Co.) Valleyfield, Quebec, Canada
JY	The Lee Tire & Rubber Company (Neumaaticos Goodyear S.A.) Hurlingham, F.C.N.S.M., Argentina
J1	Phillips Petroleum Company Bartlesville, OK, USA
J2	Bridgestone Singapore Company Jurong Town, Singapore
J3	Gumarne 1 Maja Puchov, Czechoslovakia
J4	Rubena, N.P. Nachod, Czechoslovakia
J5	The Lee Tire & Rubber Company Logan, OH, USA
J6	Jaroslavi Tire Company Jaroslavi, U.S.S.R.
J7	R & J Mfg. Corporation Plymouth, IN, USA

Code No.	Tire Manufacturer
J8	Da Chung Hua Rubber Ind. Co. Shanghai, China
J9	P.T. Intirub Besar, Jakarta, Indonesia
J0	Korea Inocee Kasei Co., Ltd. Masan, Korea
KA	The Lee Tire & Rubber Company (Goodyear Tire & Rubber Co.) Granville, New South Wales, Australia
KB	The Lee Tire & Rubber Company (Goodyear Tire & Rubber Co.) Thomastown, Victoria, Australia
KC	The Lee Tire & Rubber Company (Companhia Goodyear do Brasil) Sao Paulo, Brazil
KD	The Lee Tire & Rubber Company (Goodyear de Colombia S.A.) Cali, Colombia
KE	The Lee Tire & Rubber Company (Goodyear Congo) Republic of the Congo
KF	The Lee Tire & Rubber Company (Compagnie Francaise Goodyear) 80 Amiens-Somme, France
KH	The Lee Tire & Rubber Company (Deutsche Goodyear G.M.B.H.) Phillipsburg Brucksal, Germany
KJ	The Lee Tire & Rubber Company (Goodyear International) 64 Fulda, Germany
KK	The Lee Tire & Rubber Company (Goodyear Hellas S.A.I.C.) Thessaloniki, Greece
KL	The Lee Tire & Rubber Company (Goodyear International) Guatemala City, Guatemala
KM	The Lee Tire & Rubber Company Grand Duchy of Luxembourg
KN	The Lee Tire & Rubber Company (Goodyear India Ltd. Factory) District Gurgaon, India
KP	The Lee Tire & Rubber Company (The Goodyear Tire & Rubber Co.) Bogor, Republic of Indonesia
KT	The Lee Tire & Rubber Company (Goodyear Italiana) Latina, Italy
KU	The Lee Tire & Rubber Company (Goodyear Jamaica Ltd.) Jamaica, West Indies

Code No.	Tire Manufacturer
KV	The Lee Tire & Rubber Company (Compania Hulera Goodyear) Mexico City, Mexico
KW	The Lee Tire & Rubber Company (Compania Goodyear del Peru) Lima, Peru
KX	The Lee Tire & Rubber Company (The Goodyear Tire & Rubber Co.) Las Pinas, Rizal, Philippines
KY	The Lee Tire & Rubber Company (Goodyear Tire & Rubber Co. (GB)) Glasgow, Scotland
K1	Phillips Petroleum Company Stow, OH, USA
K2	The Lee Tire & Rubber Company Madisonville, KY, USA
K2	Kenda Rubber Ind. Co., Ltd. Yualin, Taiwan
K4	Uniroyal S.A. (Uniroyal Goodrich) Queretaro, Mexico
K5	VEB Reifenkombinat Furstenwalde Democratic Republic of Germany
K6	The Lee Tire & Rubber Company Lawton, OK, USA
K7	The Lee Tire & Rubber Company Santiago, Chile
K8	The Kelly-Springfield Tire Co. Selangor, Malaysia
K9	Natier Tire & Rubber Co., Ltd. Shetou, Changhua, Taiwan
K0	Michelin Korea Tire Co., Ltd. San-Kun, Kyungsangnam, Korea
LA	The Lee Tire & Rubber Company (Goodyear Tyre & Rubber Co.) Uitenhage, South Africa
LB	The Lee Tire & Rubber Company (Goodyear Gummi Fabics Aktirbolag) Norrkoping, Sweden
LC	The Lee Tire & Rubber Company (Goodyear International) Bangkok, Thailand
LD	The Lee Tire & Rubber Company (Goodyear International) Kocaeli, Turkey
LE	The Lee Tire & Rubber Company (CA Goodyear de Venezuela) Valencia, Edo Carabobo, Venezuela
LF	The Lee Tire & Rubber Company (Goodyear Tire & Rubber Co. (GB)) Wolverhampton, England

Code No.	Tire Manufacturer
LJ	Uniroyal Englebert Belgique S.A. Herstal-les-Liege, Belgium
LK	Productors Niaconal de Llantas S.A. Cali, Colombia
LL	Uniroyal Englebert France S.A. Compiegne, France
LM	Uniroyal Englebert Deutschland D.A. Aachen, Germany
LN	Uniroyal SA (Uniroyal Goodrich) Mexico, Mexico
LP	Uniroyal Ltd., Tire & Gen. Products Newbridge, Midlothian, Scotland
LT	Uniroyal Endustri Turk Anonim Adapazari, Turkey
LU	Uniroyal, C.A. Valencia, Venezuela
LV	General Tire Canada Ltd. Barrie, Ontario, Canada
LW	Trelleborg Gummifariks Akiebolag Trelleborg, Sweden
LX	Mitsuboshi Belting Ltd. Kobe, Japan
LY	Mitsuboshi Belting Ltd. Shikoku, Japan
L1	Goodyear Taiwan Ltd. Taipei, Taiwan
L2	WUON Poong Ind. Co., Ltd. Pusan, Korea
L3	Tong Shin Chemical Products Co. Seoul, Korea
L4	Centrala Ind. de Prelucrare Cauciuc Oltentei, Romania
L5	BRISA Bridgestone Sabanci P.K. Izmit, Turkey
L6	MODI Rubber Limited Meerut UP, India
L7	Intreprinderea De Anvelope Zalau Zalau, Judetul Saloj, Romania
L8	Dunlop Zimbabwe Ltd. Domington, Bulawaye, Zimbabwe
L9	Panther Tyres, Ltd. Aintree, Liverpool, United Kingdom
L0	Mfg. Francaise Des Pneumatiques Clermont Ferrand Cedex, France
MA	Goodyear Tire & Rubber Co. (Plant 1) Akron, OH, USA
MB	Goodyear Tire & Rubber Co. (Plant 2) Akron, OH, USA

Code No.	Tire Manufacturer
MC	The Goodyear Tire & Rubber Co. Danville, VA, USA
MD	The Goodyear Tire & Rubber Co. Gadsden, AL, USA
ME	The Goodyear Tire & Rubber Co. Jackson, MI, USA
MF	The Goodyear Tire & Rubber Co. Los Angeles, CA, USA
MH	The Goodyear Tire & Rubber Co. New Bedford, ME, USA
MJ	The Goodyear Tire & Rubber Co. Topeka, KS, USA
MK	The Goodyear Tire & Rubber Co. Union City, TN, USA
ML	The Goodyear Tire & Rubber Co. Cumberland, MD, USA
MM	The Goodyear Tire & Rubber Co. Fayetteville, NC, USA
MN	The Goodyear Tire & Rubber Company Freeport, IL, USA
MO	Neumaticos De Chile S.A. Coquimbo, Chile
MP	The Goodyear Tire & Rubber Company Tyler, TX, USA
MT	The Goodyear Tire & Rubber Co. Conshohocken, PA, USA
MU	Neumaticos Goodyear, S.A. Hurlingham F.C.N.G.S.M., Argentina
MV	The Goodyear Tyre & Rubber Co. Ltd. New South Wales, Australia
MW	The Goodyear Tyre & Rubber Co., Ltd. Thomastown, Victoria, Australia
MX	Companhia Goodyear do Brasil Sao Paulo, Brazil
MY	Goodyear de Colombia, S.A. Cali, Colombia
M1	Goodyear Maroc, S.A. Casablanca, Morocco
M2	The Goodyear Tire & Rubber Co. Madisonville, KY, USA
M3	Michelin Tire Corporation Greenville, SC, USA
M4	The Goodyear Tire & Rubber Co. Logan, OH, USA
M5	Michelin Tire Mfg. Co. of Canada Kentville, Nova Scotia, Canada
M6	Goodyear Tire & Rubber Co. Lawton, OK, USA

Code No.	Tire Manufacturer
M7	Goodyear de Chile, S.A.I.C. Santiago, Chile
M8	Premier Tyres Limited Kerala State, India
M9	Uniroyal Tire Corp. Middlebury, CT, USA
NA	Goodyear Congo Republic of the Congo
NB	The Goodyear Tyre & Rubber Co. Wolverhampton, England
NC	Compagnie Francaise Goodyear Amiens-Somme, France
ND	Deutsche Goodyear G.M.B.H. Phillipsburg Bruchsal, Germany
NE	Gummiwerke Fuida G.M.B.H. Fulda, Germany
NF	Goodyear Hellas S.A.I.C. Thessaloniki, Greece
NH	Gran Industria de Neumatticos Guatemala City, Guatemala
NJ	Goodyear S.A. Grand Duchy of Luxembourge
NK	Goodyear India Ltd. Distric Gurgaon, India
NL	The Goodyear Tire & Rubber Co., Ltd. Bogor, Republic of Indonesia
NM	Goodyear Italiana S.P.A. Latina, Italy
NN	Goodyear Jamaica Ltd. Jamaica, West Indies
NO	South Pacific Tyres Victoria, Australia
NP	Compania Hulera Goodyear Oxo Mexico City, Mexico
NT	Compania Goodyear del Peru Lima, Peru
NU	Goodyear Tire & Rubber Co. Las Pinas, Rizal, Philippines
NV	Goodyear Tyre & Rubber Co. Ltd. Glasgow, Scotland
NW	Goodyear Tyre & Rubber Co. Uitenhage, South Africa
NX	Goodyear Gummi Faabriks Aktirbolag Norrkoping, Sweden
NY	Goodyear (Thailand) Ltd. Bangkok, Thailand
N1	Maloja AG. Pneu-und-Gummiwerke Gelterkinden, Switzerland

Code No.	Tire Manufacturer
N2	Hurtubise Nutread Inc. Tonawanda, NY, USA
N3	Nitto Tire Co., Ltd. Tohin-Cho, Inabe-Gun Mie-Ken, Japan
N4	Centrala Ind. de Prelucrare Cauciue Republic Socialista Romania
N5	Pneumant, VEB Reifenwerk Riesa Riesa, Germany
N6	Pneumant Democratic Republic of Germany
N7	Intreprinderea De Anvelope Caracal Caracal, Judetul Olt., Romania
N8	Lee Tire & Rubber Co. (Goodyear) Selangor, Malaysia
N9	Cla Pneus Tropical Feira de Santana, Bahla, Brazil
PA	Goodyear Lastikleri Tas Kocaeli, Turkey
PB	CA Goodyear de Venezuela Valencia, Edo Carabobo, Venezuela
PC	Goodyear Tire & Rubber Co. Medicine Hat, Alberta, Canada
PD	Goodyear Tire & Rubber Co. Valleyfield, Quebec, Canada
PE	Seiberling Rubber Co. Toronto, Ontario, Canada
PF	The Goodyear Tire & Rubber Co. Toronto, Ontario, Canada
PH	Kelly-Springfield Tire Co. Cumberland, MD, USA
PI	Gislaved Daek AB Gislaved, Sweden
PJ	Kelly-Springfield Tire Co. Fayetteville, NC, USA
PK	Kelly-Springfield Tire Co. Freeport, IL, USA
PL	Kelly-Springfield Tire Co. Tyler, TX, USA
PM	Kelly-Springfield Tire Co. (Goodyear Tire & Rubber Co.) Conshohocken, PA, USA
PN	Kelly-Springfield Tire Co. (Goodyear Tire & Rubber Co.) Akron, OH, USA
PO	South Pacific Tyres New South Wales, Australia
PP	Kelly-Springfield Tire Co. (Goodyear Tire & Rubber Co.) Akron, OH, USA

Code No.	Tire Manufacturer
PT	Kelly-Springfield Tire Co. (Goodyear Tire & Rubber Co.) Danville, VA, USA
PU	Kelly-Springfield Tire Co. (Goodyear Tire & Rubber Co.) Gadsden, AL, USA
PV	Kelly-Springfield Tire Co. (Goodyear Tire & Rubber Co.) Jackson, MI, USA
PW	Kelly-Springfield Tire Co. (Goodyear Tire & Rubber Co.) Los Angeles, CA, USA
PX	Kelly-Springfield Tire Co. (Goodyear Tire & Rubber Co.) New Bedford, ME, USA
PY	Kelly-Springfield Tire Co. (Goodyear Tire & Rubber Co.) Topeka, KS, USA
P1	Gummifabriken Gislaved Aktiebolag Gislaved, Sweden
P2	Kelly-Springfield Tire Co. Madisonville, KY, USA
P3	Skepplanda Gummi, AB Alvangen, Sweden
P4	Kelly-Springfield Tire Co. Logan, OH, USA
P5	General Popo S.A. San Luis Potosi, Mexico
P6	Kelly-Springfield Tire Co. Lawton, OH, USA
P7	Kelly-Springfield Tire Co. Santiago, Chile
P8	No. 2 Rubber Plant Oingdao Quingdao, Shandong, China
P9	MRF, Ltd., PB No. 1 Ponda Goa, India
TA	Kelly-Springfield Tire Co. (Goodyear Tire & Rubber Co.) Union City, TN, USA
TB	Kelly-Springfield Tire Co. Hurlingham F.C.N.G.S.M. Argentina
TC	Kelly-Springfield Tire Co. (Goodyear Tyre & Rubber Co.) New South Wales, Australia
TD	Kelly-Springfield Tire Co. (Goodyear Tyre & Rubber Co.) Thomastown, Victoria, Australia
TE	Kelly-Springfield Tire Co. (Companhia Goodyear do Brasil) Sao Paulo, Brazil

Code No.	Tire Manufacturer
TF	Kelly-Springfield Tire Co. (Goodyear de Colombia, S.A.) Yumbo, Calle, Colombia
TH	Kelly-Springfield Tire Co. (Goodyear Congo) Republic of the Congo
TJ	Kelly-Springfield Tire Co. (Goodyear Tyre & Rubber Co.) Wolverhampton, England
TK	Kelly-Springfield Tire Co. (Compagnie Francaise Goodyear S.A.) Amiens-Somme, France
TL	Kelly-Springfield Tire Co. (Deutsche Goodyear G.M.B.H.) Bruchsal, Germany
TM	Kelly-Springfield Tire Co. (Goodyear International) Fulda, Germany
TN	Kelly-Springfield Tire Co. (Goodyear Hellas S.A.I.C.) Thessaloniki, Greece
TO	South Pacific Tyres Campbellfield, Victoria
TP	The Kelly-Springfield Tire Co. (Goodyear International) Guatemala City, Guatemala
TT	The Kelly-Springfield Tire Co. (Goodyear S.A) Grand Duchy of Luxembourg
TU	Kelly-Springfield Tire Co. (Goodyear India Ltd.) District Gurgaon, India
TV	Kelly-Springfield Tire Co. (Goodyear Tire & Rubber Co.) Bogor, Republic of Indonesia
TW	Kelly-Springfield Tire Co. (Goodyear Italiana) Latina, Italy
TX	Kelly-Springfield Tire Co. (Goodyear Jamaica Ltd.) Jamaica, West Indies
TY	Kelly-Springfield Tire Co. (Compania Hulera Goodyear) Mexico City, D.F. Mexico
T1	Hankook Tire Mfg. Co. Ltd. Seoul, Korea
T2	Ozos (Uniroyal) Olsztyn, Poland
T3	Debickle Zattldy Opon Samochodowych (Uniroyal AG) Debica, Poland

Code No.	Tire Manufacturer
T4	S.A. Carideng (Rubberfactory) Lanaken, Belgium
T5	Tigar-Pirot Pirot, Yugoslavia
T6	Hulera Tomel S.A. San Antonia, Mexico, D.F.
T7	Hankook Tire Mfg. Co., Ltd. Daejun, Korea
T8	Goodyear Malaysia Berhad Selangor, Malaysia
T9	MRF, Ltd. Arkonam, India
UA	Kelly-Springfield Tire Co. (Compania Goodyear del Peru) Lima, Peru
UB	Kelly-Springfield Tire Co. (Goodyear Tire & Rubber Co.) Las Pinas, Rizal, Philippines
UC	Kelly-Springfield Tire Co. (Goodyear Tyre & Rubber Co.) Glasgow, Scotland
UD	Kelly-Springfield Tire Co. (Goodyear Tyre & Rubber Co.) Uitenhage, South Africa
UE	Kelly-Springfield Tire Co. (Goodyear Gummi Fabriks Aktiebolag) Norrkoping, Sweden
UF	Kelly-Springfield Tire Co. (Goodyear Thailand Ltd.) Bangkok, Thailand
UH	Kelly-Springfield Tire Co. (Goodyear Lastikleri Tas Pk 2 Izmit) Kocaeli, Turkey
UJ	Kelly-Springfield Tire Co. (CA Goodyear de Venezuela) Edo Carabob, Venezuela
UK	Kelly-Springfield Tire Co. (Goodyear Tire & Rubber Co.) Medicine Hat, Alberta, Canada
UL	Kelly-Springfield Tire Co. (Goodyear Tire & Rubber Co.) Valleyfield, Quebec, Canada
UM	Kelly-Springfield Tire Co. (Seiberling Rubber Co. of Canada) Toronto, Ontario, Canada
UN	Kelly-Springfield Tire Co. (Goodyear Tire & Rubber Co.) Toronto, Ontario, Canada
UO	South Pacific Tyres Thomastown, Vic., Australia

Code No.	Tire Manufacturer	Code No.	Tire Manufacturer
UP	Cooper Tire & Rubber Co. Findlay, OH, USA	VK	The Firestone Tire & Rubber Co. Salinas, CA, USA
UT	Cooper Tire & Rubber Co. Texarkana, AR, USA	VL	The Firestone Tire & Rubber Co. Hamilton, Ontario, Canada
UU	Carlisle Tire & Rubber Division Carlisle, PA, USA	VM	The Firestone Tire & Rubber Co. Calgary, Alberta, Canada
UV	Kzowa Rubber Ind. Co., Ltd. Nishinariku, Osaka, Japan	VN	Bridgestone-Firestone Inc. Joliette, Quebec, Canada
UW	Okada Tire Industry Ltd. Katsushika-ku, Tokyo, Japan	VO	South Pacific Tyres Upper Hutt, Wellington, New Zealand
UX	Federal Corporation Taipei, Taiwan	VP	The Firestone Tire & Rubber Co. Bari, Italy
UY	Cheng Shin Rubber Ind. Co. Ltd. Meci Kong, Teipi, Taiwan	VT	The Firestone Tire & Rubber Co. Bilbao, Spain
U1	Lien Shin Tire Co. Ltd. Taipei, Taiwan	VU	Universal Tire Company Lancaster, PA, USA
U2	Sumitomo Rubber Industries, Ltd. Shirakawa City Fukushima Prefecture, Japan	VV	The Firestone Tire & Rubber Co. Viskafors, Sweden
U3	Miloje Zakic, Krusevac, Yugoslavia	VW	Ohtsu Tire & Rubber Co. Ltd. Osaka, Japan
U4	Geo. Byers Sons, Inc. Columbus, OH, USA	VX	Firestone Tyre & Rubber Co., Ltd. Brentford, Middlesex, England
U5	Farbenfabriken Bayer GmBH Leverkusen, Germany	VY	Firestone Tyre & Rubber Co. Ltd. Denbighshire, Wrexhan, Wales
U6	Pneumant Democratic Republic of Germany	V1	Livingston's Tire Shop Hubbard, OH, USA
U7	Pneumant Democratic Republic of Germany	V2	Vsesojuznoe Ojedinenic Avtoexport Volzhsk, USSR
U8	Hsin-Fung Fac. of Nankang Rubber Co. Taiwan Province, R.O.C.	V3	TA Hsin Rubber Tire Co. Ltd. Taipei Hsieng, Taiwan
U9	Cooper Tire & Rubber Co. Tupelo, MS, USA	V4	Ohtsu Tire & Rubber Co. Miyazaki Prefecture, Japan
VA	The Firestone Tire & Rubber Co. Akron, OH, USA	V5	Firestone El Centenario S.A. Mexico City, Mexico
VB	The Firestone Tire & Rubber Co. Akron, OH, USA	V6	Firestone Cuernavaca Cuernavaca, Mexico
VC	The Firestone Tire & Rubber Co. Albany, GA, USA	V7	Vsesojuznoe Ojedinenie Avtoexport Voronezh, USSR
VD	Bridgestone-Firestone Inc. Decatur, IL, USA	V8	Boras Gummifabrik AB Boras, Sweden
VE	Bridgestone-Firestone Inc. Des Moines, IA, USA	V9	M & H Tire Co. Gardner, MA, USA
VF	The Firestone Tire & Rubber Co. South Gate, CA, USA	WA	Firestone—France S.A. Bethune, France
VH	The Firestone Tire & Rubber Co. Memphis, TN, USA	WB	Industria Akron De Costa Rico San Jose, Costa Rica, S.A.
VJ	The Firestone Tire & Rubber Co. Pottstown, PA, USA	WC	Firestone Australia Pty. Ltd. Sydney, New South Wales, Australia

Code No.	Tire Manufacturer
WD	Fabrik fur Firestone Produkte A.G. Prattlen, Switzerland
WE	Nankang Rubber Tire Corp. West Taipei, Taiwan
WF	The Firestone Tire & Rubber Co. Burgos, Spain
WH	The Firestone Tire & Rubber Co. Boras, Sweden
WM	Dunlop Olympic Tyres W. Footscray, Victoria, Australia
WO	Inocce Rubber Co., Inc. Phatumtani, Thailand
WT	Madras Rubber Factory Ltd. Madras, India
WU	Ceat Typres of India Ltd. Bhandup, Bombay, India
WV	General Rubber Corporation Taipei, Taiwan
WW	Euzkadi Co. Hulera Euzkadi, S.A. Mexico City, Mexico
WX	Euzkadi Co. Hulera Euzkadi, A.A. La Presa, Edo de, Mexico
WY	Euzkadi Co. Hulera Euzkadi, S.A. Guadalijara, Jalisco, Mexico
W1	Bridgestone/Firestone, Inc. LaVergne, TN, USA
W2	Bridgestone/Firestone Inc. Wilson, NC, USA
W3	Vredestein Doetinchem B. V. Doetinehem, The Netherlands
W4	Dunlop Olympic Tyres Somerton, Vic., Australia
W5	Firestone de la Argentina Province de Buenos Aires, Argentina
W6	Philtread Tire & Rubber Corp. Makati, Rizal Philippines
W7	Firestone Portuguesa, S.A.R.L. Alcochete, Portugal
W8	Siam Tyre Co., Ltd. Bangkok, Thailand
W9	Ind. de Pneumaticos Firestone S.A. Rio de Janeiro, Brazil
XA	Industrie Pirelli S.p.A., V. le Milan, Italy
XB	Industri Pirelli S.p.A. Settimo, Torinese, Torino, Italy
XC	Industrie Pirelli S.p.A. Tivoil-Roma, Italy

Code No.	Tire Manufacturer
XD	Industrie Pirelli S.p.A. Messina, Italy
XE	Industrie Pirelli S.p.A. Ferrandina, Italy
XF	Productos Pirelli S.A. Barcelona, Spain
XH	Pirelli Hallas, S.A. Patrasso, Greece
XJ	Turk Pirelli Lastilkeri, S.A. Istanbul, Turkey
XK	Pirelli, S.A. Sao Paulo, Brazil
XL	Pirelli, S.A. Campinas, Brazil
XM	Pirelli Co. Platennse de Neumaticos, Merio, Buenos Aires, Argentino
XN	Pirelli, Ltd. Carlisle, England
XP	Pirelli Ltd. Burton-on-Trent, England
XT	Veith-Pirelli, A.G. Sandbach, Germany
XV	Dayton Tire & Rubber Co. (Firestone Tire & Rubber Co.) Hamilton, Ontario, Canada
XW	Dayton Tire & Rubber Co. (Firestone Tire & Rubber Co.) Calgary, Alberta, Canada
XX	Bandag, Inc. Muscatine, IA, USA
XY	Dayton Tire & Rubber Co. (Firestone Tire & Rubber Co.) Joliette, Quebec, Canada
X1	Tong Shin Chemical Prod. Co., Ltd. Seoul, Korea
X2	Hwa Fong Rubber Ind. Co. Ltd. Yualin, Taiwan
X3	Vaesojuznoe Ojedinenic Avtoexport Belaya Tserkov, USSR
X4	Pars Tyre Co. Saveh, Iran
X5	J.K. Industries Ltds. Kankroli, Rajasthan, India
X6	Vsesojuznoe Ojedinenie Avtoexport Bobruysk, USSR
X7	Vsesojuznoe Ojedinenie Avtoexport Chimkentsky, USSR
X8	Vsesojuznoe Ojedinenie Avtoexport Dnepropetrovski, USSR

Code No.	Tire Manufacturer
X9	Vsesojuznoe Ojedinenie Avtoexport Moscow, USSR
X0	Vsesojuznoe Ojedinenie Avtoexport Mizhenkamsk, USSR
YA	The Dayton Tire & Rubber Co. (Firestone Tire & Rubber Co.) Akron, OH, USA
YB	The Dayton Tire & Rubber Co. (Firestone Tire & Rubber Co.) Akron, OH, USA
YC	The Dayton Tire & Rubber Co. (Firestone Tire & Rubber Co.) Albany, GA, USA
YD	The Dayton Tire & Rubber Co. (Firestone Tire & Rubber Co.) Decatur, IL, USA
YE	The Dayton Tire & Rubber Co. (Firestone Tire & Rubber Co.) Des Moines, IA, USA
YF	The Dayton Tire & Rubber Co. (Firestone Tire & Rubber Co.) South Gate, CA, USA
YH	The Dayton Tire & Rubber Co. (Firestone Tire & Rubber Co.) Memphis, TN, USA
YJ	The Dayton Tire & Rubber Co. (Firestone Tire & Rubber Co.) Pottstown, PA, USA
YK	The Dayton Tire & Rubber Co. (Firestone Tire & Rubber Co.) Salinas, CA, USA
YL	Oy Nokia A.B. Nokia, Finland
YM	Seiberling Tire & Rubber Co. (Firestone Tire & Rubber Co.) Akron, OH, USA
YN	Seiberling Tire & Rubber Co. (Firestone Tire & Rubber Co.) Akron, OH, USA
YO	Kumho and Co., Inc. Chunnam, Korea
YP	Seiberling Tire & Rubber Co. (Firestone Tire & Rubber Co.) Albany, GA, USA
YT	Seiberling Tire & Rubber Co. (Firestone Tire & Rubber Co.) Decatur, IL, USA
YU	Seiberling Tire & Rubber Co. (Firestone Tire & Rubber Co.) Des Moines, IA, USA

Code No.	Tire Manufacturer
YV	Seiberling Tire & Rubber Co. (Firestone Tire & Rubber Co.) South Gate, CA, USA
YW	Seiberling Tire & Rubber Co. (Firestone Tire & Rubber Co.) Memphis, TN, USA
YX	Seiberling Tire & Rubber Co. (Firestone Tire & Rubber Co.) Pottstown, PA, USA
YY	Seiberling Tire & Rubber Co. (Firestone Tire & Rubber Co.) Salinas, CA, USA
Y1	Companhia Goodyear Do Brasil San Paulo, Brazil
Y2	Dayton Tire & Rubber Co. Wilson, NC, USA
Y3	Seiberling Tire & Rubber Co. Wilson, NC, USA
Y4	Dayton Tire & Rubber Co. Barberton, OH, USA
Y5	Shanghai Tsen Tai Rubber Factory Shanghai, People's Republic of China
Y6	Sime Tyres Inter. Sdn. Bhd. Kedah Darulaman, Malaysia
Y7	Bridgestone/Firestone, Inc. La Vergne, TN, USA
Y8	Bombay Tyres International Ltd. Bombay, India
Y9	P.T. Gadjah Tungual Jawa Barat, Rep. of Indonesia
1A	Union Rubber Ind. Co., Ltd. Taipei, Taiwan, R.O.C.
2A	Jiuh Shuenn Enterprises Co., Ltd. Wufeng, Taichung, Taiwan, R.O.C.
3A	Hualin Rubber Plant People's Republic of China
4A	Vee Rubber Co., Ltd. Samutsakaom Prov., Thailand
5A	Vee Rubber Inter. Co., Ltd. Bangkok, Thailand
6A	Roadstone Tyre & Rubber Co., Ltd. Nontaburi Bangkok, Thailand
7A	Kings Tire Ind. Co., Ltd. Taiwan, R.O.C.
8A	Zapater, Diaz I.C.S.A. Tire Co. Buenos Aires, Argentina
9A	Siamese Rubber Co., Ltd. Bangkok, Thailand

Code No.	Tire Manufacturer
0A	Siam Rubber Ltd. Part. Samutsakhon, Thailand
1B	Neumaticos De Venezuela C.A. Estadro Carabobo, Venezuela
2B	Deestone Limited Samutsakhom, Thailand
3B	Tianjin/United Tire & Rubber Tianjin, P.R. China
4B	Goodyear Canada, Inc. Napanee, Ontario, Canada
5B	The Kelly-Springfield Tire Co. Goodyear Canada, Inc. Napanee, Ontario, Canada
6B	Mt. Vernon Plt. of General Tire Mt. Vernon, IL, USA
7B	Bridgestone—Firestone Inc. Decatur, IL, USA
8B	Bridgestone—Firestone Inc. Des Moines, IA, USA
9B	Bridgestone—Firestone, Inc. Joliette, Quebec, Canada
0B	Bridgestone—Firestone Inc. Wilson, NC, USA
1C	Bridgestone—Firestone Inc. Oklahoma City, OK, USA
2C	Bridgestone/Firestone Inc. Morrison, TN, USA
3C	Mt. Vernon Plt. of General Tire Mt. Vernon, IL, USA
4C	South Pacific Tyres Somerton, Vic., Australia

Code No.	Tire Manufacturer
5C	The Firestone Tire & Rubber Co. Bridgestone Brand Bilbao, Spain
6C	The Firestone Tire & Rubber Co. Bridgestone Brand Burgos, Spain
7C	The Firestone Tire & Rubber Co. Papanui, Christ Church, New Zealand
8C	Firestone—France S.A. Bethune, France
9C	The Firestone Tire & Rubber Co. Bridgestone Brand Bari, Italy
0C	Michelin Siam Co., Ltd. Chonburi, Thailand
1D	Bridgestone Shimonoseki Plant Yamaguchi-ken, Japan
2D	Silverstone Tire & Rubber Co. Taiping, Perak, Malaysia
3D	Cooper Tire and Rubber Co. Albany, GA, USA
4D	Bridgestone/Firestone, Inc. Firestone Brand Morrison, TN, USA
5D	Bridgestone/Firestone, Inc. Dayton Brand Morrison, TN, USA

English–Metric (SI) Conversion*

Inches to Millimeters						Millimeters to Inches		
Inches			Inches				Inches	
Fraction	Decimal	mm	Fraction	Decimal	mm	mm	Decimal	Fraction
1/64	0.016	0.40	17/32	0.531	13.49	0.5	0.020	1/64
1/32	0.031	0.79	9/16	0.563	14.29	1	0.039	3/64
3/64	0.047	1.19	19/32	0.594	15.08	2	0.079	5/64
1/16	0.063	1.59	5/8	0.625	15.88	3	0.118	1/8
5/64	0.078	1.98	21/32	0.656	16.67	4	0.157	5/32
3/32	0.094	2.38	11/16	0.688	17.46	5	0.197	13/64
7/64	0.109	2.78	23/32	0.719	18.26	6	0.236	15/64
1/8	0.125	3.18	3/4	0.750	19.05	7	0.276	9/32
5/32	0.156	3.97	25/32	0.781	19.84	8	0.315	5/16
3/16	0.188	4.76	13/16	0.813	20.64	9	0.354	23/64
7/32	0.219	5.56	27/32	0.844	21.43	10	0.394	25/64
1/4	0.250	6.35	7/8	0.875	22.23	11	0.433	7/16
9/32	0.281	7.14	29/32	0.906	23.02	12	0.472	15/32
5/16	0.313	7.94	15/16	0.938	23.81	13	0.512	33/64
11/32	0.344	8.73	31/32	0.969	24.61	14	0.551	35/64
3/8	0.375	9.53	1	1.000	25.4	15	0.591	19/32
13/32	0.406	10.32				16	0.630	5/8
7/16	0.438	11.11				17	0.669	43/64
15/32	0.469	11.91				18	0.709	45/64
1/2	0.500	12.70				19	0.748	3/4
						20	0.787	25/32
						21	0.827	53/64
						22	0.866	55/64
						23	0.906	29/32
						24	0.945	15/16
						25	0.984	63/64

*Courtesy of FMC.

Fraction/Decimal/Millimeter Conversion Chart

1mm = .03937"   .001" = .0254mm

Fraction	Decimal	Millimeters
1/64	0.015625	0.397
1/32	.03125	0.794
3/64	.046875	1.191
1/16	.0625	1.588
5/64	.078125	1.984
3/32	.09375	2.381
7/64	.109375	2.778
1/8	.1250	3.175
9/64	.140625	3.572
5/32	.15625	3.969
11/64	.171875	4.366
3/16	.1875	4.763
13/64	.203125	5.159
7/32	.21875	5.556
15/64	.234375	5.953
1/4	.2500	6.350
17/64	.265625	6.747
9/32	.28125	7.144
19/64	.296875	7.541
5/16	.3125	7.938
21/64	.328125	8.334
11/32	.34375	8.731
23/64	.359375	9.128
3/8	.3750	9.525
25/64	.390625	9.922
13/32	.40625	10.319
27/64	.421875	10.716
7/16	.4375	11.113
29/64	.453125	11.509
15/32	.46875	11.906
31/64	.484375	12.303
1/2	.5000	12.700

Fraction	Decimals	Millimeters
33/64	0.515625	13.097
17/32	.53125	13.494
35/64	.546875	13.891
9/16	.5625	14.288
37/64	.578125	14.684
19/32	.59375	15.081
39/64	.609375	15.478
5/8	.6250	15.875
41/64	.640625	16.272
21/32	.65625	16.669
43/64	.671875	17.066
11/16	.6875	17.463
45/64	.703125	17.859
23/32	.71875	18.256
47/64	.734375	18.653
3/4	.7500	19.050
49/64	.765625	19.447
25/32	.78125	19.844
51/64	.796875	20.241
13/16	.8125	20.638
53/64	.828125	21.034
27/32	.84375	21.431
55/64	.859375	21.828
7/8	.8750	22.225
57/64	.890625	22.622
29/32	.90625	23.019
59/64	.921875	23.416
15/16	.9375	23.813
61/64	.953125	24.209
31/32	.96875	24.606
63/64	.984375	25.003
1	1.000	25.400

MM	INCHES	MM	INCHES
.1	.0039	46	1.8110
.2	.0079	47	1.8504
.3	.0118	48	1.8898
.4	.0157	49	1.9291
.5	.0197	50	1.9685
.6	.0236	51	2.0079
.7	.0276	52	2.0472
.8	.0315	53	2.0866
.9	.0354	54	2.1260
1	.0394	55	2.1654
2	.0787	56	2.2047
3	.1181	57	2.2441
4	.1575	58	2.2835
5	.1969	59	2.3228
6	.2362	60	2.3622
7	.2756	61	2.4016
8	.3150	62	2.4409
9	.3543	63	2.4803
10	.3937	64	2.5197
11	.4331	65	2.5591
12	.4724	66	2.5984
13	.5118	67	2.6378
14	.5512	68	2.6772
15	.5906	69	2.7165
16	.6299	70	2.7559
17	.6693	71	2.7953
18	.7087	72	2.8346
19	.7480	73	2.8740
20	.7874	74	2.9134
21	.8268	75	2.9528
22	.8661	76	2.9921
23	.9055	77	3.0315
24	.9449	78	3.0709
25	.9843	79	3.1102
26	1.0236	80	3.1496
27	1.0630	81	3.1890
28	1.1024	82	3.2283
29	1.1417	83	3.2677
30	1.1811	84	3.3071
31	1.2205	85	3.3465
32	1.2598	86	3.3858
33	1.2992	87	3.4252
34	1.3386	88	3.4646
35	1.3780	89	3.5039
36	1.4173	90	3.5433
37	1.4567	91	3.5827
38	1.4961	92	3.6220
39	1.5354	93	3.6614
40	1.5748	94	3.7008
41	1.6142	95	3.7402
42	1.6535	96	3.7795
43	1.6929	97	3.8189
44	1.7323	98	3.8583
45	1.7717	99	3.8976
		100	3.9370

Decimal Equivalents

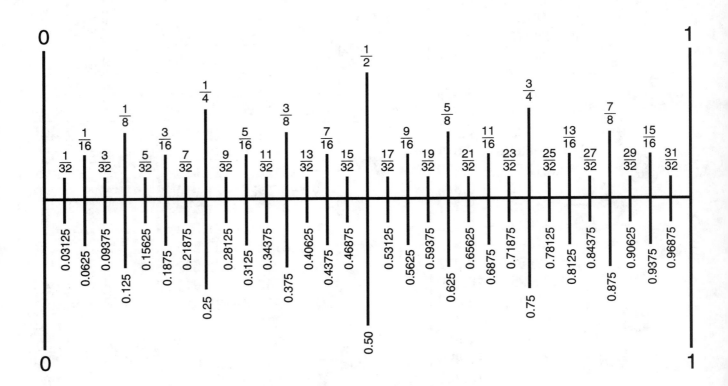

FRACTION/DECIMAL CONVERSION CHART

Alignment Angle Conversions

Angle Conversions from Degrees and Minutes

Degrees and Minutes	Decimal Degrees	Fractional Degrees
0°05′	0.08°	1/12°
0°10′	0.16°	1/6°
0°15′	0.25°	1/4°
0°20′	0.33°	1/3°
0°25′	0.42°	5/12°
0°30′	0.50°	1/2°
0°35′	0.58°	7/12°
0°40′	0.67°	2/3°
0°45′	0.75°	3/4°
0°50′	0.83°	5/6°
0°55′	0.92°	11/12°
1°00′	1.00°	1°
(minutes of an hour)	(cents of a dollar)	

1 minute = 01′

60 minutes = 1°

Angle Conversions from Fractional Degrees

Fractional Degrees	Degrees and Minutes	Decimal Degrees
1/8°	0°08′	0.125°
1/4°	0°15′	0.25°
3/8°	0°23′	0.375°
1/2°	0°30′	0.50°
5/8°	0°38′	0.625°
3/4°	0°45′	0.75°
7/8°	0°53′	0.875°
1°	1°00′	1.00°
	(minutes of an hour)	(cents of a dollar)

Angle Conversion from Decimal Degrees

Decimal Degrees	Degrees and Minutes	Fractional Degrees
0.05°	0°03′	1/20°
0.10°	0°06′	1/10°
0.15°	0°09′	3/20°
0.20°	0°12′	1/5°
0.25°	0°15′	1/4°
0.30°	0°18′	3/10°
0.35°	0°21′	7/20°
0.40°	0°24′	2/5°
0.45°	0°27′	9/20°
0.50°	0°30′	1/2°
0.55°	0°33′	11/20°
0.60°	0°36′	3/5°
0.65°	0°39′	13/20°
0.70°	0°42′	7/10°
0.75°	0°45′	3/4°
0.80°	0°48′	4/5°
0.85°	0°51′	17/20°
0.90°	0°54′	9/10°
0.95°	0°57′	19/20°
1.00°	1°00′	1°
(cents of a dollar)	(minutes of an hour)	

Conversions from Fractional Inches

Fractional Inches	Decimal Inches	Millimeters
1/32″	0.03″	0.79 mm
1/16″	0.06″	1.59 mm
3/32″	0.09″	2.38 mm
1/8″	0.13″	3.18 mm
5/32″	0.16″	3.97 mm
3/16″	0.19″	4.76 mm
7/32″	0.22″	5.56 mm
1/4″	0.25″	6.35 mm
9/32″	0.28″	7.14 mm
5/16″	0.31″	7.94 mm
11/32″	0.34″	8.73 mm
3/8″	0.38″	9.53 mm
13/32″	0.41″	10.32 mm
7/16″	0.44″	11.11 mm
15/32″	0.47″	11.91 mm
1/2″	0.50″	12.70 mm

Automotive Names and Addresses

A TO Z TOOL CO., INC.
446 W. St. Charles Road
Villa Park, IL 60181
(708) 279-4118

ALIGN-TECH
3520 Beltline Blvd.
Minneapolis, MN 55416
(800) 328-5203

AMERICAN AUTOMOBILE ASSOCIATION (AAA)
8111 Gatehouse Road
Falls Church, VA 22047
(703) 222-6000

AMERICAN PETROLEUM INSTITUTE (API)
1220 L. Street, N.W.
Washington, D.C. 20005
(202) 682-8000

AMERICAN RACING EQUIPMENT
17600 S. Santa Fe Ave.
Rancho Dominguez, CA 90221
(213) 635-7806

ARN-WOOD COMPANY, INC.
2360 West Bates
Englewood, CO 80110
(303) 761-5650

AUTO DIMENSIONS, INC.
P.O. Box 442057
Eden Prairie, MN 55347
(800) 627-0050

AUTO SAFETY HOTLINE
(800) 424-9393

AUTOMOTIVE DISMANTLERS AND RECYCLERS
ASSOCIATION (ADRA)
Formerly the NATIONAL AUTO AND TRUCK
WRECKERS ASSOC. (NATWA)
1133 15th Street, N.W.
Washington, D.C. 20005
(202) 293-2372

AUTOMOTIVE HALL OF FAME, INC.
P.O. Box 1727
Midland, MI 48641-1727
(517) 631-5760

AUTOMOTIVE PARTS AND ACCESSORIES ASSOC.,
INC. (APCS)
5100 Forbes Boulevard
Lanham, MD 20706
(301) 459-9110

AUTOMOTIVE PARTS REBUILDERS ASSOCIATION,
INC. (APRA)
6849 Old Dominion Drive
McLean, VA 22101
(703) 790-1050

AUTOMOTIVE SERVICE ASSOCIATION (ASA)
1901 Airport Freeway
P.O. Box 929
Bedford, TX 76095-0929
(817) 644-6190

AUTOMOTIVE SERVICE COUNCILS, INC. (ASC)
188 Industrial Drive, Suite 112
Elmhurst, IL 60126
(312) 530-2330

AUTOMOTIVE SERVICE INDUSTRY ASSOCIATION
(ASIA)
25 Northwest Point Blvd.
Elk Grove Village, IL 60007-1030
(708) 228-1310

AUTOMOTIVE TECHNICIANS ASSOCIATION
INTERNATIONAL (ATAI)
139 W. Maple
Birmingham, MI 48009
(313) 433-0136

AUTOMOTIVE UNDERCAR TRADE ORGANIZATION
(AUTO)
750 Lake Cook Rd., Suite 410
Buffalo Grove, IL 60089
(800) 582-1359

AUTOMOTIVE WAREHOUSE DISTRIBUTORS
ASSOCIATION, INC. (AWDA)
9140 Ward Pkwy.
Kansas City, MO 64114
(816) 444-3500

BBS OF AMERICA, INC.
33 Murray Hill Drive
Spring Valley, NY 10977
(914) 425-3900

BEAR AUTOMOTIVE SERVICE EQUIPMENT CO.
P.O. Box 880
New Berlin, WI 53151
(414) 786-5212

BEE-LINE CO.
P.O. Box 709
Bettendorf, IA 52722
(319) 332-4066

BISHMAN DIV.
Rugby Industries
West Hwy. 212,
P.O. Box 1270
Watertown, SD 57201

BLACK & DECKER (U.S.) INC.
10 North Park Drive
Hunt Valley, MD 21030
(301) 683-7000

BLACKHAWK HAND TOOLS
Automotive Div. of National Hand Tool Corp.
12828 Valley Branch Lane
P.O. Box 59857 (75229)
Dallas, TX 75234
(214) 247-6838

BORROUGHS DIV., SEALED POWER CORP.
2429 N. Burdick St.
Kalamazoo, MI 49007
(616) 345-2700

BRANICK INDUSTRIES, INC.
P.O. Box 1937
Fargo, ND 58107
(701) 235-4446

BRIDGESTONE USA, INC.
1 Bridgestone Pk.
Nashville, TN 37214
(615) 391-0088

CAL-VAN TOOLS
Div. of Chemi-Trol Chemicals Co.
1500 Walter Avenue
Fremont, OH 43420
(419) 334-2692

CAMPBELL HAUSFELD
100 Production Drive
Harrison, OH 45030
(513) 367-4811

CAR CARE COUNCIL (CCC)
One Grand Lake Drive
Port Clinton, OH 43452
(419) 734-5343

CAR-O-LINER CO.
29900 Anthony Drive
Wixom, MI 48096
(313) 624-5900

CENTER LINE TOOL CORP.
13521 Freeway Drive
Santa Fe Springs, CA 90670
(213) 921-9637
(714) 994-0500

CHAMP SERVICE LINE
Div. of Standard Motor Products
845 S. 9th Street
Edwardsville, KS 66111
(913) 441-6500

CHAMPION PARTS REBUILDERS, INC.
2525 22nd St.
Oak Brook, IL 60521
(312) 986-6100

CHICAGO PNEUMATIC TOOL CO.
2200 Bleecker St.
Utica, NY 13501
(315) 792-2800

CHIEF AUTOMOTIVE SYSTEMS, INC.
1924 E. 4th St., Box 1368
Grand Island, NE 68801
(800) 445-9262
(308) 384-9747

CHRYSLER CORPORATION
26001 Lawrence Ave.
Center Line, MI 48015

COKER TIRE
1317 Chestnut St.
P.O. Box 72554
Chattanooga, TN 37407
(800) 251-6336
(615) 265-6368

CORNWELL QUALITY TOOLS
667 Seville Road
Wadsworth, OH 44281
(216) 336-3506

CRAGAR INDUSTRIES
19007 S. Reyes Avenue
Compton, CA 90221
(213) 639-6211

DANA CORP.
P.O. Box 455
Toledo, OH 43692
(419) 866-7800

DORMAN PRODUCTS
1000 Alliance Road
Cincinnati, OH 45242
(513) 984-1000

DUNLOP TIRE CORP.
P.O. Box 1109
Buffalo, NY 14240-1109
(716) 773-8253

EASCO/K-D TOOLS
3575 Hempland Road
Lancaster, PA 17604
(717) 285-4581

EFFICIENT SYSTEMS, INC.
P.O. Box 478
N. Ridgeville, OH 44039
(800) 752-3264

ENVIRONMENTAL PROTECTION AGENCY
401 M St., S.W.
Washington, DC 20460
(202) 828-3535

EQUIPMENT & TOOL INSTITUTE (ETI)
1545 Waukegan Road
Glenview, IL 60025-2187
(708) 729-8550

FACOM HAND TOOLS
3535 W. 47th Street
Chicago, IL 60632
(312) 523-1300

FEDERATION OF AUTOMOTIVE QUALIFIED
TECHNICIANS (FAQT)
251 W. Franklin Avenue
Reed City, MI 49677
(616) 832-3399
(800) 866-3278

FIRESTONE TIRE & RUBBER CO.
1200 Firestone Pky.
Akron, OH 44317
(216) 379-7218

FMC CORP.
Auto. Ser. Equip. Div.
Exchange Avenue
Conway, AR 72032
(501) 327-4433

FORD MOTOR COMPANY
P.O. Box 07150
Detroit, MI 48207

FRICTION MATERIALS STANDARDS INSTITUTE,
INC. (FMSI)
E-210, Route 4
Paramus, NJ 07652
(201) 845-0440

GENERAL TIRE
1 General St.
Akron, OH 44329
(216) 798-3500

GOODYEAR TIRE & RUBBER CO.
1144 E. Market St.
Akron, OH 44316
(216) 796-6827

GP TOOLS
Kloster Research and Development
333 14th Street
Toledo, OH 43624
(419) 241-5555

HAND TOOLS INSTITUTE (HTI)
25 N. Broadway
Tarrytown, NY 10591
(914) 322-0040

HOFMANN CORP.
1520 Commerce Dr.
Elgin, IL 60123

HOLLANDER PUBLISHING COMPANY, INC.
P.O. Box 9405
Minneapolis, MN 55440

HOOSIER TIRE MIDWEST
1028 North St.
Springfield, IL 62704
(217) 546-4024

HUNTER ENGINEERING CO.
11250 Hunter Dr.
Bridgeton, MO 63044
(314) 731-3020

INGALLS ENGINEERING
34 Boston Ct., Unit A
Longmont, CO 80501
(303) 651-1297

INGERSOLL-RAND CO.
Power Tool Div.
Allen & Martinsville Rds.
P.O. Box 1776
Liberty Corner, NJ 07938
(201) 647-6000

INTER-INDUSTRY CONFERENCE ON AUTO
COLLISION REPAIR (I-CAR)
3701 Algonquin Road, Suite 400
Rolling Meadows, IL 60008-3118
(708) 590-1191

INTERNATIONAL TRUCK PARTS ASSOCIATION
(ITPA)
7127 Braeburn Place
Bethesda, MD 20817
(202) 544-3090

KAL-EQUIP TOOLS
10011 Walford Ave.
Cleveland, OH 44102
(216) 561-2233

KASTAR, INC.
P.O. Box 1616
Racine, WI 53401
(414) 554-7500

KELLY-SPRINGFIELD TIRE CO.
Willowbrook Rd.
Cumberland, MD 21502-2599
(301) 777-6000

KENT-MOORE TOOL GROUP
Sealed Power Corp.
28635 Mound Rd.
Warren, MI 48092
(313) 574-2332

KEN-TOOL DIVISION
Warren Tool
768 E. North Street
Akron, OH 44305
(216) 535-7177

KWIK-EZEE, INC.
54 Brooklyn Avenue
Westbury, NY 11590
(516) 333-3120

LISLE CORP.
807 E. Main St.
Clarinda, IA 51632
(712) 542-5101

LOCTITE CORP.
Automotive & Consumer Group
4450 Cranwood Ct.
Cleveland, OH 44128
(216) 475-3600

M & H TIRE CO.
P.O. Box 6
77 Industrial Row
Gardner, MA 01440
(508) 632-9501

MAC TOOLS, INC.
P.O. Box 370
Washington Court House, OH 43160
(614) 335-4112

MAREMONT CORPORATION
250 East Kehoe Blvd.
Carol Stream, IL 60188
(312) 462-8500

MATCO TOOLS CORP.
4403 Allen Rd.
Stow, OH 44224
(216) 929-4949

McCREARY TIRE & RUBBER CO.
P.O. Box 749
Indiana, PA 15701
(412) 349-9010

MICHELIN TIRE CORP.
P.O. Box 19001
Greenville, SC 29602
(803) 458-5000

MICKEY THOMPSON PERFORMANCE TIRES
P.O. Box 227
Cuyahoga Falls, OH 44222
(800) 222-9092 (U.S.)
(216) 928-9092 (OH)

MONROE SHOCKS AND STRUTS
One International Drive
Monroe, MI 48161
(313) 243-8239

MOOG AUTOMOTIVE INC.
6565 Wells Avenue
St. Louis, MO 63133
(314) 385-3400

NATIONAL AUTOMOBILE DEALERS ASSOCIATION
(NADA)
8400 Westpark Drive
McLean, VA 22102
(703) 821-7000

NATIONAL AUTOMOTIVE TECHNICIANS
EDUCATION FOUNDATION (NATEF)
13505 Dulles Technology Drive
Herndon, VA 22071-3415
(703) 713-0100

NATIONAL HIGHWAY TRAFFIC SAFETY
ADMINISTRATION
U.S. Department of Transportation
400 Seventh St., S.W.
Washington, DC 20590
(800) 424-9393
(202) 366-0123

NATIONAL INSTITUTE FOR AUTOMOTIVE SERVICE
EXCELLENCE (ASE)
13505 Dulles Technology Drive
Herndon, VA 22071-3415
(703) 713-3800

NATIONAL LUBRICATION GREASE INSTITUTE
(NLGI)
4635 Wyandotte Street
Kansas City, Missouri 64112
(816) 931-9480

NATIONAL SAFETY COUNCIL
(Promotes research and education of the public)
444 North Michigan Avenue
Chicago, IL 60611
(312) 527-4800

NATIONAL TIRE DEALERS & RETREADERS
ASSOCIATION (NTDRA)
1250 I Street, N.W., Suite 400
Washington, D.C. 20005
(800) 876-8372

NATIONAL WHEEL AND RIM ASSOCIATION (NWRA)
5121 Bowden Road, Suite 303
Jacksonville, FL 32216-5950
(904) 737-2900

NORTH AMERICAN COUNCIL OF AUTOMOTIVE
TEACHERS (NACAT)
P.O. Box 3568
Glen Ellyn, IL 60138-3568
(708) 932-1937

NORTHSTAR MANUFACTURING COMPANY, INC.
6100 Baker Road
Minnetonka, MN 55345-5909
(800) 828-0255

OLD FORGE TOOLS
7750 King Rd.
Spring Arbor, MI 49283
(517) 750-1840
(800) 338-3360

OTC
Div. of Sealed Power Corp.
655 Eisenhower Dr.
Owatonna, MN 55060
(507) 455-7102

PIRELLI-ARMSTRONG TIRE CORP.
500 Sargent Drive
New Haven, CT 06536-0201
(203) 784-2200

RIKEN-AMERICA, INC.
1113 E. 230th St.
Carson, CA 90745
(800) 421-1838 (U.S.)
(800) 262-1144 (CA)

ROTARY LIFT
P.O. Box 30205
Airport Station
Memphis, TN 38130
(901) 345-2900

RUBBER MANUFACTURERS ASSOCIATION (RMA)
1400 K St. NW, Suite 900
Washington, DC 20005
(202) 682-4800

SAFETY-KLEEN CORP.
777 Big Timber Road
Elgin, IL 60123
(312) 697-8460

SCHLEY PRODUCTS, INC.
5350 E. Hunter Avenue
Anaheim Hills, CA 92807
(714) 693-7666

SCHRADER AUTOMOTIVE GROUP
566 Mainstream Drive
P.O. Box 675
Nashville, TN 37202-0675
(615) 256-3400

SHIM•A•LINE INC.
3520 Beltline Blvd.
Minneapolis, MN 55416
(800) 328-5203

SHIMCO PRODUCTS, INC.
204 Commerce Drive
Suite G
Fort Collins, CO 80524

SIOUX TOOLS, INC.
2901 Floyd Blvd.
P.O. Box 507
Sioux City, IA 51102
(712) 252-0525

SKF AUTOMOTIVE PRODUCTS
1100 First
King of Prussia, PA 19406
(215) 265-1900

SNAP-ON TOOL CO.
2801 80th St.
Kenosha, WI 53141
(414) 656-5200

SOCIETY OF AUTOMOTIVE ENGINEERS (SAE)
400 Commonwealth Drive
Warrendale, PA 15096
(412) 776-4841

SOCIETY OF COLLISION REPAIR SPECIALISTS
(SCRS)
P.O. Box 3765
Tustin, CA 92680
(714) 838-3115

SPECIALTY EQUIPMENT MARKET ASSOCIATION
(SEMA)
1575 S. Valley Vista Drive
Diamond Bar, CA 92765
(714) 396-0289

SPECIALTY PRODUCTS CORP.
P.O. Box 923
Longmont, CO 80502-0923
(303) 772-2103
(800) 525-6505

SPRING SERVICE ASSOCIATION (SSA)
4015 Marks Road, Suite 2B
Medina, OH 44256
(216) 725-7160

STANLEY TOOLS
Div. of the Stanley Works
600 Myrtle St.
New Britain, CT 06050
(203) 225-5111

STEMPF
3206 Bloomington Avenue South
Minneapolis, MN 55407
(800) 328-4460

STEWART-WARNER CORP.
Alemite & Instrument Div.
1826 Diversey Pkwy.
Chicago, IL 60614
(312) 883-7662

SUNEX INTERNATIONAL
P.O. Box 4215
Highway 250 Bypass
Greenville, SC 29608
(803) 834-8759

SUPERIOR PNEUMATIC & MFG., INC.
855 Canterbury Rd.
P.O. Box 40420
Cleveland, OH 44140
(216) 871-8780

SUSPENSION SPECIALISTS (SSA)
4015 Marks Rd., Suite 2B
Medina, OH 44256
(216) 725-7260

THEXTON MFG. CO.
7685 Parklawn Ave.
P.O. Box 35008
Minneapolis, MN 55435
(612) 831-4171

THE TIRE AND RIM ASSOCIATION, INC.
3200 W. Market Street
Akron, OH 44313
(216) 666-8121

TIRE INDUSTRY SAFETY COUNCIL (TISC)
National Press Building, Suite 844
Washington, D.C. 20045
(202) 783-1022

TOYO TIRE (U.S.A.) CORP.
300 W. Artesia Blvd.
Compton, CA 90220
(213) 537-2820

TRUCK FRAME & AXLE REPAIR ASSOCIATION
(TFARA)
915 E. 99th St.
Brooklyn, NY 11236-4011
(718) 257-6133

TRW INC.
Replacement Parts Div.
8001 Pleasant Valley Rd.
Cleveland, OH 44131-5582
(216) 447-1879

UNIROYAL GOODRICH TIRE COMPANY
600 S. Main
Akron, OH 44397
(216) 374-3000

UNIVERSAL RODAC CORP.
1928 Starr-Batt, Unit E
Rochester Hills, MI 48309
(313) 853-2323

U.S. DEPARTMENT OF TRANSPORTATION
National Highway Traffic Safety Administration
400 Seventh St., S.W.
Washington, D.C. 20590

YOKOHAMA TIRE CORPORATION
601 S. Acacia Ave.
Fullerton, CA 92631
(714) 870-3800
(800) 423-4544 ext. 3951

Answers to Even-Number ASE Certification-Type Questions

Chapter 1	2-d, 4-b, 6-d, 8-b, 10-c
Chapter 2	2-b, 4-c, 6-b, 8-d, 10-b
Chapter 3	2-b, 4-c, 6-d, 8-d, 10-d
Chapter 4	2-c, 4-d, 6-b, 8-c, 10-b
Chapter 5	2-c, 4-b, 6-b, 8-d, 10-d
Chapter 6	2-b, 4-a, 6-a, 8-a, 10-a
Chapter 7	2-b, 4-b, 6-b, 8-c, 10-a

Chapter 8	2-c, 4-a, 6-c, 8-c, 10-a
Chapter 9	2-c, 4-c, 6-c, 8-b, 10-a
Chapter 10	2-b, 4-c, 6-b, 8-b, 10-a
Chapter 11	2-a, 4-c, 6-c, 8-b, 10-a
Chapter 12	2-b, 4-a, 6-c, 8-b, 10-a
Chapter 13	2-b, 4-a, 6-c, 8-c, 10-d

Glossary

ABS Antilock brakes.

ACI Automotive Components Inc.

Ackerman effect The angle of the steering arms causes the inside wheel to turn more sharply than the outer wheel when making a turn. This produces toe-out on turns (TOOT).

Alamite fitting *See* Zerk.

Alloy A metal that contains one or more other elements, usually added to increase strength or give the base metal important properties.

Anti-dive A term used to describe the geometry of the suspension that controls the movement of the vehicle during braking. It is normal for a vehicle to nose-dive slightly during braking and is designed into most vehicles.

Anti-roll bar *See* Stabilizer bar.

Anti-squat A term used to describe the geometry of the suspension that controls the movement of the vehicle body during acceleration. 100 percent anti-squat means that the body remains level during acceleration. Less than 100 percent indicates that the body "squats down," or lowers in the rear during acceleration.

Anti-sway bar *See* Stabilizer bar.

API American Petroleum Institute.

APRA Automotive Parts Rebuilders Association.

Articulation test A test specified by some vehicle manufacturers that tests the amount of force necessary to move the inner tie rod end in the ball socket assembly. The usual specification for this test is greater than 1 lb. (4 N) and less than 6 lb. (26 N) of force.

ASE Abbreviation for the National Institute for Automotive Service Excellence, a nonprofit organization for the testing and certification of vehicle service technicians.

Aspect ratio The ratio of height to width of a tire. A tire with an aspect ratio of 60 (a 60 series tire) has a height (from rim to tread) of 60 percent of its cross-sectional width.

ASTM American Society for Testing Materials.

Axial In line along with the axis or centerline of a part or component. Axial play in a ball joint means looseness in the same axis as the ball joint stud.

Back spacing The distance between the back rim edge and the center section mounting pad of a wheel.

Backside setting *See* Back spacing.

Ball socket assembly An inner tie rod end assembly that contains a ball and socket joint at the point where the assembly is threaded on to the end of the steering rack.

Bias belted A bias ply tire with additional belt material just under the tread area.

Bias ply The body plies cover the entire tire and are angled as they cross from bead to bead.

Bolt circle The diameter (in inches or millimeters) of a circle drawn through the center of the bolt holes in a wheel.

Bounce test A test used to check the condition of shock absorbers.

Breather tube A tube that connects the left and right bellows of a rack and pinion steering gear.

Brinelling A type of mechanical failure used to describe a dent in metal such as that which occurs when a shock load is applied to a bearing. Named after Johann A. Brinell, a Swedish engineer.

Bump steer Used to describe what occurs when the steering linkage is not level, causing the front tires to turn inward or outward as the wheels and suspension move up and down. Automotive chassis engineers call it *roll steer.*

Bump stop *See* Jounce bumper.

Camber The inward or outward tilt of the wheels from true vertical as viewed from the front or rear of the vehicle. Positive camber means the top of the wheel is out from center of the vehicle more than the bottom of the wheel.

Cardan joint A type of universal joint named for a 16th-century Italian mathematician.

Caster The forward or backward tilt of an imaginary line drawn through the steering axis as viewed from the side of the vehicle. Positive caster is where an imaginary line would contact the road surface in front of the contact path of the tire.

Caster sweep   A process used to measure caster during a wheel alignment procedure where the front wheels are rotated first inward, then outward, a specified amount.

Center bolt   A bolt used to hold the leaves of a leaf spring together in the center. Also called a *centering pin.*

Center support bearing   A bearing used to support the center of a long driveshaft on a rear-wheel-drive vehicle. Also called a *steady bearing.*

Centering pin   *See* Center bolt.

Centerline steering   Used to describe the position of the steering wheel while driving on a straight, level road. The steering wheel should be centered or within plus or minus 3 degrees, as specified by many vehicle manufacturers.

Chassis   The frame, suspension, steering, and machinery of a motor vehicle.

Coefficient of friction   A measure of the amount of friction, usually measured from 0 to 1. A low number (0.3) indicates low friction, and a high number (0.9) indicates high friction.

Coil spring   A spring steel rod wound in a spiral (helix) shape. Used in both front and rear suspension systems.

Compensation   A process used during a wheel alignment procedure where the sensors are calibrated to eliminate errors in the alignment readings that may be the result of a bent wheel or unequal installation of the sensor on the wheel of the vehicle.

Composite   A term used to describe the combining of individual parts into a larger component. For example, a composite leaf spring is constructed of fiberglass and epoxy; a composite master brake cylinder contains both plastic parts (reservoir) and metal parts (cylinder housing).

Compression bumper   *See* Jounce bumper.

Compression rod   *See* Strut rod.

Cone   The inner race or ring of a bearing.

Constant velocity joint   Commonly called *CV joints.* CV joints are driveline joints that can transmit engine power through relatively large angles without a change in the velocity, as is usually the case with conventional Cardan-type U-joints.

Controller   Commonly used to describe a computer or an electronic control module.

Cotter key   A metal loop used to retain castle nuts by being installed through a hole. Size is measured by diameter and length (for example, 1/8″ × 1 1/2″). Also called a *cotter pin.* Named for the Old English verb meaning "to close or fasten."

Coupling disc   *See* Flexible coupling.

Cow catcher   A large spring seat used on many General Motors MacPherson strut units. If the coil spring breaks, the cow catcher prevents one end of the spring from moving outward and cutting a tire.

Cross camber/caster   The difference of angle from one side of the vehicle to the other. Most manufacturers recommend a maximum difference side to side of 1/2 degree for camber and caster.

Cross steer   A type of steering linkage commonly used on light and medium trucks.

Cup   The outer race or ring of a bearing.

CV joints   Constant velocity joints.

Diff   An abbreviation or slang for *differential.*

Differential   A mechanical unit containing gears that provides gear reduction and a change of direction of engine power and permits the drive wheels to rotate at different speeds, as required when turning a corner.

Directional stability   Ability of a vehicle to move forward in a straight line with a minimum of driver control. Crosswinds and road irregularities will have little effect if directional stability is good.

Dog tracking   The condition where the rear wheels do not follow directly behind the front wheels. Named for dogs that run with their rear paws offset toward one side so that their rear paws will not hit their front paws.

DOT   Department of Transportation.

Double Cardan   A universal joint that uses two conventional Cardan joints together, to allow the joint to operate at greater angles.

Drag link   Used to describe a link in the center of the steering linkage; usually called a *center link.*

Drag rod   *See* Strut rod.

Drift   A mild pull that does not cause a force on the steering wheel that the driver must counteract (also known as *lead*). Also refers to a tapered tool used to center a component in a bolt hole prior to installing the bolt.

Dropping point   The temperature at which a grease passes from a semisolid to a liquid state under conditions specified by ASTM.

Dry park test   A test of steering and/or suspension components. With the wheels in the straight-ahead position and the vehicle on flat level ground, have an assistant turn the steering wheel while looking and touching all steering and suspension components, checking for any looseness.

Durometer   The hardness rating of rubber products, named for an instrument used to measure hardness that was developed about 1890.

Dust cap   A functional metal cap that keeps grease in and dirt out of wheel bearings. Also called a *grease cap.*

EPA   Environmental Protection Agency.

Flexible coupling   A part of the steering mechanism between the steering column and the steering gear or rack and pinion assembly. Also called a *rag joint* or *steering coupling disc.* The purpose of the flexible coupling is to keep noise vibration and harshness from being transmitted from the road and steering to the steering wheel.

Flow control valve   Regulates and controls the flow of power steering pump hydraulic fluids to the steering gear or rack and pinion assembly. The flow control valve is usually part of the power steering pump assembly.

Follower ball joint   A ball joint used in a suspension system to provide support and control without having the weight of the vehicle or the action of the springs transferred through the joint itself. Also called a *friction ball joint.*

Forward steer   *See* Front steer.

4 DR   Four-door model.

4 × 4   The term used to describe a four-wheel-drive vehicle. The first "4" indicates the number of wheels of the vehicle; the second "4" indicates the number of wheels that are driven by the engine.

4 × 2   The term used to describe a two-wheel-drive truck. The "4" indicates the number of wheels of the vehicle; the "2" indicates the number of wheels that are driven by the engine.

Free play   The amount that the steering wheel can move without moving the front wheels. The maximum allowable amount of free play is less than 2″ for a parallelogram-type steering system, and 3/8″ for a rack and pinion steering system.

Frequency   The number of times a complete motion cycle takes place during a period of time (usually measured in seconds).

Friction   The resistance to sliding of two bodies in contact with each other.

Friction ball joint   Outer suspension pivot that does not support the weight of the vehicle. Also called a *follower ball joint.*

Front steer   A construction design of a vehicle that places the steering gear and steering linkage in front of the centerline of the front wheels. Also called *forward steer.*

FWD   Front-wheel drive.

Galvanized steel   Steel with a zinc coating to help protect it from rust and corrosion.

Garter spring   A spring used in a seal to help keep the lip of the seal in contact with the moving part.

GKN   Guest, Keene, and Nettelfolds.

Gland nut   The name commonly used to describe the large nut at the top of a MacPherson strut housing. This gland nut must be removed to replace a strut cartridge.

Gram   A metric unit of weight equal to 1/1000 kilogram (1 oz × 28 = 1 gram). An American dollar bill or paper clip weighs about 1 gram.

Grease cap   A functional metal cap that keeps grease in and dirt out of wheel bearings. Also called a *dust cap.*

Grease retainer   *See* Grease seal.

Grease seal   A seal used to prevent grease from escaping and to prevent dirt and moisture from entering.

Green tire   An uncured assembled tire. After the green tire is placed in a mold under heat and pressure, the rubber changes chemically and comes out of the mold formed and cured.

Grommet   An eyelet (usually made from rubber) used to protect, strengthen, or insulate around a hole or passage.

GVW   Gross vehicle weight. GVW is the weight of the vehicle plus the weight of all passengers and cargo up to the limit specified by the manufacturer.

Half shaft   Drive axles on a front-wheel-drive vehicle or from a stationary differential to the drive wheels.

Halogenated compounds   Chemicals containing chlorine, fluorine, bromine, or iodine. These chemicals are generally considered to be hazardous, and any product containing these chemicals should be disposed of using approved procedures.

Haltenberger linkage   A type of steering linkage commonly used on light trucks.

HD   Heavy duty.

Hanger bearing   *See* Center support bearing.

Helper springs   Auxiliary or extra springs used in addition to the vehicle's original springs to restore proper ride height or to increase the load-carrying capacity of the vehicle.

Hertz   A unit of measurement of frequency. One Hertz is one cycle per second, abbreviated Hz. Named for Heinrich R. Hertz, a 19th-century German physicist.

Hooke's Law   The force characteristics of a spring discovered by Robert Hooke (1635–1703), an English physicist. Hooke's Law states that "the deflection (movement or deformation) of a spring is directly proportional to the applied force."

HSS   High-strength steel. A low-carbon alloy steel that uses various amounts of silicon, phosphorus, and manganese.

Hub cap   A functional and decorative cover over the lug nut portion of the wheel. *Also see* Wheel cover.

Hydrophilic   A term used to describe a type of rubber used in many all-season tires where the rubber has an affinity for water (rather than repelling it).

Hydrophobic   A term used to describe the repelling of water.

Hydroplaning   Condition that occurs when driving too fast on wet roads. The water on the road gets trapped between the tire and the road, forcing the tire onto a layer of water and off the road surface. All traction between the tire and the road is lost.

Hypoid gear set   A ring gear and pinion gear set that mesh below the centerline of the ring gear. This type of gear set allows the driveshaft to be lower in the vehicle, yet requires special hypoid gear lubricant.

Included angle   SAI angle added to the camber angle of the same wheel.

Independent suspension   A suspension system that allows a wheel to move up and down without undue effect on the opposite side.

Iron Refined metal from iron ore (ferrous oxide) in a furnace. *Also see* Steel.

IRS Independent rear suspension.

ISO International Standards Organization.

Isolator bushing Rubber bushing used between the frame and the stabilizer bar. Also known as a *stabilizer bar bushing.*

Jounce Used to describe up-and-down movement.

Jounce bumper A rubber or urethane stop to limit upward suspension travel. Also called a *bump stop, strikeout bumper, suspension bumper,* or *compression bumper.*

Kerf Large water grooves in the tread of a tire.

King pin A pivot pin commonly used on solid axles or early model twin I-beam axles that rotate in bushings and allow the front wheels to rotate. The knuckle pivots about the king pin.

King pin inclination Inclining the tops of the king pins toward each other creates a force stabilizing to the vehicle.

KPI King pin inclination (also known as *steering axis inclination—SAI*). The angle formed between true vertical and a line drawn between the upper and lower pivot points of the spindle.

Ladder frame A steel frame for a vehicle that uses cross braces along the length, similar to the rungs of a ladder.

Lateral runout A measure of the amount a tire or wheel is moving side to side while being rotated. Excessive lateral runout can cause a shimmy-type vibration if the wheels are on the front axle.

Lead A mild pull that does not cause a force on the steering wheel that the driver must counteract (also known as *drift*).

Leaf spring A spring made of several pieces of flat spring steel.

Live axle A solid axle used on the drive wheels; it contains the drive axles that propel the vehicle.

Load index An abbreviated method that uses a number to indicate the load-carrying capabilities of a tire.

Load-carrying ball joint Used in a suspension system to provide support and control, and through which the weight (load) of the vehicle is transferred to the frame.

LOBRO joint A brand name of CV joint.

Lock nut *See* Prevailing torque nut.

LSD An abbreviation commonly used for limited-slip differentials.

LT Light truck.

M&S Mud and snow.

MacPherson strut A type of front suspension with the shock absorber and coil spring in one unit, which rotates when the wheels are turned. Assembly mounts to the vehicle body at the top and to one ball joint and control arm at the lower end. It is named for its inventor, Earle S. MacPherson.

Match mount The process of mounting a tire on a wheel and aligning the valve stem with a mark on the tire. The mark on the tire represents the high point of the tire; the valve stem location represents the smallest diameter of the wheel.

Memory steer A lead or pull of a vehicle caused by faults in the steering or suspension system. If after making a turn, the vehicle tends to pull in the same direction as the last turn, then the vehicle has memory steer.

Millisecond One-thousandth of one second (1/1000).

Minutes A unit of measure of an angle. Sixty minutes equal 1 degree.

Moly grease Grease containing molybdenum disulfide.

Morning sickness A slang term used to describe temporary loss of power steering assist when cold caused by wear in the control valve area of a power rack and pinion unit.

MSDS Material safety data sheets.

NHTSA National Highway Traffic Safety Administration.

NLGI National Lubricating Grease Institute. Usually associated with grease. The higher the NLGI number, the firmer the grease. #000 is very fluid, whereas #5 is very firm. The consistency most recommended is NLGI #2 (soft).

Noise Noise is the vibration of air caused by a body in motion.

NVH Noise, vibration, and harshness.

OE Original equipment.

OEM Original equipment manufacturer.

Offset The distance the center section (mounting pad) is offset from the centerline of the wheel.

Orbital steer *See* Bump steer.

Overcenter adjustment An adjustment made to a steering gear while the steering is turned through its center straight-ahead position. Also known as a *sector lash adjustment.*

Overinflation Used to describe a tire with too much tire pressure (greater than maximum allowable pressure).

Oversteer A term used to describe the handling of a vehicle where the driver must move the steering wheel in the opposite direction from normal while turning a corner. Oversteer handling is very dangerous. Most vehicle manufacturers design their vehicles to understeer rather than oversteer.

Oz-in. Measurement of imbalance. 3 oz-in. means that an object is out of balance; it would require a 1-oz. weight placed 3″ from the center of the rotating object or a 3-oz. weight placed 1″ from the center or any other combination that when multiplied equals 3 oz-in.

Panhard rod A horizontal steel rod or bar attached to the rear axle housing at one end and the frame at the other to keep the center of the body directly above the center of the rear axle during cornering and suspension motions. Also called a *track rod*.

Parallelogram A geometric box shape where opposite sides are parallel (equal distance apart). A type of steering linkage used with a conventional steering gear that uses a pitman arm, center link, idler arms and tie rods.

Penetration A test for grease where the depth of a standard cone is dropped into a grease sample and its depth is measured.

Perimeter frame A steel structure for a vehicle that supports the body of the vehicle under the sides, as well as the front and rear.

Pickle fork A tapered fork used to separate chassis parts that are held together by a nut and a taper. Hitting the end of the pickle fork forces the wedge portion of the tool between the parts to be separated and "breaks the taper." A pickle fork tool is generally *not* recommended because the tool can tear or rip the grease boot of the part being separated.

Pitch The pitch of a threaded fastener refers to the number of threads per inch.

Pitman arm A short lever arm that is splined to the steering gear cross shaft. It transmits the steering force from the cross shaft to the steering linkage.

Pitman shaft *See* Sector shaft.

Platform The platform of a vehicle includes the basic structure (frame and/or major body panels), as well as the basic steering and suspension components. One platform may be the basis for several different brand vehicles.

Pound foot (lb. ft.) A measurement of torque. A 1-pound pull, 1 foot from the center of an object.

PPM Parts per million.

Prevailing torque nut A special design of nut fastener that is deformed slightly or has other properties that permit the nut to remain attached to the fastener without loosening.

Prop shaft An abbreviation for propeller shaft.

Propeller shaft A term used by many manufacturers for a driveshaft.

PSI Pounds per square inch.

Pull Vehicle tends to go left or right while traveling on a straight, level road.

Rack and pinion A type of lightweight steering unit that connects the front wheels through tie rods to the end of a long shaft called a *rack*. When the driver moves the steering wheel, the force is transferred to the rack and pinion assembly. Inside the rack housing is a small pinion gear that meshes with gear teeth, which are cut into the rack.

Radial runout A measure of the amount a tire or wheel is out of round. Excessive radial runout can cause a tramp-type vibration.

Radial tire A tire whose carcass plies run straight across (or almost straight across) from bead to bead.

Radius rod A suspension component to control longitudinal (front-to-back) support; it is usually attached with rubber bushings to the frame at one end and the axle or control arm at the other end. *Also see* Strut rod.

Rag joint *See* Flexible coupling.

Ratio The expression for proportion. For example, in a typical rear axle assembly, the driveshaft rotates three times faster than the rear axles. It is expressed as a ratio of 3:1 and read as "three to one." Power train ratios are always expressed as driving divided by driven gears.

RBS Rubber bonded socket.

Rear spacing *See* Back spacing.

Rear steer A construction design of a vehicle that places the steering gear and steering linkage behind the centerline of the front wheels.

Rebuilt *See* Remanufactured.

Remanufactured A term used to describe a process where a component is disassembled, cleaned, inspected, and reassembled using new or reconditioned parts. According to the Automotive Parts Rebuilders Association (APRA), this same procedure is also called *rebuilt*.

RIM Reaction injection molded.

RMP Reaction moldable polymer.

Road crown A roadway where the center is higher than the outside edges. Road crown is designed into many roads to drain water off the road surface.

Roll bar *See* Stabilizer bar.

Roll steer *See* Bump steer.

Run-flat tires Tires specially designed to operate for reasonable distances and speeds without air inside to support the weight of the vehicle. Run-flat tires usually require the use of special rims designed to prevent the flat tire from coming off the wheel.

RWD Rear-wheel drive.

SAE Society of Automotive Engineers.

Saginaw Brand name of steering components manufactured in Saginaw, Michigan, USA.

SAI Steering axis inclination (same as KPI).

Schrader valve A type of valve used in tires, air-conditioning, and fuel injection systems. Invented in 1844 by August Schrader.

Scrub radius Refers to where an imaginary line drawn through the steering axis intersects the ground compared to the centerline of the tire. *Zero* scrub radius means the line intersects at the center line of the tire. *Positive* scrub radius means that the line intersects below the road surface. *Negative* scrub radius means the line intersects above the road surface. It is also called *steering offset* by some vehicle manufacturers.

Sector lash Refers to clearance (lash) between a section of gear (sector) on the pitman shaft in a steering gear. *Also see* Overcenter adjustment.

Sector shaft The output shaft of a conventional steering gear. It is a part of the sector shaft in a section of a gear that meshes with the worm gear and is rotated by the driver when the steering wheel is turned. It is also called a *pitman shaft*.

SED   Sedan.

SEMA   Specialty Equipment Manufacturers Association.

Setback   The amount the front wheels are set back from true parallel with the rear wheels. Positive setback means the right front wheel is set back farther than the left. Setback can be measured as an angle formed by a line perpendicular (90°) to the front axles.

Shackle   A mounting that allows the end of a leaf spring to move forward and backward as the spring moves up and down during normal operation of the suspension.

Shim   A thin metal spacer.

Shimmy   A vibration that results in a rapid back-and-forth motion of the steering wheel. A bent wheel or a wheel assembly that is not correctly balanced dynamically is a common cause of shimmy.

Short/long arm suspension   Abbreviated SLA. A suspension system with a short upper control arm and a long lower control arm. The wheel changes very little in camber with a vertical deflection. Also called *double-wishbone-type suspension*.

Sipes   Small traction-improving slits in the tread of a tire.

SLA   Short/long arm suspension.

Slip angle   The angle between the true centerline of the tire and the actual path followed by the tire while turning.

SMC   Sheet molding compound.

Solid axle   A solid supporting axle for both front or both rear wheels. Also referred to as a straight axle or nonindependent axle.

Space frame construction   A type of vehicle construction that uses the structure of the body to support the engine and drivetrain, as well as the steering and suspension. The outside body panels are nonstructure.

Spalling   A term used to describe a type of mechanical failure caused by metal fatigue. Metal cracks then break out into small chips, slabs, or scales of metal.

Spider   Center part of a wheel. Also known as the *center section*.

Spindle nut   Nut used to retain and adjust the bearing clearance of the hub to the spindle.

Sprung weight   The weight of a vehicle that is supported by the suspension.

Stabilizer bar   A hardened steel bar connected to the frame and both lower control arms to prevent excessive body roll. Also called an *anti-sway* or *anti-roll bar*.

Stabilizer links   Usually consists of a bolt, spacer, and nut to connect (link) the end of the stabilizer bar to the lower control arm.

Steady bearing   *See* Center support bearing.

Steel   Refined iron metal with most of the carbon removed.

Steering arms   Arms bolted to or forged as a part of the steering knuckles. They transmit the steering force from the tie rods to the knuckles, causing the wheels to pivot.

Steering coupling disc   *See* Flexible coupling.

Steering gear   Gears on the end of the steering column that multiply the driver's force to turn the front wheels.

Steering knuckle   The inner portion of the spindle that pivots on the king pin or ball joints.

Steering offset   *See* Scrub radius.

Straight axle   *See* Solid axle.

Strikeout bumper   *See* Jounce bumper.

Strut rod   Suspension member used to control forward/backward support to the control arms. Also called *tension* or *compression rod* (*TC rod*) or *drag rod*.

Strut rod bushing   A rubber component used to insulate the attachment of the strut rod to the frame on the body of the vehicle.

Stud   A short rod with threads on both ends.

Suspension   Parts or linkages by which the wheels are attached to the frame or body of a vehicle. These parts or linkages support the vehicle and keep the wheels in proper alignment.

Suspension bumper   *See* Jounce bumper.

Sway bar   Shortened name for anti-sway bar. *See* Stabilizer bar.

TC rod   *See* Strut rod.

Tension rod   *See* Strut rod.

Thrust angle   The angle between the geometric centerline of the vehicle and the thrust line.

Thrust line   The direction the rear wheels are pointed as determined by the rear-wheel toe.

Tie rod   A rod connecting the steering arms together.

Toe-in   The difference in measurement between the front of the wheels and the back of the wheels (the front are closer than the back).

Toe-out   The back of the tires are closer than the front.

Torque   A twisting force that may or may not result in motion. Measured in lb. ft. or Newton-meters.

Torque steer   Torque steer occurs in front-wheel-drive vehicles when engine torque causes a front wheel to change its angle (toe) from straight ahead. The resulting pulling effect of the vehicle is most noticeable during rapid acceleration, especially whenever upshifting of the transmission creates a sudden change in torque.

Torsion bar   A type of spring in the shape of a straight bar. One end is attached to the frame of the vehicle, and the opposite end is attached to a control arm of the suspension. When the wheels hit a bump, the bar twists and then untwists.

Torx   A type of fastener which features a star-shaped indentation for a tool. A registered trademark of the Camcar Division of Textron.

Total toe   The total (combined) toe of both wheels, either front or rear.

Track   The distance between the centerline of the wheels as viewed from the front or rear.

Track rod   A horizontal steel rod or bar attached to rear axle housing at one end and the frame at the other to keep the center of the rear axle centered on the body. Also known as a *panhard rod.*

Tracking   Used to describe the fact that the rear wheels should track directly behind the front wheels.

Tramp   A vibration usually caused by up-and-down motion of an out-of-balance or out-of-round wheel assembly.

Turning radius   Refers to the angle of the steering knuckles that allow the inside wheel to turn at a sharper angle than the outside wheel whenever turning a corner. Also known as toe-out on turns (TOOT) or the Ackerman effect.

2 DR   Two-door model.

UNC   Unified national coarse.

Underinflation   A term used to describe a tire with too little tire pressure (less than minimum allowable pressure).

Understeer   A term used to describe the handling of a vehicle where the driver must turn the steering wheel more and more while turning a corner.

UNF   Unified national fine.

Unit body   A type of vehicle construction first used by the Budd Company of Troy, Michigan, that does not use a separate frame. The body is built strong enough to support the engine and the power train, as well as the suspension and steering system. The outside body panels are part of the structure. *Also see* Space frame construction.

Unsprung weight   The parts of a vehicle not supported by the suspension system. Examples of items that are typical unsprung weight include wheels, tires, and brakes.

Vibration   An oscillation, shake, or movement that alternates in opposite directions.

VOC   Volatile organic compounds.

Vulcanization   A process where heat and pressure combine to change the chemistry of rubber.

Wander   A type of handling which requires constant steering wheel correction to keep the vehicle going straight.

Watt's link   A type of track rod which uses two horizontal rods pivoting at the center of the rear axle.

Wear bars   *See* Wear indicators.

Wear indicators   Bald area across the tread of a tire when 2/32″ or less of tread depth remains.

Wear indicator ball joint   A ball joint design with a raised area around the grease fitting. If the raised area is flush or recessed with the surrounding area of the ball joint, the joint is worn and must be replaced.

Weight-carrying ball joint   *See* Load-carrying ball joint.

Wheel cover   A functional and decorative cover over the entire wheel. *Also see* Hub cap.

Wheelbase   The distance between the centerline of the two wheels as viewed from the side.

Wishbone suspension   *See* Short/long arm suspension.

W/O   Without.

Worm and roller   A steering gear that uses a worm gear on the steering shaft. A roller on one end of the cross shaft engages the worm.

Worm and sector   A steering gear that uses a worm gear that engages a sector gear on the cross shaft.

Zerk   A name commonly used for a grease fitting. Named for its developer, Oscar U. Zerk, in 1922, an employee of the Alamite Corporation. Besides a Zerk fitting, a grease fitting is also called an *Alamite fitting.*

Index